**METHODS IN MOLECULAR BIOLOGY**

*Series Editor*
John M. Walker
School of Life and Medical Sciences
University of Hertfordshire
Hatfield, Hertfordshire, UK

For further volumes:
http://www.springer.com/series/7651

For over 35 years, biological scientists have come to rely on the research protocols and methodologies in the critically acclaimed *Methods in Molecular Biology* series. The series was the first to introduce the step-by-step protocols approach that has become the standard in all biomedical protocol publishing. Each protocol is provided in readily-reproducible step-by step fashion, opening with an introductory overview, a list of the materials and reagents needed to complete the experiment, and followed by a detailed procedure that is supported with a helpful notes section offering tips and tricks of the trade as well as troubleshooting advice. These hallmark features were introduced by series editor Dr. John Walker and constitute the key ingredient in each and every volume of the *Methods in Molecular Biology* series. Tested and trusted, comprehensive and reliable, all protocols from the series are indexed in PubMed.

# Cancer Systems and Integrative Biology

Edited by

**Usha N. Kasid**

*Lombardi Comprehensive Cancer Center, Georgetown University, Washington, DC, USA*

**Robert Clarke**

*The Hormel Institute, University of Minnesota, Austin, MN, USA*

Humana Press

*Editors*
Usha N. Kasid
Lombardi Comprehensive
Cancer Center
Georgetown University
Washington, DC, USA

Robert Clarke
The Hormel Institute
University of Minnesota
Austin, MN, USA

ISSN 1064-3745 ISSN 1940-6029 (electronic)
Methods in Molecular Biology
ISBN 978-1-0716-3162-1 ISBN 978-1-0716-3163-8 (eBook)
https://doi.org/10.1007/978-1-0716-3163-8



This Humana imprint is published by the registered company Springer Science+Business Media, LLC, part of Springer Nature.
The registered company address is: 1 New York Plaza, New York, NY 10004, U.S.A.

# Preface

Recent advances in the integration of multi-omics and multiparametric strategies, high-throughput analysis of large datasets, and application of state-of-the-art methods at the single-cell resolution level and in 3D organoid models have revolutionized the field of precision oncology. The overarching goal of these efforts is to develop tools to monitor spatial and temporal changes in tumors, track tumor markers in blood, and ultimately develop precision medicine to combat the disease in real time. Various methodologies and reviews in this book, contributed by leading experts in cancer systems and integrative biology, will be an invaluable resource for researchers, pharmaceutical scientists, and oncologists interested in expanding their knowledge base in the current developments in cancer research.

*Washington, DC, USA* — *Usha N. Kasid*
*Austin, MN, USA* — *Robert Clarke*

# Contents

# Contributors

MAYAR ALLAM • *Wallace H. Coulter Department of Biomedical Engineering, Georgia Institute of Technology and Emory University, Atlanta, GA, USA*
SHARAD AWASTHI • *Department of Epigenetics and Molecular Carcinogenesis, The University of Texas MD Anderson Cancer Center, Houston, TX, USA*
ALAKESH BERA • *Department of Anatomy, Physiology and Genetics, and Institute for Molecular Medicine, Uniformed Services University School of Medicine (USUHS), Bethesda, MD, USA; Surgery Branch, Center for Cancer Research, National Cancer Institute, National Institutes of Health, Bethesda, MD, USA*
PAWEL BIALK • *Gene Editing Institute, ChristianaCare Health System, Newark, DE, USA*
DEEPA BISHT • *Department of Epigenetics and Molecular Carcinogenesis, The University of Texas MD Anderson Cancer Center, Houston, TX, USA*
BENJAMIN C. BLUM • *Center for Network Systems Biology, Boston University, Boston, MA, USA; Department of Biochemistry, Boston University School of Medicine, Boston, MA, USA*
HOWARD E. BOUDREAU • *Georgetown Lombardi Comprehensive Cancer Center, Georgetown University Medical Center, Washington, USAD.C.*
SHUANGYI CAI • *Wallace H. Coulter Department of Biomedical Engineering, Georgia Institute of Technology and Emory University, Atlanta, GA, USA*
PATRICK CHANG • *Department of Research Pathology, Genentech, Inc., South San Francisco, CA, USA*
DIGONTO CHATTERJEE • *Department of Anatomy, Physiology and Genetics, and Institute for Molecular Medicine, Uniformed Services University School of Medicine (USUHS), Bethesda, MD, USA*
ROBERT CLARKE • *The Hormel Institute, University of Minnesota, Austin, MN, USA; I. J. Holton Chair in Cancer Research, Austin, MN, USA*
ZEYNEP H. COBAN AKDEMIR • *Human Genetics Center, Department of Epidemiology, Human Genetics, and Environmental Sciences, School of Public Health, The University of Texas Health Science Center at Houston, Houston, TX, USA*
AHMET F. COSKUN • *Wallace H. Coulter Department of Biomedical Engineering, Georgia Institute of Technology and Emory University, Atlanta, GA, USA; Interdisciplinary Bioengineering Graduate Program, Georgia Institute of Technology, Atlanta, GA, USA; Parker H. Petit Institute for Bioengineering and Bioscience, , Georgia Institute of Technology, Atlanta, GA, USA*
TINGWEI DENG • *Department of Materials Science and Engineering, University of California Irvine, Irvine, CA, USA*
ZHIYONG DING • *Mills Institute for Personalized Cancer Care, Fynn Biotechnologies Ltd, Jinan City, Shandong Province, P. R. China*
JACOB J. EDWARDS • *Department of Surgery, University of Michigan, Ann Arbor, MI, USA*
OFER EIDELMAN • *Department of Anatomy, Physiology and Genetics, and Institute for Molecular Medicine, Uniformed Services University School of Medicine (USUHS), Bethesda, MD, USA*

ANDREW EMILI • *Center for Network Systems Biology, Boston University, Boston, MA, USA; Department of Biochemistry, Boston University School of Medicine, Boston, MA, USA; Division of Oncological Sciences, Knight Cancer Institute, Oregon Health & Science University, Portland, OR, USA*

TIMOTHY L. FRANKEL • *Department of Surgery, University of Michigan, Ann Arbor, MI, USA*

SUMANTA GHOSH • *Department of Epigenetics and Molecular Carcinogenesis, The University of Texas MD Anderson Cancer Center, Houston, TX, USA*

ARMANDO E. GIULIANO • *Department of Surgery, Cedars-Sinai Medical Center, West Hollywood, CA, USA; Department of Surgery, Samuel Oschin Comprehensive Cancer Institute, Cedars-Sinai Medical Center, West Hollywood, CA, USA*

RAJGOPAL GOVINDARAJAN • *Division of Pharmaceutics and Pharmacology, The Ohio State University College of Pharmacy, Columbus, OH, USA; Translational Therapeutics, The Ohio State University Comprehensive Cancer Center, Columbus, OH, USA*

BRIAN D. GRIFFITH • *Department of Surgery, University of Michigan, Ann Arbor, MI, USA*

MARCO HADISURYA • *Department of Biochemistry, Purdue University, West Lafayette, IN, USA*

ARTUR HAHN • *Hamburg, Germany*

JERED B. HAUN • *Department of Biomedical Engineering, University of California Irvine, Irvine, CA, USA; Department of Materials Science and Engineering, University of California Irvine, Irvine, CA, USA; Department of Chemical and Biomolecular Engineering, University of California Irvine, Irvine, CA, USA; Chao Family Comprehensive Cancer Center, University of California Irvine, Irvine, CA, USA; Center for Advanced Design and Manufacturing of Integrated Microfluidics, University of California, Irvine, Irvine, CA, USA*

CHRISTIAN HECKENDORF • *Center for Network Systems Biology, Boston University, Boston, MA, USA; Department of Biochemistry, Boston University School of Medicine, Boston, MA, USA*

SAMUEL HENDERSON • *Wallace H. Coulter Department of Biomedical Engineering, Georgia Institute of Technology and Emory University, Atlanta, GA, USA; George W. Woodruff School of Mechanical Engineering, Georgia Institute of Technology, Atlanta, GA, USA*

VANESSA HERRERA • *Department of Biomedical Engineering, University of California Irvine, Irvine, CA, USA*

JACK HESTER • *Department of Anatomy, Physiology and Genetics, and Institute for Molecular Medicine, Uniformed Services University School of Medicine (USUHS), Bethesda, MD, USA*

SACHI HORIBATA • *Precision Health Program, Michigan State University, East Lansing, MI, USA; Department of Pharmacology and Toxicology, College of Human Medicine, Michigan State University, East Lansing, MI, USA*

ANTON ILIUK • *Department of Biochemistry, Purdue University, West Lafayette, IN, USA; Tymora Analytical Operations, West Lafayette, IN, USA*

ERINA IWABUCHI • *Department of Pathology and Histotechnology, Tohoku University Graduate School of Medicine, Sendai, Japan*

HAO JIANG • *Department of Biochemistry and Molecular Genetics, University of Virginia School of Medicine, Charlottesville, VA, USA*

SEUNG-GI JIN • *Department of Epigenetics, Van Andel Institute, Grand Rapids, MI, USA*

JENNIFER JOHNSON • *Department of Epigenetics, Van Andel Institute, Grand Rapids, MI, USA*

CATHERINE JOZWIK • *Department of Anatomy, Physiology and Genetics, and Institute for Molecular Medicine, Uniformed Services University School of Medicine (USUHS), Bethesda, MD, USA*

USHA N. KASID • *Georgetown Lombardi Comprehensive Cancer Center, Georgetown University Medical Center, Washington, USAD.C.*

EVAN T. KELLER • *Biointerfaces Institute, University of Michigan, Ann Arbor, MI, USA; Single Cell Spatial Analysis Program, University of Michigan, Ann Arbor, MI, USA; Department of Urology, School of Medicine, University of Michigan, Ann Arbor, MI, USA; Unit for Laboratory Animal Medicine, University of Michigan, Ann Arbor, MI, USA; Computational Medicine and Bioinformatics, School of Medicine, University of Michigan, Ann Arbor, MI, USA*

JILL M. KELLER • *Biointerfaces Institute, University of Michigan, Ann Arbor, MI, USA; Department of Urology, School of Medicine, University of Michigan, Ann Arbor, MI, USA; Unit for Laboratory Animal Medicine, University of Michigan, Ann Arbor, MI, USA*

ERIC B. KMIEC • *Gene Editing Institute, ChristianaCare Health System, Newark, DE, USA*

ADITI KUMAR • *Wallace H. Coulter Department of Biomedical Engineering, Georgia Institute of Technology and Emory University, Atlanta, GA, USA*

FELIX T. KURZ • *Division of Radiology, German Cancer Research Center, Heidelberg, Germany*

MELISSA A. LARSON • *Transgenic and Gene-Targeting Institutional Facility, The Kansas University Medical Center, Kansas City, KS, USA*

MATTHEW L. LAWTON • *Center for Network Systems Biology, Boston University, Boston, MA, USA; Department of Biochemistry, Boston University School of Medicine, Boston, MA, USA*

DMITRY LESHCHINER • *Precision Health Program, Michigan State University, East Lansing, MI, USA; Department of Pharmacology and Toxicology, College of Human Medicine, Michigan State University, East Lansing, MI, USA*

MARK M. LI • *Livestrong Cancer Institutes, Department of Oncology, Dell Medical School, The University of Texas at Austin, Austin, TX, USA*

WEI LI • *Hunan Provincial Key Laboratory of Medical Virology, College of Biology, Hunan University, Changsha, Hunan, China; Department of Biochemistry and Molecular Genetics, University of Virginia School of Medicine, Charlottesville, VA, USA*

XIA LI • *Mills Institute for Personalized Cancer Care, Fynn Biotechnologies Ltd, Jinan City, Shandong Province, P. R. China*

MICHELLE V. LIHON • *Department of Biochemistry, Purdue University, West Lafayette, IN, USA*

WEIWEI LIN • *Center for Network Systems Biology, Boston University, Boston, MA, USA; Department of Biochemistry, Boston University School of Medicine, Boston, MA, USA*

XIAOJUN MA • *Department of Biomedical Informatics, School of Medicine, University of Pittsburgh, Pittsburgh, PA, USA; UPMC Hillman Cancer Center, University of Pittsburgh, Pittsburgh, PA, USA*

JAKE J. MCGUE • *Department of Surgery, University of Michigan, Ann Arbor, MI, USA*

YASUHIRO MIKI • *Department of Nursing, Faculty of Medical Science and Welfare, Tohoku Bunka Gakuen University, Sendai, Japan*

MICHAEL M. MURATA • *Department of Surgery, Cedars-Sinai Medical Center, West Hollywood, CA, USA*

VEENA Y. NAVEEN • *Department of Biomedical Engineering, University of California Irvine, Irvine, CA, USA*

DEBASIS NAYAK • *Division of Pharmaceutics and Pharmacology, The Ohio State University College of Pharmacy, Columbus, OH, USA*

Hatice Ulku Osmanbeyoglu • *Department of Biomedical Informatics, School of Medicine, University of Pittsburgh, Pittsburgh, PA, USA; UPMC Hillman Cancer Center, University of Pittsburgh, Pittsburgh, PA, USA; Department of Bioengineering, School of Engineering, University of Pittsburgh, Pittsburgh, PA, USA; Department of Biostatistics, University of Pittsburgh, Pittsburgh, PA, USA*

Melissa Ozbeyler • *Wallace H. Coulter Department of Biomedical Engineering, Georgia Institute of Technology and Emory University, Atlanta, GA, USA; George W. Woodruff School of Mechanical Engineering, Georgia Institute of Technology, Atlanta, GA, USA*

Gerd P. Pfeifer • *Department of Epigenetics, Van Andel Institute, Grand Rapids, MI, USA*

Koushul Ramjattun • *Department of Biomedical Informatics, School of Medicine, University of Pittsburgh, Pittsburgh, PA, USA; UPMC Hillman Cancer Center, University of Pittsburgh, Pittsburgh, PA, USA*

Arvind Rao • *Single Cell Spatial Analysis Program, University of Michigan, Ann Arbor, MI, USA; Department of Biostatistics, School of Public Health, University of Michigan, Ann Arbor, MI, USA; Computational Medicine and Bioinformatics, School of Medicine, University of Michigan, Ann Arbor, MI, USA*

Tyler Risom • *Department of Research Pathology, Genentech, Inc., South San Francisco, CA, USA*

Natalia Rivera-Torres • *Gene Editing Institute, ChristianaCare Health System, Newark, DE, USA*

Jennifer Robinson • *Georgetown Lombardi Comprehensive Cancer Center, Georgetown University Medical Center, Washington, USAD.C.*

Sandra Rost • *Department of Research Pathology, Genentech, Inc., South San Francisco, CA, USA*

April Sagan • *Department of Biomedical Informatics, School of Medicine, University of Pittsburgh, Pittsburgh, PA, USA; UPMC Hillman Cancer Center, University of Pittsburgh, Pittsburgh, PA, USA*

Nidhi Sahni • *Department of Epigenetics and Molecular Carcinogenesis, The University of Texas MD Anderson Cancer Center, Houston, TX, USA; Department of Bioinformatics and Computational Biology, The University of Texas MD Anderson Cancer Center, Houston, TX, USA; Quantitative and Computational Biosciences Program, Baylor College of Medicine, Houston, TX, USA*

Lilly Saiontz • *Wallace H. Coulter Department of Biomedical Engineering, Georgia Institute of Technology and Emory University, Atlanta, GA, USA*

Patricia M. Schnepp • *Biointerfaces Institute, University of Michigan, Ann Arbor, MI, USA; Department of Urology, School of Medicine, University of Michigan, Ann Arbor, MI, USA*

Greg Shelley • *Biointerfaces Institute, University of Michigan, Ann Arbor, MI, USA; Department of Urology, School of Medicine, University of Michigan, Ann Arbor, MI, USA*

Gloria M. Sheynkman • *Department of Molecular Physiology and Biological Physics, University of Virginia, Charlottesville, VA, USA; Department of Biochemistry and Molecular Genetics, School of Medicine, University of Virginia, Charlottesville, VA, USA; Center for Public Health Genomics, and UVA Comprehensive Cancer Center, University of Virginia, Charlottesville, VA, USA*

Meera Srivastava • *Department of Anatomy, Physiology and Genetics, and Institute for Molecular Medicine, Uniformed Services University School of Medicine (USUHS), Bethesda, MD, USA*

TAKASHI SUZUKI • *Department of Pathology and Histotechnology, Tohoku University Graduate School of Medicine, Sendai, Japan; Department of Anatomic Pathology, Tohoku University Graduate School of Medicine, Sendai, Japan*

HISASHI TANAKA • *Department of Surgery, Cedars-Sinai Medical Center, West Hollywood, CA, USA; Department of Surgery, Samuel Oschin Comprehensive Cancer Institute, Cedars-Sinai Medical Center, West Hollywood, CA, USA; Department of Surgery and Biomedical Sciences, Samuel Oschin Comprehensive Cancer Institute, Cedars-Sinai Medical Center, West Hollywood, CA, USA*

YAO TANG • *Greenebaum Cancer Center, University of Maryland School of Medicine, Baltimore, MD, USA*

W. ANDY TAO • *Department of Biochemistry, Purdue University, West Lafayette, IN, USA; Department of Chemistry, Purdue University, West Lafayette, IN, USA; Tymora Analytical Operations, West Lafayette, IN, USA; Department of Medicinal Chemistry and Molecular Pharmacology, Purdue University, West Lafayette, IN, USA; Purdue Institute for Cancer Research, Purdue University, West Lafayette, IN, USA*

STEPHANIE THE • *Biointerfaces Institute, University of Michigan, Ann Arbor, MI, USA; Department of Biostatistics, School of Public Health, University of Michigan, Ann Arbor, MI, USA*

CAROLYN J. VIVIAN • *Department of Cancer Biology, The Kansas University Medical Center, Kansas City, KS, USA*

JAY L. VIVIAN • *Transgenic and Gene-Targeting Institutional Facility, The Kansas University Medical Center, Kansas City, KS, USA*

TOMMY V. VO • *Department of Biochemistry and Molecular Biology, College of Human Medicine, Michigan State University, East Lansing, MI, USA*

NAN WANG • *Mills Institute for Personalized Cancer Care, Fynn Biotechnologies Ltd, Jinan City, Shandong Province, P. R. China*

BRENNA WEADICK • *Division of Pharmaceutics and Pharmacology, The Ohio State University College of Pharmacy, Columbus, OH, USA*

DANNY R. WELCH • *Departments of Cancer Biology, Internal Medicine (Hematology/Oncology), Molecular and Integrative Physiology, and Pathology and Laboratory Medicine, The Kansas University Medical Center and The University of Kansas Comprehensive Cancer Center, Kansas City, KS, USA*

XIAOFENG WU • *Department of Chemistry, Purdue University, West Lafayette, IN, USA*

QIAN XU • *Guangdong Provincial Key Laboratory of Infectious Diseases and Molecular Immunopathology, Department of Pathology, Shantou University Medical College, Shantou, China*

MEILING YAN • *Department of Pathology, Union Hospital, Tongji Medical College, Huazhong University of Science and Technology, Wuhan, China*

S. STEPHEN YI • *Livestrong Cancer Institutes, Department of Oncology, Dell Medical School, The University of Texas at Austin, Austin, TX, USA; Oden Institute for Computational Engineering and Sciences (ICES), The University of Texas at Austin, Austin, TX, USA; Department of Biomedical Engineering, Cockrell School of Engineering, The University of Texas at Austin, Austin, TX, USA; Interdisciplinary Life Sciences Graduate Programs (ILSGP), College of Natural Sciences, The University of Texas at Austin, Austin, TX, USA*

JAMES ZIAI • *Department of Research Pathology, Genentech, Inc., South San Francisco, CA, USA*

# Chapter 1

# Introduction: Cancer Systems and Integrative Biology


## Robert Clarke



## Abstract

The insights provided by the holistic approaches of systems and integrative biology offer a means to address the multiple levels of complexity evident in cancer biology. Often focused on the use of large-scale, high-dimensional omics data for discovery in silico, integration with lower-dimensional data and lower-throughput wet laboratory studies allows for the development of a more mechanistic understanding of the control, execution, and operation of complex biological systems. While no single volume can cover all of the advances across this broad and rapidly developing field, we here provide reviews, methods, and detailed protocols for several state-of-the-art approaches to probe cancer biology from an integrative systems perspective. The protocols presented are intended for easy implementation in the laboratory and often offer a clear rationale for their development and application. This introduction provides a very brief description of systems and integrative biology as context for the chapters that follow, with a short overview of each chapter to allow the reader to easily and quickly find those protocols of most interest.



**Key words** Cancer, Data analysis, Integrative biology, Systems biology, Methods


Systems and integrative biology combines, in a holistic manner, data obtained from different technological platforms—often across very high-dimensional omic scale measurements of the genome, transcriptome, proteome, and metabolome—with knowledge of time, space, and mechanism to create computational and mathematical models that explain and/or predict the function(s) of living systems. Modeling and model interpretation require the multidisciplinary integration of critical concepts from computer science, experimental biology and medicine, informatics, and statistics [1]. Thus, systems approaches rarely (if ever) reduce the understanding or interpretation of a system or function to a single gene or small topological feature comprised of several interacting nodes. It is the more widely applied reductionist approaches that focus on single nodes (genes, proteins, metabolites) or relatively small, low-dimensional signaling features or pathways (nodes and the edges that connect them).


Usha N. Kasid and Robert Clarke (eds.), *Cancer Systems and Integrative Biology*, Methods in Molecular Biology, vol. 2660, https://doi.org/10.1007/978-1-0716-3163-8_1,

Major differences between systems and reductionist approaches include the scale and scope of perspective and analysis. Systems analyses place a central emphasis on the integration and coordination of actions across multiple, high-dimensional data spaces (computational modeling) that can often inform or guide modeling lower-dimensional scale data (mechanistic or semi-mechanistic mathematical modeling) to better understand or predict the control and/or execution of either a function within a complex system, or the entire system itself. Indeed, one of the more rapidly developing systems approaches is the iterative integration of computational analyses of high-dimensional data to identify tractable parameter spaces for mathematical modeling of low-dimensional data. Modeling of these lower-dimensional spaces can provide new insights on node-edge activities that can then refine the design or constrain features of the search space of a high-dimensional analysis. Quantitative data on a node, as may be obtained from an in silico model(s) or wet laboratory reductionist experiment(s), can inform its priors in the Bayesian models often applied in high-dimensional data analysis, or its parameter weight in a mathematical model in lower-dimensional analysis.

Valid and insightful, the focused data and mechanistic insights obtained from reductionist approaches often provide the building blocks for modeling, validating, and/or interpreting key aspects of the in silico models built by systems scientists [1, 2]. For example, predictions from an in silico model may require mechanistic validation in wet laboratory experiments, such as when a model predicts that a specific gene is a central hub (a node with multiple edges connecting it with other nodes) that controls aspects of a specific cellular function. To understand how the node operates within the biological system, mechanistic validation of this prediction may require overexpressing and/or knocking down the expression of this node (hub gene) and then measuring how these perturbations affect the target cellular function and perhaps also the expression and/or function of other nodes to which it is connected. These wet laboratory experiments are often designed and interpreted from a reductionist perspective to assist in accomplishing the broader systems modeling goals.

Chapters in this volume address, directly or indirectly, critical steps in integrative systems analyses. The volume includes some reviews, primary observations, and descriptions of tools or workflows for use or adaptation to meet several goals for systems scientists. The wet laboratory experimental methods described include are detailed protocols that can be easily applied or adapted for use; some chapters reflect the authors' experience with a method, along with guides or tips for method implementation or optimization. For in silico tools or workflows, readers will find detailed methods to allow for ease of adoption. High-dimensional data spaces have unique properties and challenges [2, 3]; several of the informatics workflows provided here account for these issues. Beginning with Chapter 2, what follows below is a brief chapter-by-chapter overview to help readers make the most of this volume, which extends and adds substantially to our previously edited volume of cancer systems biology methods focused on gene networks [4].

Large palindromic DNA sequences occur when nucleotide sequences are identical to their reverse complements; such palindromic sequences are often associated with tumorigenesis [5]. In Chapter 2, Murata et al. describe a modified genome-wide analysis of palindrome (GAPF) protocol that includes an experimental approach and bioinformatics tool for isolating and amplifying DNA palindromes. Prior to applying PCR, denaturing and renaturing converts the DNA palindrome into dsDNA that can enrich palindromes with amplification of the target signal while concurrently reducing noise by S1 nuclease digestion. The modified approach allows for the use of low-input DNA concentrations and efficiently presents palindromes without the need for ultra-deep sequencing [5].

Transcriptome analysis has been greatly facilitated by the advent of RNA-seq, an approach that has begun to replace some of the earlier chip-based microarray platforms. The method requires the extraction of high-quality RNA, the preparation of mRNA libraries for next-generation sequencing, and bioinformatics analysis of the resulting sequences that are often compared across different experimental conditions or phenotypes. Leshchiner et al. describe an experimental and bioinformatics pipeline to perform RNA-seq in Chapter 3 [6]. Using RNA extraction and analysis from cancer cell lines as an example, a detailed experimental pipeline is described that includes the preparation of sequencing libraries without the need for either polyA-tail capture or rRNA depletion. The bioinformatics workflow for data analysis describes mapping reads to a reference genome, accurately quantifying individual transcript expression values, and comparing transcript levels among different conditions of phenotypes [6].

Metastasis is a complex process that is a major driver of cancer mortality, yet much about the process and its regulation remains to be discovered [2]. A poorly understood area is the contribution to the metastatic cascade from the mitochondrial genome, a challenge that often reflects a limited choice of appropriate animal models, the multiple roles played by mitochondria, and the dynamical features of mitochondrial biology. In Chapter 4, Welch et al. describe the rationale, development, and uses for a powerful and innovative mitochondrial-nuclear exchange (MNX) mouse model [7]. MNX mice contain the mitochondrial DNA from one strain and the nuclear DNA from another, generated by physically extracting and replacing pronuclei (one-cell embryos) from one mouse strain with those from another. Notably, the process eliminates the need for either back-crossing strains or the use of mutagens. MNX mice facilitate analysis of the differences in mtDNA of mouse strains that exhibit differing intrinsic and extrinsic effects on metastasis. Since other diseases also exhibit differences in mitochondrial function, the MNX approach may have utility well beyond the study of metastasis in cancer [7].

While established cancer cell line models from patient tumors have generated significant insights, studies with patient material xenografted directly into mice (patient-derived xenografts, PDX) or maintained in relatively short-term organoid culture (patient-derived organoids, PDO) have shown their ability to retain important features of the tumors from which they were derived [8]. Xu et al. describe in Chapter 5 a novel adaptation of the PDO method where a three-dimensional autologous culture method uses a patient's own body fluid or serum to craft both a scaffold for organoid growth and to supplement the culture medium for PDO growth [9]. Two models are described where the method supported cells isolated either from pleural effusions or malignant ascites, or from solid tissue biopsies or surgically excised tumors. The 3D autologous culture method (3D-ACM) helps to preserve critical features of the microenvironment from the original tumor and to preserve key biological properties [9]. While this chapter provides methods and protocols, Kumar et al. provide a review of single cell and spatial analysis of organoid platforms in Chapter 22 [10].

A further recent advance in the omics revolution is reflected in the development of spatial omics technologies that map the spatial distribution of protein or RNA expression in cells and tissues. Wang et al. describe their experience with the nanoString Digital Spatial Profiler (DSP) in Chapter 6. An optimized DSP protocol is described that uses next-generation-based barcoding that enables formalin-fixed paraffin-embedded (FFPE) tissues to be subjected to high-plex RNA and protein profiling [11]. Techniques for sample preparation, DSP, and post-collection procedures are described in detail, enabling readers to more effectively initiate a DSP-based workflow for RNA and protein profiling of FFPE specimens.

Single-cell RNA sequencing is one approach to deal with some forms of tissue heterogeneity in cancer specimens [12]. One major systems goal of interpreting single-cell RNA-seq data is the ability to build network models of a system of interest. A powerful approach for integrating single-cell RNA-seq data with network analysis is presented by The et al. in Chapter 7. Based on the existing method called Passing Attributes between Networks for Data Assimilation (PANDA), the authors clearly present a computational analysis pipeline that integrates gene expression data with both gene-coregulation, protein–protein interaction (PPI), and transcription factor binding motif data [13]. Application of the PANDA-based analytical pipeline is presented in the context of studying cancer drug resistance mechanisms.

Mapping spatial protein expression in heterogeneous specimens can also be accomplished using immunohistochemistry (IHC) when appropriate antibodies are available. When faced with multiple protein candidates, individual target protein analysis in single samples is often low throughput and can be time- and resource-demanding to screen a sufficient number of specimens to meet statistical power requirements. Nayak et al. [14] present an effective solution in Chapter 8 that applies the combined use of tissue microarrays (TMA) with multiplexed immunohistochemistry. The method is described in the context of studying nucleoside analog transport mechanisms in drug resistance in pancreatic cancer. The authors present a stepwise approach for performing multiplexed immunohistochemistry, TMA imaging, and the quantitative measurement of select marker expression, accompanied by a discussion of central experimental design considerations [14].

Measuring spatial protein co-expression is a central feature of mapping PPIs. While IHC analysis is useful, inadequate resolution and other limitations can make definitive PPI visualization difficult. In Chapter 9, Miki et al. describe an in situ proximity ligation assay (PLA) that allows for a more direct visualization of PPIs in FFPE or frozen sections and in cultured cells [15]. Primary antibodies to each of the PPI components first bind their antigens, followed by binding to the respective primary antibodies of secondary antibodies each of which bears an oligonucleotide. Treatment with a DNA ligase joins one oligonucleotide on one secondary antibody to another when the probes are within 40 nm of each other. Amplification of the unique ligated reporter strand by PCR is performed such that a fluorescence-labeled probe that recognizes only the amplified product can then bind and enable visualization of the linked proteins. Hence, localization of an active PPI is visualized directly, providing a spatial analysis of proteins in a PPI [15].

Mass spectrometry-based platforms can provide primary data for multiomic proteome and metabolome analyses. However, downstream analysis, as required to map a peptide sequence to the correct protein or a molecular weight or other measured property to the correct metabolite, can represent an analytical bottleneck. In Chapter 10, Heckendorf et al. describe an Omics Notebook open-source framework that supports initial exploratory analysis that can integrate and report MS-based multiomic data [16]. Omics Notebook is customizable and offers an automated and reproducible pipeline for rapid identification and visualization of functional patterns that are statistically significant and biologically interesting. The authors present a protocol for integrative analysis of multiomics data using Omics Notebook that can be adopted by high volume core laboratories and research centers.

Extrinsic signals can be sensed by membrane proteins that then initiate and propagate intrinsic signaling to control or execute specific cellular functions. Single-cell data is often available but can be sparse, with single-cell RNA-seq methods often missing low abundant transcription factor targets of extrinsic signals. Sagan et al., writing in Chapter 11, provide one solution in the form of the computational framework SPaRTAN (Single-cell Proteomic and RNA-based Transcription factor Activity Network) [17]. The novel SPaRTAN framework uses cellular indexing of transcriptome and epitope by sequencing (CITE-seq) data and information on cis-regulatory sites to capture the interactions between cell surface receptors and transcription factors that regulate expression of specific target genes. The analytical pipeline is presented in the context of applying SPaRTAN to CITE-seq data from peripheral blood mononuclear cells to connect cell surface molecules to downstream transcriptional events that control transcriptomic phenotypes [17].

Highly multiplexed antibody staining of FFPE or frozen tissues can be accomplished using multiplex ion beam imaging (MIBI) and imaging mass cytometry (IMC), where metal ions released from primary antibodies are detected by time-of-flight mass spectrometry (ToF). This workflow retains spatial orientation of the primary antibody binding [18]. In Chapter 12, Risom et al. present clear and detailed protocols for the application of this workflow that includes antibody conjugation and validation, staining, and the collection of preliminary data using either MIBI or IMC. The protocol is potentially capable of detecting up to 50 targets and their spatial orientation in tissue sections. The approach is presented in the context of studies of mouse and human pancreatic adenocarcinoma [18].

Analysis of secreted proteins is central to understanding cell–cell communication and to define specific cellular phenotypes—particularly for immune cells. Current immunofluorescence-based approaches have limited detection sensitivity, which can be problematic for single cell-based approaches since cells must secrete thousands of molecules to exceed the detection limit [19]. Naveen et al. have developed a quantum dot-based method that, with near single-molecule resolution, can monitor proteins secreted from single cells. Chapter 13 describes the quantum dot-based single-cell secretion analysis platform and its expansion to include multiplexing capabilities for cytokines, in the context of single-cell macrophage polarization changes in response to different stimuli [19].

Cells also communicate through paracrine and endocrine signaling facilitated by the production of extracellular vesicles (EVs). EVs can be disease biomarkers or provide a means to deliver drugs. Central to the study or use of EVs is the application of appropriate means for their isolation and identification. For example, the isolation and characterization of EVs from human plasma can be used for diagnostic purposes. An effective pipeline for the characterization and evaluation of plasma-derived EVs is described in Chapter 14 by Lihon et al. [20]. The pipeline includes a high recovery EVtrap method for EV isolation, protein extraction by a phase-transfer surfactant method, and both qualitative and quantitative MS strategies for proteome characterization. EVtrap uses a novel magnetic bead-based approach fully compatible with LC–MS strategies for EV proteome analysis. Data-dependent (DDA) and data-independent (DIA) acquisition protocols are presented for proteomic data acquisition and quantification by MS [20].

Posttranslational protein modification is a dynamic process that is difficult to study at the proteomic scale in part because the steady-state measurements recovered reflect the sum of each protein's rates of biosynthesis and degradation. In Chapter 15, Jozwik et al. describe the use of a novel, large-scale, targeted antibody microarray time-resolved platform that can resolve total protein changes and the rates of biosynthesis of low-abundance proteins [21]. The study is presented in the context of exploring the proinflammatory signaling proteome in lung epithelial cells from cystic fibrosis. Feasibility of the approach, which is based on protein labeling with $^{35}S$ methionine and $^{32}P$, is exemplified by the estimated proteomic kinetics of over 500 low-abundance proteins. Hence, the method is sensitive and flexible, permitting identification of alterations in posttranslational modification at a massively parallel scale [21].

Many standard approaches to study the spatial nature of cell interactions in the tumor microenvironment (TME) are often limited to either co-localization measures of a small number of antigens or require the destruction of the TME architecture. A powerful multiplex fluorescent immunohistochemistry approach (mfIHC) for use with FFPE sections is described by McGue et al. in Chapter 16 [22]. The mfIHC approach preserves critical features of the TME architecture in the embedded tissue and can maintain the spatial relationship of cells. Antigen retrieval is used to allow application of primary and secondary antibodies that is then followed by covalent binding of a fluorophore by a tyramide-based reaction. Stripping of the antibodies allows for several rounds of antibody application, largely eliminating species cross-reactivity. Signal amplification is also applied, reducing the autofluorescence often encountered in fixed tissues. The authors carefully discuss experimental design, staining, and imaging strategies for the implementation of mfIHC [23].

DNA damage, whether resulting from endogenous or exogenous stressors, is an early driver of carcinogenesis. Methods to identify DNA damage rarely achieve single base pair resolution level across the entire genome. In Chapter 17, Jin et al. describe a circle DNA sequencing method (CD-seq) that directly addresses this limitation [23]. Following extraction of damaged/modified DNA, DNA is cleaved to produce fragment sizes of ~300–400 bases, sonicated and cleaned to produce blunt ends and then ligated into circles. The modified base in the circle is cleaved, creating a double-strand break at the lesion sites, and the opened rings are ligated to Illumina sequencing linkers and used for library preparation and sequencing using standard methods. CD-seq is applicable to the study of several different types of DNA damage [23].

Developing novel strategies to eliminate drug-resistant cancer cells remains a major goal. In Chapter 18, Rivera–Torres et al. describe a CRISPR-directed gene editing approach to disrupt specific drug resistance genes [24]. The approach is presented in the context of targeting NRF2, a gene that regulates cellular stress response genes, in drug-resistant lung cancer cells. A nonviral delivery (CRISPRMax) is used to introduce CRISPR/Cas for gene disruption. The authors provide a detailed description of the methods and protocols [24].

Finding novel biomarkers and potential targets for cancer drug discovery and development remains a timely and important goal. In the context of racial disparities in pancreatic cancer, Bera et al. describe an integrative workflow for analyzing proteogenomic data and drug targets in Chapter 19 [25]. The authors make good use of publicly accessible data and tools including the cBioPortal genomics platform, TCGA data, the MD Anderson Cell Lines Project (MCLP), and information on specific genes as captured in GeneCards (www.genecards.org). Illustrating the feasibility of their approach, results from application of the workflow are presented for White, African American/Black, and Asian patients [25].

Vascular heterogeneity is a central feature of the TME of many solid tumors, and changes in tumor vasculature can affect critical functions of cancer cells including responses to altered nutrient and drug perfusion [12]. In Chapter 20, Kurz and Hahn present a protocol that uses morphological and topological data to study vascular heterogeneity and its role in therapy response. The protocol was applied to single plane illumination microscopy (SPIM) images of brain tissues where the vessels had been labeled with fluorescent lectin. SPIM images of tissue-cleared mouse brains were visualized and the vascular geometry and network topology modeled in silico to explore changes within the vasculotome [26].

Control of a cancer cell's entry into mitosis can affect the rate of proliferation and responsiveness to mitotic poisons such as the taxanes and Vinca alkaloids. Boudreau et al. explore the role of DEPDC1B (BRCC3, XTP8, XTP1) in controlling entry into mitosis in Chapter 21. The authors propose that DEPDC1B affects AKT1 and ERK signaling and show that expression of DEPDC1B is a strong candidate biomarker for breast, lung, pancreas, and renal cell cancers [27]. These initial studies implicate DEPDC1B as a protein partner of the p856 subunit of PIK3, as a regulator of ERK and AKT cross talk, and as a cell de-adhesion mitotic checkpoint. Moreover, DEPDC1B is regulated by Raf1 and is a direct target of SOX10 and so may be a central promoter of angiogenesis and metastasis [27].

In Chapter 22, Kumar et al. provide a detailed perspective on the development and application of organoids. Starting with organoid emergence and construction, the authors then discuss organoid structural and molecular analysis, application of organoid platforms, challenges, and practical issues and conclude their perspective with a discussion on general use and future prospects [10]. Above in Chapter 5, Xu et al. describe a three-dimensional autologous organoid culture method [9].

Biomolecular condensates are micro-compartments that lack a membrane and can include proteins that can form weak and multivalent intermolecular interactions separate from the solvent phase (liquid–liquid phase separation; LLPS). These condensates are key spatiotemporal regulators of their component parts, and changes in their biomolecular phase separation can cause dysregulation of the cellular processes controlled by condensate proteins. In Chapter 23, Li and Jiang review the role of phase-separated biomolecular condensates in cancer and the methods used for their study [28]. The authors start with a discussion of the analysis phase separation of a protein of interest, expand their discussion into functional analyses in cancer, and then mechanistic studies of how phase-separated biomolecular condensates regulate cancer. In addition to citations to the primary literature, the text notes useful reviews by others to help readers find more expansive discussions of methods and guidelines for LLPS study [28]. The following chapter (Chapter 24) includes a discussion of the role of gain-of-function (GOF) mutations in protein aggregation and LLPS [29].

While loss−/disruption-of-function mutations have received significant attention, in Chapter 24, Li et al. review the role of gain-of-function (GOF) mutations in precision medicine [29]. The authors provide a broad discussion of GOF mutations, how these have been characterized in multiomic studies, and the mechanistic effects of GOF mutations in signaling networks. How more recent advances in bioinformatics and computations tools and resources can aid the study of GOF and the consequences of their acquisition also is discussed. The review begins with a discussion of GOF and continues with their epigenetic regulation and consideration of the role of GOF mutations on transcription factor function, lncRNAs, pseudogenes, PPIs, and posttranslational regulation. While these sections deal primarily with GOFs and cancer, additional sections consider neurodegenerative and inflammatory diseases. Adding to the material presented above in Chapter 23 by Li and Jiang [28] on phase-separated bimolecular condensates in cancer, the chapter here explores how GOFs may affect LLPS function. Finally, the authors describe a series of key computational resources for predicting GOF mutations and end with a brief concluding section and comments on the potential for future studies of GOF mutations [29].

The contents of this volume cover a broad area of topics related to cancer systems and integrative biology. The brief introduction to each chapter provided above is intended primarily as a guide and to help readers quickly and easily find the material of most interest. While it is not possible to cover all areas with the depth evident in the chapters here and in our prior volume on cancer gene network methods [4], it is hoped that readers will find the insights from several chapters relevant and useful in their own cancer systems and integrative biology studies.

## References


1. Clarke R, Tyson JJ, Tan M, Baumann WT, Xuan J, Wang Y (2019) Systems biology: perspectives on multiscale modeling in research on endocrine-related cancers. Endoc Relat Cancer 26:R345–R368
2. Clarke R, Kraikivski P, Jones BC, Sevigny CM, Sengupta S, Wang Y (2020) A systems biology approach to discovering pathway signaling dysregulation in metastasis. Cancer Metastasis Rev 39:903–918
3. Clarke R, Ressom HW, Wang A, Xuan J, Liu MC, Gehan EA, Wang Y (2008) The properties of very high dimensional data spaces: implications for exploring gene and protein expression data. Nat Rev Cancer 8:37–49
4. Clarke R (2016) Introduction – cancer gene networks. In: Kasid UN, Clarke R (eds) Cancer gene networks. Springer, New York, pp 1–9
5. Murata MM, Giuliano AE, Tanaka H (2023) Genome-wide analysis of palindrome formation with next generation sequencing (GAPF-seq) and a bioinformatics pipelline for assessing *de novo* palindromes in cancer genomes. In: Kasid UN, Clarke R (eds) Cancer systems and integrative biology. Springer, New York
6. Leshchiner D, Vo TV, Horibata S (2023) Sample preparation and differential gene expression analysis of human cancer cell lines by RNA sequencing. In: Kasid UN, Clarke R (eds) Cancer systems and integrative biology. Springer, New York
7. Welch DR, Larson MA, Vivian CJ, Vivian JL (2023) Generating mitochondrial-nuclear exchange (MNX) mice to identify mitochondrial determinants of cancer metastasis. In: Kasid UN, Clarke R (eds) Cancer systems and integrative biology. Springer, New York
8. Clarke R, Jones BC, Sevigny CM, Hilakivi-Clarke LA, Sengupta S (2021) Experimental models of endocrine responsive breast cancer: strengths and limitations. Cancer Drug Resist 4:762–783
9. Xu Q, Yan ML, Tang Y (2023) 3-dimensional autologous culture method for precision oncology. In: Kasid UN, Clarke R (eds) Cancer systems and integrative biology. Springer, New York
10. Kumar A, Cai S, Allam M, Henderson S, Ozbeyler M, Saiontz L, Coskun AF (2023) Single cell spatial analysis of emergent organoid platforms. In: Kasid UN, Clarke R (eds) Cancer systems and integrative biology. Springer, New York
11. Wang N, Li X, Ding Z (2023) High-plex spatial profiling of RNA and protein using digital spacial profiler. In: Kasid UN, Clarke R (eds) Cancer systems and integrative biology. Springer, New York
12. Andrade de Oliviera K, Sengupta S, Yadav A, Clarke R (2023) The complex nature of heterogeneity and its roles in breast cancer biology and therapeutic responsiveness. Front Endocrinol. (in press)
13. The S, Schnepp PM, Shelley G, Keller JM, Rao A, Keller ET (2023) Integration of single cell RNA-sequencing and network analysis to investigate mechanisms of drug resistance. In: Kasid UN, Clarke R (eds) Cancer systems and integrative biology. Springer, New York
14. Nayak D, Weadick B, Govindarajan R (2023) Combination of tissue microarray profiling and multiplexed immunohistochemistry approaches to investigate transport mechanisms of nucleoside analog resistance. In: Kasid UN, Clarke R (eds) Cancer systems and integrative biology. Springer, New York
15. Miki Y, Iwabuchi E, Suzuki T (2023) *In situ* proximity ligation assay to visualize protein-protein interactions in tumor specimens. In: Kasid UN, Clarke R (eds) Cancer systems and integrative biology. Springer, New York
16. Heckendorf C, Blum BJ, Lin W, Lawton ML, Emili A (2023) Integration of metabolomic and proteomic data to uncover actionable metabolic pathways. In: Kasid UN, Clarke R (eds)

Cancer systems and integrative biology. Springer, New York

17. Sagan A, Ma X, Ramjattun K, Osmanbeyoglu HU (2023) Linking expression of cell-surface receptors with transcription factors by computational analysis of paired single-cell proteomes and transcriptomes. In: Kasid UN, Clarke R (eds) Cancer systems and integrative biology. Springer, New York
18. Risom T, Chang P, Rost S, Ziai J (2023) Mass spectrometry based tissue imaging of the tumor microenvironment. In: Kasid UN, Clarke R (eds) Cancer systems and integrative biology. Springer, New York
19. Naveen VY, Deng T, Herrera V, Haun JB (2023) Multiplexed immunoassay using quantum dots to monitor proteins secreted from single cells at near-single molecule resolution. In: Kasid UN, Clarke R (eds) Cancer systems and integrative biology. Springer, New York
20. Lihon MV, Hadisurya M, Wu X, Iliuk A, Tao WA (2023) Isolation and identification of plasma extracellular vesicles protein biomarkers. In: Kasid UN, Clarke R (eds) Cancer systems and integrative biology. Springer, New York
21. Jozwik C, Eidleman O, Srivastava M (2023) Discovery of a hidden proinflammatory signaling proteome using large scale targeted antibody microarray platform. In: Kasid UN, Clarke R (eds) Cancer systems and integrative biology. Springer, New York
22. McGue JJ, Edwards JJ, Griffith BD, Frankel TL (2023) Multiplex fluorescent immunohistochemistry for preservation of tumor microenvironment architecture and spatial relationship of cells in tumor tissues. In: Kasid UN, Clarke R (eds) Cancer systems and integrative biology. Springer, New York
23. Jin S-G, Johnson J, Pfeifer GP (2023) Circle damage sequencing for whole-genome analysis of DNA damage. In: Kasid UN, Clarke R (eds) Cancer systems and integrative biology. Springer, New York
24. Rivera-Torres N, Bialik P, Kmiec EB (2023) CRISPR-directed gene editing as a method to induce chemoresistance in lung cancer cells. In: Kasid UN, Clarke R (eds) Cancer systems and integrative biology. Springer, New York
25. Bera A, Chatterjee D, Hester J, Srivastava M (2023) Integrated *in silico* analysis of proteogenomic and drug targets for pancreatic cancer survival. In: Kasid UN, Clarke R (eds) Cancer systems and integrative biology. Springer, New York
26. Kurz FT, Hahn A (2023) Advanced computational methods to evaluate vascular heterogeneity in tumor tissue based on single plane Ilumination microscopy. In: Kasid UN, Clarke R (eds) Cancer systems and integrative biology. Springer, New York
27. Boudreau HE, Robinson J, Kasid UN (2023) Illuminating DEPDC1B in multi-pronged regualtion of tumor progression. In: Kasid UN, Clarke R (eds) Cancer systems and integrative biology. Springer, New York
28. Li W, Jiang H (2023) Analysis of phase-separated biomolecualr condensates in cancer. In: Kasid UN, Clarke R (eds) Cancer systems and integrative biology. Springer, New York
29. Li MM, Awasthi S, Ghosh S, Sahni N, Yi SS (2023) Gain-of-function variomics and multi-omics network biology for precision medicine. In: Kasid UN, Clarke R (eds) Cancer systems and integrative biology. Springer, New York

# Chapter 2

# Genome-Wide Analysis of Palindrome Formation with Next-Generation Sequencing (GAPF-Seq) and a Bioinformatics Pipeline for Assessing De Novo Palindromes in Cancer Genomes

**Michael M. Murata, Armando E. Giuliano, and Hisashi Tanaka**

## Abstract

DNA palindromes are a type of chromosomal aberration that appears frequently during tumorigenesis. They are characterized by sequences of nucleotides that are identical to their reverse complements and often arise due to illegitimate repair of DNA double-strand breaks, fusion of telomeres, or stalled replication forks, all of which are common adverse early events in cancer. Here, we describe the protocol for enriching palindromes from genomic DNA sources with low-input DNA amounts and detail a bioinformatics tool for assessing the enrichment and location of de novo palindrome formation from low-coverage whole-genome sequencing data.

**Key words** DNA palindromes, Fold-back inversions, Inverted repeats, Breakage–fusion–bridge (BFB) cycles, Large chromosomal aberrations, Genomic amplification, Genomic instability, Next-generation sequencing, Bioinformatics

## 1 Introduction

Large structural rearrangements of chromosomes are common in human cancers, and they often enable gene amplification and subsequent tumor progression [1–3]. This type of genomic instability is described by breakage–fusion–bridge (BFB) cycles in which broken telomeres or other regions of DNA allow chromosome fusion and the formation of dicentric chromosomes that break unevenly during anaphase (Fig. 1) [4–8]. Several cycles of this breakage–fusion–bridge create large palindromic sequences in the DNA, also called fold-back inversions, as well as copy number gains and losses [6, 7, 9]. A microarray-based approach for the genome-wide analysis of palindrome formation called GAPF was developed in 2005 to assess palindrome-containing regions and to identify

Usha N. Kasid and Robert Clarke (eds.), *Cancer Systems and Integrative Biology*, Methods in Molecular Biology, vol. 2660,
https://doi.org/10.1007/978-1-0716-3163-8_2,

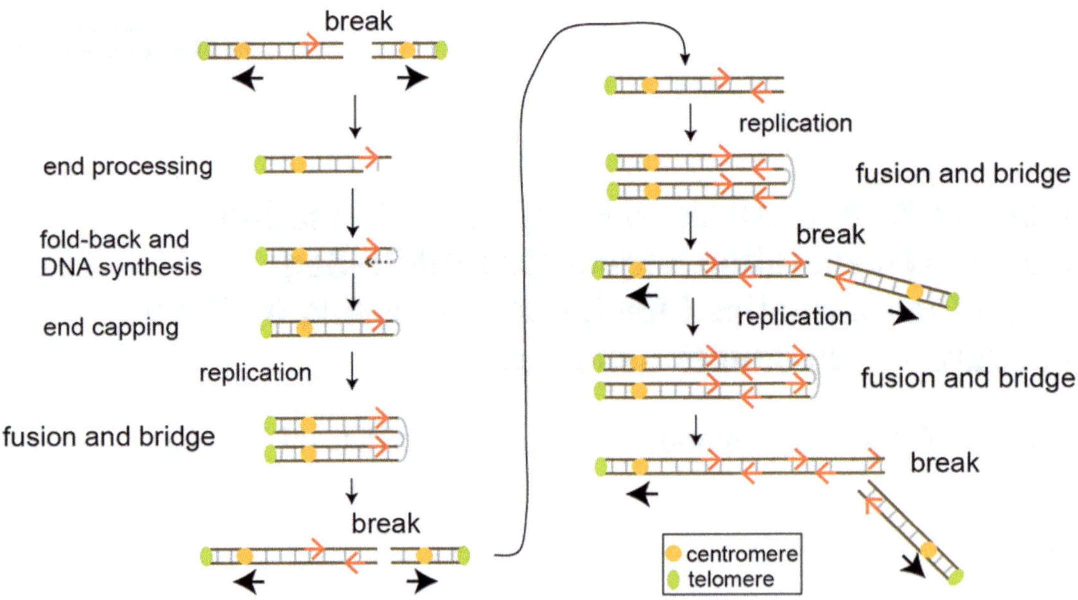


**Fig. 1** Breakage–fusion–bridge (BFB) cycles. Double-stranded DNA breaks can create DNA palindromes via BFB cycles. Subsequent BFB cycles can create copy number variation. These processes are common in human cancers and facilitate genome instability. (Modified from [9])

potential loci for subsequent gene amplification [10–12]. Briefly, input genomic DNA is digested with KpnI or SbfI, two relatively infrequently cutting restriction enzymes. The reaction products are then combined, and the mixture is boiled to denature double-stranded DNA (dsDNA). The solution is immediately quenched in ice water to rapidly anneal palindromic sequences, which are physically tethered together and can easily align complimentary sequences, whereas non-palindromic sequences are disassociated during boiling and remain as single-stranded DNA. The DNA solution is digested using the single-strand specific nuclease S1 to isolate the DNA palindromes. The remaining dsDNA is used to create libraries for next-generation sequencing (Fig. 2). This strategy was modified in 2014 for high-throughput sequencing to pinpoint palindromic junctions [4, 13].

Here, we describe a modified GAPF protocol for isolating and amplifying DNA palindromes from genomic DNA sources with low-input DNA amounts and a bioinformatics pipeline for assessing the enrichment and location of de novo palindrome formation (Fig. 3). Native DNA palindromes typically represent a structural challenge for genomic studies because the Taq polymerase involved in PCR and library construction for whole-genome sequencing cannot navigate the secondary structure of self-annealed palindromes [14]. Therefore, these technologies may underestimate palindromes and fold-back inversions. With GAPF, the denaturing and renaturing step prior to any PCR steps converts the DNA

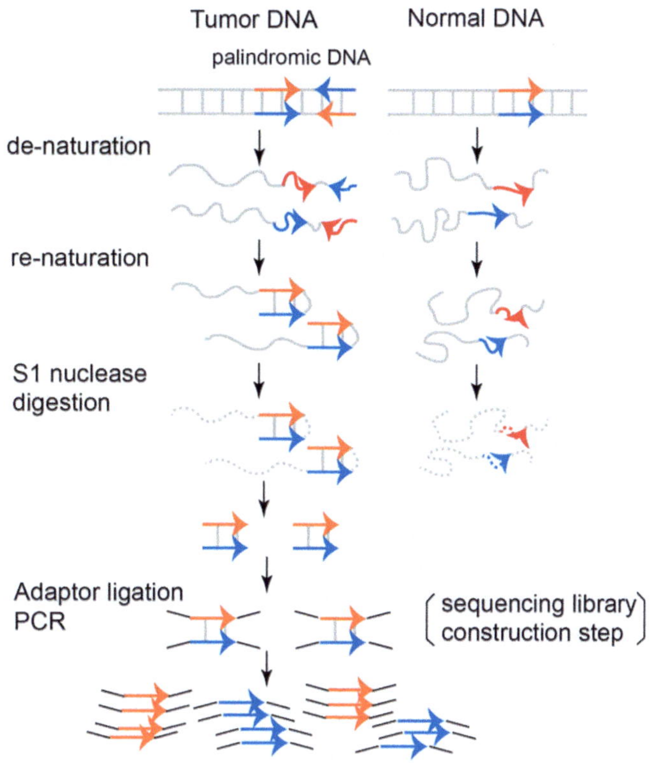


**Fig. 2** Principles of genome-wide analysis of palindrome formation (GAPF). DNA containing palindromic sequences can be denatured and rapidly renatured via intrastrand self-annealing of complimentary sequences that share a DNA backbone to form double-stranded DNA. Non-palindromic DNA sequences remain single-stranded and are digested by a single-strand specific nuclease. The double-stranded DNA can be used to create DNA libraries for sequencing during which palindromic signal is amplified by PCR

palindrome into dsDNA amenable to amplification by PCR. Furthermore, this procedure for enriching palindromes confers the advantages of simultaneously amplifying target signal (via PCR) and reducing background noise (via S1 nuclease digestion) without targeted analysis and, thus, can efficiently present palindromes in sequencing data without ultra-deep sequencing [15].

## 2 Materials

### 2.1 GAPF

1. SbfI-HF and CutSmart buffer (New England Biolabs, R3642S).
2. KpnI-HF and NEBuffer 1.1 (New England Biolabs, R3142S).
3. S1 nuclease and buffer (Invitrogen, 18001016).

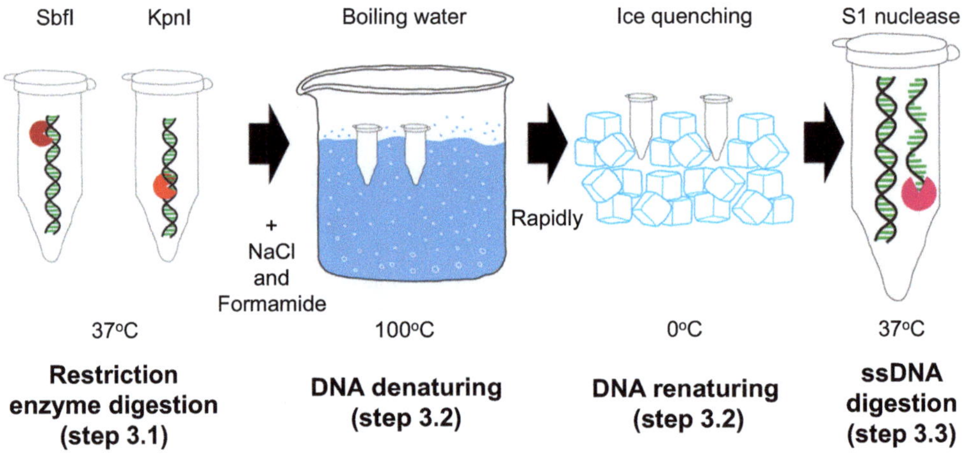


**Fig. 3** Laboratory procedure for GAPF. Genomic DNA is digested by infrequently cutting restriction enzymes SbfI or KpnI. With the addition of NaCl and formamide to help facilitate denaturing, the DNA solutions are mixed and boiled. Then, the DNA solution is rapidly quenched in ice water to renature DNA palindromes into double-stranded DNA. Single-strand specific S1 nuclease is added to digest any non-palindromic DNA that remains single-stranded

4. Monarch PCR and DNA Cleanup Kit (New England Biolabs, T1030S).
5. Formamide.
6. 5 M NaCl.
7. Nuclease-free $H_2O$.
8. Microcentrifuge tubes.
9. Thin-wall microcentrifuge tubes.
10. Microcentrifuge tube cap locks.

### *2.2 Library Construction*

1. NEBNext Ultra II FS DNA Library Prep Kit for Illumina (New England Biolabs, E7805S).
2. AMPure XP Beads (Beckman Coulter Inc., A63881).
3. 10 mM Tris–HCl, pH 7.5–8.0 with 10 mM NaCl (for adapter dilution).
4. 80% ethanol (freshly prepared).
5. 1× TE: 10 mM Tris–HCl, pH 8.0, 1 mM EDTA.
6. 0.1× TE: 1:10 dilution of 1× TE in water.
7. Qubit Assay Kit (Invitrogen, Q32851).
8. Thin-wall PCR tube strips.
9. Magnetic stand/rack.

## 3 Methods

Prepare all solutions using analytical grade reagents, and store them at room temperature unless indicated otherwise. Carry out all procedures at room temperature unless specified otherwise. Follow waste disposal regulations when disposing waste materials.

### 3.1 DNA Fragmentation (Restriction Enzyme Digestion)

1. Mix 30–1000 ng of DNA with nuclease-free $H_2O$ to a total volume of 34 μL in a 1.7 mL microcentrifuge tube (*see* **Note 1**).
2. In a new 1.7 mL microcentrifuge tube, mix 17 μL of the DNA solution with 1 μL KpnI (10 U) and 2 μL 10× NEBuffer 1.1 for a total volume of 20 μL (*see* **Note 2**) (Fig. 3).
3. In a new 1.7 mL microcentrifuge tube, mix the remaining 17 μL of the DNA solution with 1 μL SbfI (10 U) and 2 μL CutSmart buffer for a total volume of 20 μL (*see* **Note 3**).
4. Incubate at 37 °C in a water bath overnight (>16 h).
5. Briefly spin in a microcentrifuge to bring the liquid to the bottom.
6. Heat at 65 °C for 20 min to inactivate restriction enzymes (*see* **Note 4**).

### 3.2 Snap-Back

1. Briefly spin in a microcentrifuge to bring the liquid to the bottom.
2. Mix the 20 μL of KpnI-digested DNA and 20 μL of SbfI-digested DNA with 1.8 μL 5 M NaCl, 45 μL formamide, and 3.2 μL nuclease-free $H_2O$ in a thin-wall PCR tube (*see* **Note 5**).
3. Apply a cap lock to prevent the tube from opening during DNA denaturing.
4. Heat the DNA mixture in boiling water for 7 min to denature DNA (*see* **Note 6**).
5. Immediately quench the DNA mixture in ice water for 5 min to rapidly renature DNA (*see* **Note 7**).

### 3.3 S1 Digestion

1. Briefly spin in a microcentrifuge to bring the liquid to the bottom.
2. Add 4.8 μL 5 M NaCl, 12 μL 10× S1 nuclease buffer, 2 μL S1 nuclease (20 U/μL), and 11.2 μL nuclease-free $H_2O$ to the DNA mixture (*see* **Note 8**).
3. Incubate at 37 °C in a water bath for 1 h.

### *3.4 Purify DNA (Monarch PCR and DNA Cleanup Kit)*

1. Centrifugation should be carried out at 16,000 × g (~13,000 rpm) at room temperature.
2. Add 240 μL DNA Cleanup Binding Buffer to the S1 digested-DNA sample.
3. Mix well by pipetting ten times.
4. Briefly spin in a microcentrifuge to bring the liquid to the bottom.
5. Move liquid to a column, insert column into a 2 mL collection tube, and close the cap.
6. Centrifuge for 1 min and then discard the flow-through.
7. Add 200 μL DNA Wash Buffer, centrifuge for 1 min, and then discard the flow-through.
8. Repeat **step** 7 once.
9. Insert the empty column into the collection tube and centrifuge for 1 min.
10. Transfer the column to a new collection tube.
11. Add 15 μL DNA Elution Buffer and incubate for 1 min at room temperature.
12. Centrifuge for 1 min.
13. Add 10 μL DNA Elution Buffer and incubate for 1 min at room temperature.
14. Centrifuge for 1 min and save the sample (*see* **Note 9**).

### *3.5 Library Construction (NEBNext Ultra II FS DNA Library Prep Kit for Illumina)*

1. Mix 22 μL of DNA, 4 μL nuclease-free $H_2O$, 7 μL NEBNext Ultra II FS Reaction Buffer and 2 μL NEBNext Ultra II FS Enzyme Mix in a PCR tube.
2. Vortex reaction for 5 s, and briefly spin in a centrifuge to bring the liquid to the bottom.
3. In a thermocycler with the lid heated to 75 °C, incubate the reaction for 15 min at 37 °C followed by 30 min at 65 °C and then hold at 4 °C.
4. Add to the reaction mixture 1 μL Ligation Enhancer, 2.5 μL diluted NEBNext Adaptor, and 30 μL Ligation Master Mix.
5. Mix well by pipetting ten times set to 50 μL, and briefly spin in a microcentrifuge to bring the liquid to the bottom.
6. In a thermocycler with no heated lid, incubate the reaction for 15 min at 20 °C, and then hold at 4 °C.
7. Add to the reaction mixture 3 μL USER Enzyme.
8. Mix well by pipetting ten times set to 50 μL, and briefly spin in a microcentrifuge to bring the liquid to the bottom.
9. In a thermocycler with the lid heated to at least 47 °C, incubate the reaction for 15 min at 37 °C, and then hold at 4 °C.

10. Vortex magnetic beads.
11. Add 57 μL magnetic beads to adaptor-ligated DNA.
12. Incubate at room temperature for 5 min.
13. Place magnetic bead DNA mixture on magnet for 5 min.
14. Remove supernatant.
15. On magnet, add 200 μL 80% ethanol, wait for 30 s, and then remove supernatant.
16. Repeat **step 15** once.
17. Air-dry the magnetic beads for 3 min.
18. Off magnet, add 17 μL 0.1× low TE buffer.
19. Mix well by pipetting ten times.
20. Incubate at room temperature for 5 min.
21. Place magnetic bead DNA mixture on magnet for 5 min.
22. Remove 15 μL supernatant and put into a new PCR tube.
23. Add 5 μL Universal PCR Primer, 5 μL Index Primer, and 25 μL NEBNext Q5 Master Mix.
24. Mix well by pipetting ten times set to 40 μL, and briefly spin in a microcentrifuge to bring the liquid to the bottom.
25. In a thermocycler with the lid heated to at least 103 °C, incubate the reaction for 30 s at 98 °C followed by 20 cycles of 10 s at 98 °C and 75 s at 65 °C, then 5 min at 65 °C, and hold at 4 °C.
26. Vortex magnetic beads.
27. Add 45 μL magnetic beads to adaptor-ligated DNA.
28. Incubate at room temperature for 5 min.
29. Place magnetic bead DNA mixture on magnet for 5 min.
30. Remove supernatant.
31. On magnet, add 200 μL 80% ethanol, wait for 30 s, and then remove supernatant.
32. Repeat **step 31** once.
33. Air-dry the magnetic beads for 3 min.
34. Off magnet, add 33 μL 0.1× low TE buffer.
35. Mix well by pipetting ten times.
36. Incubate at room temperature for 5 min.
37. Place magnetic bead DNA mixture on magnet for 5 min.
38. Remove 30 μL supernatant and store in a DNA LoBind tube.
39. Measure concentration of DNA with High Sensitivity Qubit Fluorometer for dsDNA using 2 μL of sample.
40. Check size distribution with Agilent Bioanalyzer High Sensitivity DNA chip.

41. Sequence samples using an Illumina-based sequencing platform with low sequencing depth (0.5–1.0× coverage is sufficient).

### 3.6 Data Analysis

1. Trim raw *.fastq data with Trim_galore (v0.6.1) and Cutadapt (v2.3) with parameters "--length 55."
2. Align trimmed *.fastq data to hg38 reference genome using Bowtie2 (v2.3.5) with unpaired alignment.
3. Convert *.sam alignment file using Samtools (v1.9) to binary format and sort the subsequent *.bam files.
4. Filter uniquely mapped reads by applying a mapping quality filter of 40 using the "samtools view" command with parameters "-b -q 40" (*see* **Note 10**).
5. Extract the number of sequencing reads after applying the mapped quality filter to determine the per-million scaling factor to normalize for mapping depth (*see* **Note 11**).
6. Sort *.bam file using Samtools and convert to *.bed format using Bedtools (v2.28.0).
7. Sort *.bed files using the "sort" command with parameters "-k1,1 -k2,2n."
8. Use Bedtools2 to take an alignment of reads as input and generate a coverage track as output in 1 kb nonoverlapping bins with parameters "-sorted -counts."
9. Use the scaling factor to normalize the coverage in 1 kb bins for the mapping depth.
10. Locate regions of high coverage bins to identify de novo DNA palindromes (*see* **Note 12**).

## 4 Notes

1. This protocol has been optimized to efficiently enrich palindromes from low-input DNA sources.
2. Digestion by restriction enzymes is necessary to cut DNA into fragments that can be effectively denatured by boiling in later steps. It is recommended to create master mixes of the restriction enzyme solution when handling multiple samples.
3. In order to capture large DNA palindromes (~5 kb), DNA needs to be cut infrequently by KpnI or SbfI, and so, GAPF performs best when these restriction enzymes are used separately.
4. During heat inactivation, begin heating water so that it is boiling by the time the heat inactivation step is completed.

This minimizes the amount of time that DNA digested by KpnI or SbfI is exposed to the other restriction enzyme when these solutions are later combined.

5. Minimize the amount of time that KpnI and SbfI are combined prior to boiling. It is recommended to create a master mix of the snap-back solution when handling multiple samples. Formamide and NaCl are necessary additions in order to facilitate efficient DNA denaturing.
6. The timing is critical for denaturing and renaturing the DNA. Boiling for too long starts to degrade DNA, while too short does not efficiently denature dsDNA into ssDNA.
7. It is best to prepare the ice water early (e.g., during heat inactivation of the restriction enzymes) so that the water can be cooled by the ice. Occasional stirring of the ice water during quenching will also prevent local increases of temperature.
8. It is recommended to create a master mix of the nuclease solution when handling multiple samples.
9. 2 or 3 μL of sample can be used for measuring sample concentration by a Qubit 3.0 Fluorometer to assess the overall depletion of DNA by comparing the remaining DNA to the input DNA prior to GAPF.
10. The hg38 human reference genome contains palindromic sequences that will be amplified by this procedure [16]. Because alignment software will attempt to find a single point of origin, reads that can align to either arm of the palindrome will have a low mapping quality. To detect de novo palindrome formation in tumor samples, palindromes found in the reference genome can be removed using a filter for uniquely mapped reads.
11. The per-million scaling factor is calculated by dividing the total number of reads in the file by 1,000,000.
12. The threshold for what is considered "high coverage" can change depending on how efficiently GAPF enriched DNA palindromes. After the per-million scaling factor, the average coverage in 1 kb bins is approximately 0.3, so an appropriate threshold may be between 1.0 and 5.0 depending on the background signal in single-copy regions of the genome.

## Acknowledgments

This work is supported by the National Cancer Institute (2 R01 CA149385), Department of Defense (W81XWH-18-1-0058), Cedars-Sinai Medical Center (to H.T.), and the Margie and Robert E. Petersen Foundation (to A.E.G.).

## References


1. Li Y, Roberts ND, Wala JA, Shapira O, Schumacher SE, Kumar K, Khurana E, Waszak S, Korbel JO, Haber JE, Imielinski M, Akdemir KC, Alvarez EG, Baez-Ortega A, Beroukhim R, Boutros PC, Bowtell DDI, Brors B, Burns KH, Campbell PJ, Chan K, Chen K, Cortés-Ciriano I, Dueso-Barroso A, Dunford AJ, Edwards PA, Estivill X, Etemadmoghadam D, Feuerbach L, Fink JL, Frenkel-Morgenstern M, Garsed DW, Gerstein M, Gordenin DA, Haan D, Haber JE, Hess JM, Hutter B, Imielinski M, Jones DTW, Ju YS, Kazanov MD, Klimczak LJ, Koh Y, Korbel JO, Kumar K, Lee EA, Lee JJK, Li Y, Lynch AG, Macintyre G, Markowetz F, Martincorena I, Martinez-Fundichely A, Meyerson M, Miyano S, Nakagawa H, Navarro FCP, Ossowski S, Park PJ, Pearson JV, Puiggròs M, Rippe K, Roberts ND, Roberts SA, Rodriguez-Martin B, Schumacher SE, Scully R, Shackleton M, Sidiropoulos N, Sieverling L, Stewart C, Torrents D, Tubio JMC, Villasante I, Waddell N, Wala JA, Weischenfeldt J, Yang L, Yao X, Yoon SS, Zamora J, Zhang CZ, Weischenfeldt J, Beroukhim R, Campbell PJ, P.S.V.W. Group, and P. Consortium (2020) Patterns of somatic structural variation in human cancer genomes. Nature 578(7793): 112–121
2. Hellman A, Zlotorynski E, Scherer SW, Cheung J, Vincent JB, Smith DI, Trakhtenbrot L, Kerem B (2002) A role for common fragile site induction in amplification of human oncogenes. Cancer Cell 1(1):89–97
3. Tanaka H, Watanabe T (2020) Mechanisms underlying recurrent genomic amplification in human cancers. Trends Cancer 6(6):462–477
4. Marotta M, Chen X, Watanabe T, Faber PW, Diede SJ, Tapscott S, Tubbs R, Kondratova A, Stephens R, Tanaka H (2013) Homology-mediated end-capping as a primary step of sister chromatid fusion in the breakage-fusion-bridge cycles. Nucleic Acids Res 41(21): 9732–9740
5. Rudolph KL, Millard M, Bosenberg MW, DePinho RA (2001) Telomere dysfunction and evolution of intestinal carcinoma in mice and humans. Nat Genet 28(2):155–159
6. Coquelle A, Pipiras E, Toledo F, Buttin G, Debatisse M (1997) Expression of fragile sites triggers intrachromosomal mammalian gene amplification and sets boundaries to early amplicons. Cell 89(2):215–225
7. Tanaka H, Tapscott SJ, Trask BJ, Yao MC (2002) Short inverted repeats initiate gene amplification through the formation of a large DNA palindrome in mammalian cells. Proc Natl Acad Sci U S A 99(13):8772–8777
8. Maser RS, DePinho RA (2002) Connecting chromosomes, crisis, and cancer. Science 297(5581):565–569
9. Marotta M, Onodera T, Johnson J, Budd GT, Watanabe T, Cui X, Giuliano AE, Niida A, Tanaka H (2017) Palindromic amplification of the ERBB2 oncogene in primary HER2-positive breast tumors. Sci Rep 7:41921
10. Tanaka H, Bergstrom DA, Yao MC, Tapscott SJ (2005) Widespread and nonrandom distribution of DNA palindromes in cancer cells provides a structural platform for subsequent gene amplification. Nat Genet 37(3):320–327
11. Guenthoer J, Diede SJ, Tanaka H, Chai X, Hsu L, Tapscott SJ, Porter PL (2012) Assessment of palindromes as platforms for DNA amplification in breast cancer. Genome Res 22(2):232–245
12. Diede SJ, Guenthoer J, Geng LN, Mahoney SE, Marotta M, Olson JM, Tanaka H, Sj T (2010) DNA methylation of developmental genes in pediatric medulloblastomas identified by denaturation analysis of methylation differences. Proc Natl Acad Sci U S A 107(1): 234–239
13. Yang H, Volfovsky N, Rattray A, Chen X, Tanaka H, Strathern J (2014) GAP-Seq: a method for identification of DNA palindromes. BMC Genomics 15:394
14. Rattray AJ (2004) A method for cloning and sequencing long palindromic DNA junctions. Nucleic Acids Res 32(19):e155
15. Papageorgiou L, Eleni P, Raftopoulou S, Mantaiou M, Megalooikonomou V, Vlachakis D (2018) Genomic big data hitting the storage bottleneck. EMBnet J 24:e910
16. Warburton PE, Giordano J, Cheung F, Gelfand Y, Benson G (2004) Inverted repeat structure of the human genome: the X-chromosome contains a preponderance of large, highly homologous inverted repeats that contain testes genes. Genome Res 14(10A):1861–1869

# Chapter 3

# Sample Preparation and Differential Gene Expression Analysis of Human Cancer Cell Lines by RNA Sequencing

**Dmitry Leshchiner, Tommy V. Vo, and Sachi Horibata**

## Abstract

RNA sequencing (RNA-seq) is a method used for the high-throughput quantification of mRNA in a biological sample. It is widely used to investigate differential gene expression between drug-resistant and sensitive cancers to identify genetic mediators of drug resistance. Here, we describe a comprehensive experimental and bioinformatic pipeline to isolate mRNA from human cell lines, prepare mRNA libraries for next-generation sequencing, and perform post-sequencing bioinformatics analyses.

**Key words** RNA-sequencing, mRNA, Gene expression, Transcriptomics

## 1 Introduction

Regulation of RNA levels is a fundamental way by which all cells modulate gene expression. Traditionally, approaches such as Northern blotting and quantitative PCR have been widely used to measure target RNA levels and to perform comparative studies [1, 2]. The early use of DNA microarrays enabled these types of analyses on the transcriptome-wide scale, albeit with low resolution [3]. More recent developments in second- and third-generation sequencing have enabled unprecedented high-resolution studies into RNA regulation [4]. This revolution led to the advent of RNA sequencing (RNA-seq) methods [5]. The use of RNA-seq on cancer cells, cancer models, and patient samples have dramatically improved our understanding of the transcriptomic landscapes of cancers [6] and of the evolution of cancer drug resistance mechanisms [7]. Moreover, the recent applications of single cell RNA-seq [8] and spatial transcriptomics [9] have further enhanced our understanding of cancer drug resistance mechanisms [10].

RNA-seq often uses pools of high-quality RNA that are either enriched for polyA-tailed mRNAs or depleted for the highly abundant ribosomal RNAs (rRNAs). Nevertheless, both approaches

Usha N. Kasid and Robert Clarke (eds.), *Cancer Systems and Integrative Biology*, Methods in Molecular Biology, vol. 2660, https://doi.org/10.1007/978-1-0716-3163-8_3,

suffer from known challenges including low efficiency in the capture of mRNAs with short polyA tails (by the polyA tail enrichment approach) [11] or very low concentrations of remaining RNAs after rRNA depletion [12]. Moreover, downstream bioinformatic analysis procedures of RNA-seq data can vary greatly between end users depending on the packages used [13, 14], making it difficult for new researchers to know how to perform their studies. Here, we demonstrate a detailed pipeline that can be used to isolate high-quality RNA from human cancer cells, prepare sequencing libraries without preparation of polyA-tail capture or rRNA depletion, and perform typical downstream bioinformatic analyses of RNA-seq data [15]. Additionally, we will provide a general protocol addressing three main bioinformatic steps for many RNA-seq analysis: (1) mapping reads to the reference genome, (2) quantifying transcript levels for each gene, and (3) finding genes that are differentially expressed between different conditions such as drug-treated and drug-untreated cells or drug-sensitive and drug-resistant cancer cells.

## 2 Materials

### 2.1 RNA Preparation

1. Cell plates.
2. Media for cell culture.
3. PBS.
4. RNAseZap.
5. Cell lines.
6. 1.5 mL nuclease-free tubes.
7. Ice bucket.
8. 100% ethanol.
9. QIAGEN RNeasy Mini Kit.
10. DeNovix or NanoDrop.
11. Bioanalyzer system.
12. Optional: RNase-Free DNase Set Kit.

### 2.2 RNA-Seq Library Preparation

1. 0.2 mL nuclease-free PCR individual tubes or tube strips.
2. 1.5 mL nuclease-free tubes.
3. Molecular-grade, nuclease-free water.
4. PCR machine.
5. QIAseq FastSelect -rRNA HMR kit.
6. NEBNext Ultra II Directional RNA Library Prep Kit for Illumina.
7. AMPure XP beads (Beckman Coulter) (*see* **Note 1**).

8. Magnetic stand for 1.5 mL tubes.
9. 80% ethanol, freshly prepared within 24 h prior to use.
10. 0.1× TE, pH 7.4.
11. NEBNext Adaptor (*see* **Note 2**).
12. NEBNext Adaptor Dilution Buffer (10 mM NaCl, 10 mM Tris–HCl pH 7.5).
13. USER enzyme (NEB).
14. NEBNext Index primers (*see* **Notes 2** and **3**).
15. Qubit dsDNA high sensitivity kit.
16. Qubit 4 fluorometer.
17. Agilent TapeStation or Bioanalyzer system.

### 2.3 Bioinformatic Software and Tools

1. FastQC.
2. CutAdapt.
3. A Linux/Unix-based machine with a at least 30 GB of RAM and 50 GB of disk space (*see* **Note 4**).
4. Aligner software such as STAR [16], TopHat2 [17], RSEM [18], Salmon [19], or Kallisto [20] (*see* **Note 5**).

## 3 Methods

### 3.1 RNA Isolation (see Note 6)

1. Spray the lab bench table with RNAseZap to decontaminate the area with any RNases.
2. For adherent cancer cells (*see* **Notes 7** and **8**), rinse with PBS and place the 6-well plate on ice bucket.
3. Add 350 μL of Buffer RLT (lysis buffer) to the cells according to manufacturer's protocol.
4. Use cell scraper or pipette tips to scrape off the lysed cells and place them into 1.5 mL nuclease-free tubes.
5. For nonadherent cancer cells, rinse the cells with PBS and add 350 μL of Buffer RLT to the cell pellet.
6. Add 350 μL of 70% ethanol to the lysate and mix the samples by pipetting. Transfer 700 μL of sample mixture to RNeasy mini spin columns placed in a 2 mL collection tubes that are provided in the QIAGEN RNeasy Mini kit.
7. Centrifuge the columns for 15 s at ≥8000 × g. Discard the flow-through that is now in the collection tubes (*see* **Note 9**).
8. Following this step, there is an option to perform a DNase digestion using the RNase-Free DNase Set Kit to remove any DNA contamination in the samples.

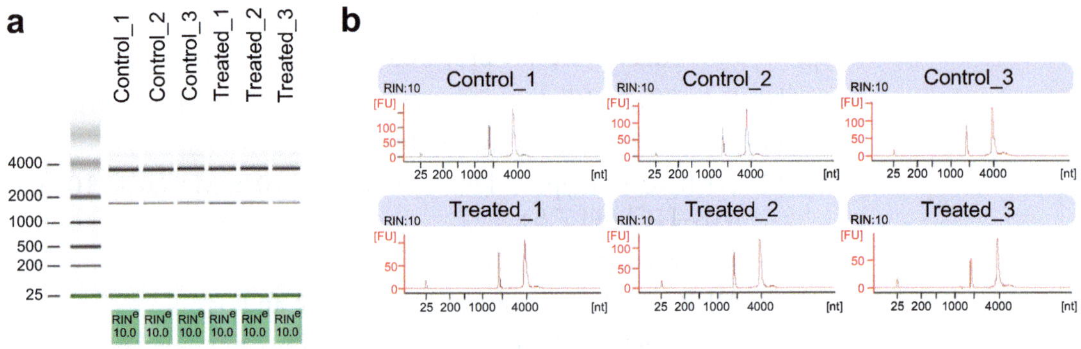


**Fig. 1** Bioanalyzer tracers of typical high-quality total RNAs from untreated control and drug treated cancer cells that are used for RNA-seq. The tracers were generated by analyzing on RNA Bioanalyzer chips. (**a**) Gel representation of the major rRNA species in total RNA samples, with RNA integrity number (RIN) provided per sample at the bottom. RIN >8 indicates high-quality RNA preps. (**b**) Plots of isolated RNA sizes (bp). Generally, only the highly abundant rRNAs are observed, as indicated by the two major peaks

9. Add 700 μL of Buffer RW1 (a stringent washing buffer) to the RNeasy spin column to wash the samples. Close the lid, and centrifuge the samples for 15 s at ≥8000 × g. Discard the flow-through.
10. Add 4 volumes of 100% ethanol to Buffer RPE.
11. Add 500 μL of Buffer RPE containing ethanol (a mild washing buffer) to the RNease spin column to wash the samples. Close the lid and centrifuge the samples for 15 s at ≥8000 × g. Discard the flow-through.
12. Repeat **step** 7 but perform a 2-min spin instead of 15 s.
13. Transfer the RNeasy spin column into a new 2 mL collection tube provided by the kit to increase the purity of the samples. Close the lid and centrifuge the samples for 1 min at ≥8000 × g to dry the membrane on the RNeasy spin column.
14. Place the RNeasy spin column into a new 1.5 mL tubes. Add 30 μL of RNase-free water and centrifuge for 15 s at ≥8000 × g. Take the collected 30 μL samples, and let it run through the RNeasy spin column again to get a higher RNA yield.
15. Check the RNA yield using DeNovix or NanoDrop. If the yield is high, dilute the samples with RNase-free water, and re-check for RNA yield and RNA quality.
16. Send samples for further RNA quality control using RNA Bioanalyzer. RNA integrity number (RIN) of at least 8 (representing high-quality minimally fragmented RNA) is recommended for RNA-seq library construction (Fig. 1a,b).

***3.2 rRNA Hybridization***

1. For each unique sample, pre-chill a 0.2 mL PCR tube on ice for at least 2 min.
2. Separately, thaw QIAseq FastSelect-rRNA reagent (from QIAseq FastSelect -rRNA HMR kit), First-Strand Synthesis buffer (from NEBNext Ultra II Directional Library prep kit), and Random Primers (from NEBNext Ultra II Directional Library prep kit) by holding tubes briefly in hands. Then, quickly vortex the tubes and centrifuge to collect all liquids to the tube bottom.
3. Add the following reagents into each pre-chilled sample tube: 1 μg of DNase I-pretreated total RNA in 4 μL volume, 4 μL of First-Strand Synthesis buffer, 1 μL of Random Primers, and 1 μL of QIAseq FastSelect -rRNA reagent. Total volume should be 10 μL per tube (*see* **Note 10**).
4. Mix by pipetting up and down five times, and then briefly centrifuge to collect all liquids to the tube bottom.
5. Place sample tube into a PCR machine with hot-lid temperature at 105 °C, and perform the following sequential incubation steps: 15 min at 94 °C, 2 min at 75 °C, 2 min at 70 °C, 2 min at 65 °C, 2 min at 60 °C, 2 min at 55 °C, 2 min at 37 °C, and 2 min at 25 °C, then hold at 4 °C (*see* **Note 11**).

***3.3 First-Strand cDNA Synthesis***

1. Prepare the next First-Strand cDNA synthesis step by gently flicking the tube bottom of the NEBNext First-Strand Synthesis Enzyme Mix three to five times, then quickly centrifuging to collect liquids, and then placing tube on ice. Briefly thaw the NEBNext Strand Specificity Reagent by holding tube in hand, then vortexing and centrifuging to collect liquids, and then placing tube on ice.
2. To each 0.2 mL PCR tube containing captured rRNA, add 8 μL NEBNext Strand Specificity Reagent, 2 μL NEBNext First-Strand Synthesis Enzyme Mix. Total volume should be 20 μL per tube.
3. Mix each tube by pipetting up and down five times and briefly centrifuging to collect liquids to the tube bottom.
4. Place sample tube into a PCR machine with hot-lid temperature at 80 °C and perform the following sequential incubation steps: 10 min at 25 °C, 15 min at 42 °C, and 15 min at 70 °C, then hold at 4 °C.

***3.4 Second-Strand cDNA Synthesis***

1. Prepare the next Second-Strand cDNA synthesis step by gently flicking the tube bottom of the NEBNext Second Strand Synthesis Enzyme Mix three to five times, then quickly centrifuging to collect liquids, and then placing tube on ice. Briefly thaw the NEBNext Second Strand Synthesis Reaction Buffer

with dUTP Mix (10×) by holding tube in hand, then vortex and centrifuge to collect liquids. Place tube on ice.

2. To each 20 μL First-Strand Synthesis product in 0.2 mL PCR tubes, add 8 μL of NEBNext Second Strand Synthesis Reaction Buffer with dUTP Mix (10×), 4 μL of NEBNext Second Strand Synthesis Enzyme, and 48 μL of molecular-grade nuclease-free water. Total volume should be 80 μL per tube.
3. Mix each tube by pipetting up and down five times and briefly centrifuging to collect liquids to the tube bottom.
4. Place sample tube into a PCR machine with hot-lid temperature at 40 °C, and incubate for 1 h at 16 °C.

### 3.5 cDNA Cleanup

1. Pre-warm AMPure XP beads by leaving aliquots of the beads at room temperature for at least 30 min prior to use.
2. For each sample, thoroughly homogenize the AMPure XP bead aliquots by vortexing, then add 144 μL of beads (1.8×) to a new 1.5 mL tube.
3. Next, add 80 μL of the corresponding sample double-stranded cDNA product to the 144 μL of AMPure XP beads in the 1.5 mL tube. Immediately mix by pipetting up and down ten times.
4. Allow the cDNA-beads mixture to sit at room temperature for 5 min.
5. Apply the 1.5 mL sample tube to a magnetic stand. Keep the tube on the stand until all AMPure XP beads have separated to the magnetized tube wall and the solution appears clear (usually takes 1–2 min).
6. Use a P1000 pipette tip to carefully remove and discard the aqueous clear solution without disturbing the AMPure XP beads (*see* **Note 12**).
7. Without removing the tube from the magnetic stand, gently add 200 μL of 80% ethanol to the tube to wash the beads. Do not resuspend the beads. Allow the tube to remain on the magnetic stand, unperturbed, for at least 30 s.
8. Use a P1000 pipette tip to carefully remove all 80% ethanol, leaving behind the beads that are still magnetized to the tube wall.
9. Repeat the wash procedure once more with new 200 μL of 80% ethanol.
10. Be sure to remove all traces of ethanol solution after the second wash by using a P10 pipette tip to remove residual solution from the tube bottom.
11. Keep tube cap open to air-dry the beads for 5 min at room temperature (*see* **Note 13**).

12. Transfer the tube from the magnetic stand to a tube holder.
13. Use a P200 pipette tip to add 53 μL of 0.1× TE, pH 7.4 directly to the AMPure XP beads. Be sure to pipette up and down five to ten times to get all the beads into solution and fully resuspended.
14. Let the resuspended beads incubate at room temperature for 2 min.
15. Apply the tube back onto the magnetic stand, and wait for all beads to separate to the tube wall. The solution should appear clear (usually takes 2 min).
16. Carefully transfer 50 μL of clear solution to a new 0.2 mL PCR tube. Completely avoid taking any AMPure XP beads.

### 3.6 cDNA End Prep

1. Prepare the next Second-Strand cDNA synthesis step by gently flicking the tube bottom of the NEBNext Ultra II End Prep Enzyme Mix three to five times, then quickly centrifuging to collect liquids, and then placing tube on ice. Briefly thaw the NEBNext Ultra II End Prep Reaction Buffer by holding tube in hand, then vortex and centrifuge to collect liquids. Place tube on ice.
2. To each 50 μL purified double-stranded cDNA in 0.2 mL PCR tubes, add 7 μL of NEBNext Ultra II End Prep Reaction Buffer, 3 μL of NEBNext Ultra II End Prep Enzyme Mix. Total volume should be 60 μL.
3. Mix well by pipetting 50 μL volumes up and down five to ten times. Avoid making air bubbles.
4. Place sample tube into a PCR machine with hot-lid temperature at 75 °C, and perform the following sequential incubation steps: 30 min at 20 °C, 30 min at 65 °C, then hold at 4 °C.
5. Proceed immediately to the next adapter ligation step.

### 3.7 cDNA Adapter Ligation

1. Thaw NEBNext Adaptor by holding tube in hand. Then briefly vortex and centrifuge to collect all liquids. Finally place tube on ice.
2. Dilute the NEBNext Adaptor 25-fold in NEBNext Adaptor Dilution Buffer. For example, to make a 50 μL diluted NEBNext Adaptor aliquot, mix 2 μL of NEBNext Adaptor and 48 μL of NEBNext Adaptor Dilution Buffer. Briefly vortex and centrifuge to collect all liquids.
3. To each 60 μL purified end-prepped double-stranded cDNA in 0.2 mL PCR tubes, add 2.5 μL of 25-fold diluted NEBNext Adaptor, 1 μL of NEBNext Ligation Enhancer, and 30 μL of NEBNext Ultra II Ligation Master Mix. Total volume should be 93.5 μL (*see* **Note 14**).

4. Immediately and gently mix the viscous solution by pipetting up and down ten times, while avoiding the introduction of air bubbles.
5. Place sample tube into a PCR machine, with hot-lid temperature off, and incubate for 15 min at 20 °C.
6. Take tube out of the PCR machine and place on ice.
7. Add 3 μL of USER enzyme to each tube. Mix by pipetting up and down three to five times, avoiding the introduction of air bubbles.
8. Place tube back into PCR machine, with hot-lid temperature at 45 °C, and incubate for an additional 15 min at 37 °C.
9. Continue to the next ligation cleanup step.

### 3.8 Ligation Cleanup

1. Pre-warm AMPure XP beads by leaving aliquots of the beads at room temperature for at least 30 min prior to use.
2. For each sample, thoroughly homogenize the AMPure XP bead aliquots by vortexing, then add 87 μL of beads (0.9×) to a new 1.5 mL tube.
3. Next, add 93.5 μL of the corresponding cDNA ligation product to the 87 μL of AMPure XP beads in the 1.5 mL tube. Immediately mix by pipetting up and down ten times.
4. Allow the ligation product-beads mixture to sit at room temperature for 10 min.
5. Apply the 1.5 mL sample tube to a magnetic stand. Keep the tube on the stand until all AMPure XP beads have separated to the magnetized tube wall and the solution appears clear (usually takes 1–2 min).
6. Use a P1000 pipette tip to carefully remove and discard the aqueous clear solution without disturbing the AMPure XP beads (*see* **Note 12**).
7. Without removing the tube from the magnetic stand, gently add 200 μL of 80% ethanol to the tube to wash the beads. Do not resuspend the beads. Allow the tube to remain on the magnetic stand, unperturbed, for at least 30 s.
8. Use a P1000 pipette tip to carefully remove all 80% ethanol, leaving behind the beads that are still magnetized to the tube wall.
9. Repeat the wash procedure once more with new 200 μL of 80% ethanol.
10. Be sure to remove all traces of ethanol solution after the second wash but using a P10 pipette tip to remove residual solution from the tube bottom.

11. Keep tube cap open to air-dry the beads for 5 min at room temperature (*see* **Note 13**).
12. Transfer the tube from the magnetic stand to a test tube holder.
13. Use a P200 pipette tip to add 17 μL of 0.1× TE, pH 7.4 directly to the AMPure XP beads. Be sure to pipette up and down five to ten times to get all the beads into solution and fully resuspended.
14. Let the resuspended beads incubate at room temperature for 2 min.
15. Apply the tube back onto the magnetic stand, and wait for all beads to separate to the tube wall. The solution should appear clear (usually takes 2 min).
16. Carefully transfer 15 μL of clear solution to a new 0.2 mL PCR tube. Completely avoid taking any AMPure XP beads.

### 3.9 Library PCR

1. Prepare the next Library PCR step by gently flicking the tube bottom of the NEBNext Ultra II Q5 Master Mix three to five times, then quickly centrifuging to collect liquids, and then placing tube on ice. Briefly thaw the Universal PCR primer and NEBNext Index Primer(s) by holding tubes in hand, then vortex and centrifuge to collect liquids. Place tubes on ice.
2. To each 15 μL purified adaptor-ligated cDNA in 0.2 mL PCR tube, add 5 μL Universal PCR primer, 5 μL NEBNext Index Primer, and 25 μL NEBNext Ultra II Q5 Master Mix. Total volume should be 50 μL (*see* **Note 15**).
3. Mix by pipetting up and down five to ten times.
4. Place sample tubes into PCR machine, with hot-lid temperature at 105 °C. Perform PCR amplification using the following cycling conditions: (1 cycle) 98 °C for 30 s, (9–12 cycles) 98 °C for 10 s, 65 °C for 75 s, and (1 cycle) 65 °C for 5 min, then hold at 4 °C (*see* **Note 16**).

### 3.10 PCR Cleanup

1. Pre-warm AMPure XP beads by leaving aliquots of the beads at room temperature for at least 30 min prior to use.
2. For each sample, thoroughly homogenize the AMPure XP bead aliquots by vortexing, then add 45 μL of beads (0.9×) to a new 1.5 mL tube.
3. Next, add 50 μL of the corresponding PCR product to the 45 μL of AMPure XP beads in the 1.5 mL tube. Immediately mix by pipetting up and down ten times.
4. Allow the PCR product-beads mixture to sit at room temperature for 5 min.

5. Apply the 1.5 mL sample tube to a magnetic stand. Keep the tube on the stand until all AMPure XP beads have separated to the magnetized tube wall and the solution appears clear (usually takes 1–2 min).
6. Use a P1000 pipette tip to carefully remove and discard the aqueous clear solution without disturbing the AMPure XP beads (*see* **Note 12**).
7. Without removing the tube from the magnetic stand, gently add 200 μL of 80% ethanol to the tube to wash the beads. Do not resuspend the beads. Allow the tube to remain on the magnetic stand, unperturbed, for at least 30 s.
8. Use a P1000 pipette tip to carefully remove all 80% ethanol, leaving behind the beads that are still magnetized to the tube wall.
9. Repeat the wash procedure once more with new 200 μL of 80% ethanol.
10. Be sure to remove all traces of ethanol solution after the second wash but using a P10 pipette tip to remove residual solution from the tube bottom.
11. Keep tube cap open to air-dry the beads for 5 min at room temperature (*see* **Note 13**).
12. Transfer the tube from the magnetic stand to a test tube holder.
13. Use a P200 pipette tip to add 23 μL of 0.1× TE, pH 7.4 directly to the AMPure XP beads. Be sure to pipette up and down five to ten times to get all the beads into solution and fully resuspended.
14. Let the resuspended beads incubate at room temperature for 2 min.
15. Apply the tube back onto the magnetic stand, and wait for all beads to separate to the tube wall. The solution should appear clear (usually takes 2 min).
16. Carefully transfer 20 μL of clear solution to a new 1.5 mL tube. Completely avoid taking any AMPure XP beads. This sample is the final library, ready for RNA sequencing on an Illumina platform.

### 3.11 Library Quantification and Assessment

1. Use 2–5 μL of final library to quantify the concentration of dsDNA using the Qubit dsDNA high sensitivity system. Do not attempt quantification using NanoDrop or Denovix systems (or similar spectroscopic systems). They are not sensitive or accurate enough to measure these libraries.

2. Use at least 1 μL of final library to also assess by Agilent TapeStation or Bioanalyzer systems. These systems can reliably provide the library fragment size distributions and dsDNA concentrations.
3. Library concentrations and average fragment sizes are then used to compute the molarity of each library (in nM units). For multiplexing, equal molarity of each library is pooled together and then processed for sequencing. We do not discuss further the details of multiplexing, sample denaturation, sample dilution, and Illumina machine running because those steps are usually performed by sequencing core personnel (since it is not expected that the typical laboratory will have their own Illumina sequencer).

### 3.12 Assessing Sequencing Quality

1. The sequencer output is usually a fastq file that contains the raw sequence reads of the RNA. Raw sequencing fastq or fastq.gz data files can be opened using quality assessment software such as FastQC. This will give a broad overview of sequencing quality for each library using metrics such as number of total reads, quality scores of base-calls, and number of duplicated reads. Figure 2 shows example FastQC outputs for a successful RNA-seq library (Fig. 2a, b).
2. The user can also modify reads (i.e., read trimming) using bioinformatic tools such as CutAdapt, depending on the quality of the sequencing data and the user needs.

### 3.13 Understanding Pre-Alignment Data and Preparation

1. A typical entry in a fastq file appears as in the example below:

   @SRR1111111 HWI-ST632:1601:1735:2104 length = 56.

   CCTGCCTGGGGGACCGACTAGGGAGCGCAGGGG AACGTGTACAAGAGCGCGCGCTG.

   + SRR1111111 HWI-ST632:1601:1735:2104 length = 56.

   FCCCDC@ HACCCCDCDBCHCCDDCIFBFCCICH FICCDDCHCBHDCFCDCIHDCHFHHE.

   Each group of four lines provides information about a single read (or pair of reads if paired-end sequencing was performed).
2. The first line, starting with the @ symbol, includes information about the SRR number (which is required for downloading the entire fastq file from NCBI's Sequence Read Archive or from NCBI GEO), the ID of the instrument, the run and flow cell, and the length of the sequenced read. If pair-ended sequencing was performed, information about the paired read is also included in the first line.
3. The second line is the base-sequence of the read.

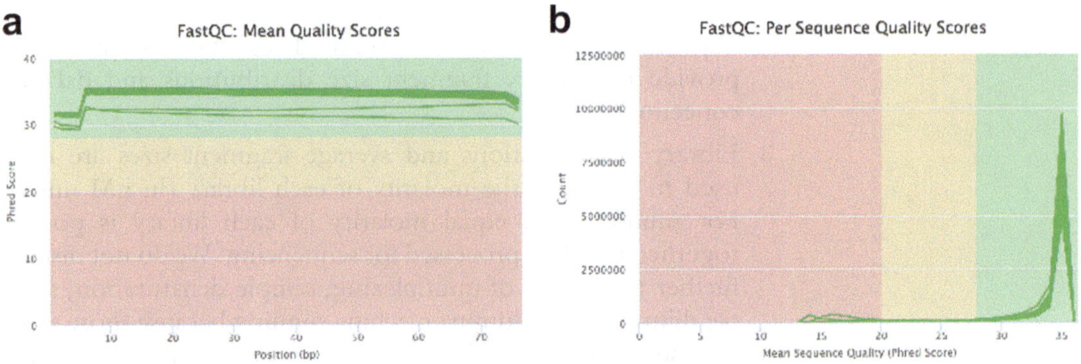


**Fig. 2** Example FastQC plots of a raw fastq data file from a typical RNA-seq experiment. In this example RNA-seq experiment, single-end reads of 75-bases were sequenced on an Illumina NextSeq500 platform. The read data in the raw fastq data file was analyzed by FastQC software. (**a**) Distribution of mean quality scores (measured as Phred scores) across all sequenced bases in a read where green indicates typically good mean quality scores. (**b**) Plot of total counts of sequenced bases (*y*-axis) at various Phred scores (*x*-axis)

4. The third line, starting with the + symbol, can be used to add any description information and usually is identical to the first line.
5. The fourth line contains the sequencing quality data for each base in the same position as in the second line.
6. The quality score is the estimated probability of a base being called incorrectly by the sequence machine. The higher the score, the higher the estimated sequencing quality.
7. Important in conducting high-quality RNA-seq analysis is deciding the initial design of the study. Comparisons should be made between biologically relevant samples, with at least three biological replicates available per condition. Sequencing depth is also important in ensuring high data quality. Depth above ten million reads is generally recommended, with around 30 million reads being the target for best-quality sequencing to allow the capture of expression changes of genes with low basal expression.

### *3.14 Building the Reference Transcriptome*

1. The first step in any RNA-seq analysis is the building of references for the alignment and quantification software. The user can find the exact command to build references in the documentation of the aligner package of choice, such as STAR, Tophat2, RSEM, Salmon, or Kallisto.
2. The genome sequence of hg38 (human) can be found using these links:
   - ftp://ftp.ncbi.nlm.nih.gov/genomes/archive/old_genbank/Eukaryotes/vertebrates_mammals/Homo_

sapiens/GRCh38/seqs_for_alignment_pipelines/GCA_000001405.15_GRCh38_no_alt_analysis_set.fna.gz

- https://www.ncbi.nlm.nih.gov/assembly/GCF_000001405.26/

3. The latest version of the reference human genome annotation can be downloaded from Gencode: https://www.gencodegenes.org/human/.
4. From here, the user needs to download a comprehensive gene annotation GTF file to prepare the reference genome by annotating where expected genes are.

### 3.15 Read Alignment

1. After running the scripts in reference building mode, the user needs to make sure that the same versions of the scripts will be used to perform the RNA-seq analysis—for some scripts, the references can be *version-specific*.
2. An example of such a script would be one that accepts a "samples.txt" file where every line is the sample name, which will be used to find the fastq file and pass it on to the aligner of choice. After that, the output of the aligner (usually SAM or BAM files) should be passed on to the feature count script. If the program performs both steps, it will only need the fastq file as the user input. The user will also need to ensure that the script can determine whether the initial sequencing was done using the single- or pair-end method.
3. It is important to perform another quality control check post-alignment.
4. This allows to check the quality of both the RNA-seq data and the alignment. This will also be able to detect contamination of samples with foreign DNA, improper references, PCR over-amplification, and poorly performed RNA-seq sample preparation.
5. Some examples of post-alignment quality assessment tools are Picard, Qualimap2, SAMTools, RSeQC, and, more recently, RNA-SeQC2.
6. Understanding the various normalized gene expression units is also very important and is required to perform correct analysis:
   (a) *CPM* (counts per million) is the ratio between number of reads mapped to a gene and the total number of reads divided by a million (scaling factor). This does not allow for comparisons between genes within the same sample, since the number of mapped reads will be affected by the gene length, with longer genes inherently having higher counts. CPM can be used to compare expression levels of the same gene across *different* samples.

(b) *RPKM* (reads per kilobase per million reads) and *FPKM* (fragments per kilobase per million reads) are calculated as the ratio of number of reads mapped to a gene, per 1000 bases of the gene, and the total number of reads (in million units). The main difference between the two metrics is that the former is used for single-end sequencing, whereas the latter is used in pair-end sequencing. Both metrics allow for within sample comparison but should not be used to compare between samples.

(c) *TPM* (transcripts per million reads) is a more advanced normalization metric which is proportional to RPKM/ FPKM were the number of reads to a gene is divided by the length of the gene in kilobases, giving reads per kilobase (RPK). Next, the sum of all RPK values in a sample is divided by a million to give the RPK scaling factor. At each base of the transcriptome, the ratio between RPK and the RPK scaling factor is the TPM. TPM is useful in comparing expression levels between genes within a sample but comparisons across different samples can be affected by library composition.

### 3.16 Differential Expression Analysis

1. This analysis will determine which genes are overexpressed or downregulated in the condition of interest, relative to a control. For example, one can perform a differential gene expression analysis to test the effect a certain drug has on the transcriptional landscape of cancer cells. Differential gene expression analysis can also be used to compare gene expression profiles between drug-sensitive and drug-resistant cancer cells to determine a target or potential marker of drug resistance. There are multiple packages available for this type of analysis such as DESeq2 [21], edgeR [22], CuffDiff2 [23], and limma [24].
2. In order to run these programs, the researcher will need to generate a read count table. Most packages accept only raw read counts and *not* normalized TPM or FPKM, since normalization removes many information that is used by the packages to make statistical calculations. Within the script, the user should identify which columns of the table represent the control conditions against which the comparison will be made.
3. It is recommended to filter out genes with low read counts to avoid any possibility of false positive differential expression calls by the program. Biological replicates can help determine whether small expression changes represent statistical artifacts or biological phenomena.

4. Another analysis that the user may want to perform is using dimension reduction techniques to test for transcriptional differences between different samples. There are several methods to perform this type of analysis such as principal component analysis (PCA) or multidimensional scaling (MDS), which will determine if any sample is a significant outlier in the data.
5. Once the preprocessing and quality control of the data are completed, the differential gene expression analysis can be performed. Every package has its own set of commands.
6. After computing the differential gene expression analysis between the control and the test group, the user should identify genes which are significantly differentially expressed by applying cutoffs to the adjusted p-values (padj) and log2 fold change values (log2FC). Commonly used cutoffs are padj $<0.05$ and $|\log2FC| > 1$.
7. Also, the user can visualize the differential gene expression using a heatmap, MA plot, or a volcano plot. This is especially useful if there are multiple control and test groups (i.e., different drugs). This will allow the user to quickly visualize the data and get an overview to begin building a scientific hypothesis.

### 3.17 Gene Set Enrichment Analysis

1. A common downstream analysis that can aid the researcher in generating scientific hypotheses based on RNA-seq data is gene set enrichment (Fig. 3).
2. Enrichment is generally defined as an over-representation of genes within a predefined set of genes, such as a biological pathway. This analysis can be performed using the gene set enrichment analysis (GSEA) tool [25] (https://www.gsea-msigdb.org/gsea/index.jsp). The data needs to be formatted specifically for this tool, as detailed at https://software.broadinstitute.org/cancer/software/gsea/wiki/index.php/Data_formats.
3. The enrichment score (ES) for a user-provided pre-defined set of genes (usually corresponding to a known pathway or gene ontology term) (*see* Fig. 3) is calculated by going down a list of genes which are ranked by the value of the differential expression. The ES for a pathway or ontology term increases when an encountered gene is present in the gene set and decreases when a gene is not in the set. The size of the increase or decrease depends on how close the gene is to the top or bottom of the ranked list.
4. A phenotype permutation test is then performed in order to determine the statistical significance of the ES. A null distribution for ES is calculated based on the permuted data, and then the ES score is compared to it in order to calculate the nominal p-value.

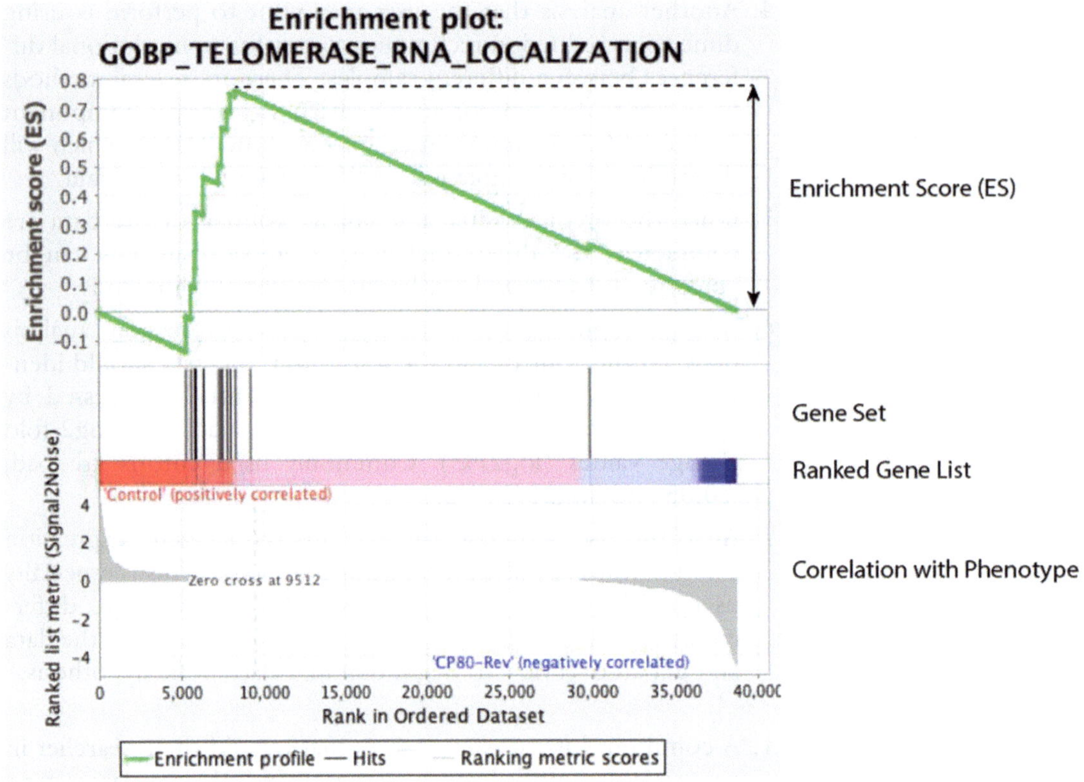


**Fig. 3** Graphical representation of enrichment scores. Enrichment score (ES) quantifies the over-representation of a user-defined gene set within a list of differentially expressed genes. Differential expression is determined from RNA-seq analyses described above. ES is calculated based on the maximum deviation value from zero during a random walk within a ranked gene list

5. The ES is then normalized by gene set size which creates the normalized enrichment score (NES). False discovery rate (FDR) is then calculated for each NES to account for random false positives that arise from the multiple hypothesis testing.

## 4 Notes

1. These beads are normally stored at 4 °C. It is best to make small aliquots of these beads (up to 1 mL). Prior to transferring to new tubes, make absolutely sure to thoroughly mix the bead solution by vortexing or shaking of the bead container.
2. Users can purchase the NEBNext Multiplex Oligos for Illumina (Index Primers Set 1 and/or 2) which contains NEBNext Adapter, USER enzyme, Universal PCR primer, and NEBNext Index primers for single-indexing of libraries (compatible with Illumina sequencing).

3. The Universal and Index primers can be found in the NEBNext Multiplex Oligos for Illumina (Index Primers Set 1 and/or 2) kits. Alternatively, they can be individually synthesized and PAGE-purified. The exact primer sequences can be found in the reference manual for the respective NEBNext Multiplex Oligos for Illumina kits.
4. Most of RNA-seq analysis is server-based due to high computational requirements. The amount of time required to run alignment of a single RNA-seq sample can range from roughly 30 min to several hours, assuming roughly 30 million reads and depending on processor speeds. Alignment is usually the most resource intensive part of RNA-seq analysis.
5. Various aligners have been developed over the years such as STAR or Tophat2, which are most commonly used. After alignment, feature counts need to be counted, RSEM is also a popular package that can perform this task.
6. Cell lysis can be done on different dates for replicates, but it is recommended that RNA isolation steps be performed together for all samples.
7. Do not overgrow the cells in the plate before collection as that can affect gene expression. Ideal would be ~70% confluency depending on the cells. Growing cells in 6-well plate gives the most optimal result. In our hands, we noticed that having more cells ($>1 \times 10^6$) clumps the RNA purification column and can result in low yield and/or poorer RNA quality. Also, while collecting the cells, the lysed samples should always be kept on ice to prevent RNA degradation.
8. For adherent cells: RNA collection should be performed between 24 and 48 h after plating to provide enough time for the cells to attach to the culture plate or flask.
9. During the centrifugation of the RNeasy spin column, the centrifugation step should be performed at room temperature and not at 4 °C to avoid any formation of precipitates that can ruin the RNA isolation. The RNA isolation step should be performed quickly but precisely. Once the RNA is purified, the samples should be kept on ice or stored at −20 °C.
10. The QIAselect reagent contains probes that will hybridize to rRNA fragments within the total RNA. Bound rRNA fragments cannot be used to generate cDNA in the subsequent steps of the library preparation procedure.
11. This step will also result in RNA fragmentation. The indicated step procedure assumes intact RNA of RIN number at least 8. For partially degraded RNA of RIN number 2–6, perform the incubation at 94 °C for 7–8 min, instead of 15 min.

12. Be careful not to allow the AMPure XP beads to dry out. Once the beads are not in liquid suspension, it can take ~5 min for them to dry. The best practice is to remove liquid from the beads, then immediately add new solution to the beads to keep them hydrated.
13. The beads should appear dark brown and a bit glossy. If the beads become light brown and flaky, they have over-dried. Over drying beads might cause unintentional loss of purified cDNA.
14. We do not advise preparing a ligation pre-mix beforehand because the mixtures may not be stable in the long term. We generally add as needed directly from stock tubes of the various reagents.
15. For each unique sample library, the same Universal primer (forward) will be used. But each unique sample library must use a different Index primer (reverse). If the user will make two or more unique libraries, we highly recommend using NEB-Next Index 6 and 12 for the first two libraries, followed by any unique indices for all remaining libraries. Having two libraries with indices 6 and 12 will ensure that the user can safely multiplex their libraries (irrespective of how many libraries they have). However, if the user will only make one library, they can use any Index primer.

    The number of PCR cycles will depend on the quality of the input RNA material and should be optimized by the user to minimize PCR overamplification, while having sufficient library for sequencing (at least 100 ng).

## Acknowledgments

This work was supported by Michigan State University startup funds to T.V.V and S.H.

### References

1. Khandjian EW, Meric C (1986) A procedure for northern blot analysis of native RNA. Anal Biochem 159:227–232
2. Streit S, Michalski CW, Erkan M, Kleeff J, Friess H (2009) Northern blot analysis for detection and quantification of RNA in pancreatic cancer cells and tissues. Nat Protoc 4:37–43
3. Trevino V, Falciani F, Barrera-Saldana HA (2007) DNA microarrays: a powerful genomic tool for biomedical and clinical research. Mol Med 13:527–541
4. Hu T, Chitnis N, Monos D, Dinh A (2021) Next-generation sequencing technologies: an overview. Hum Immunol 82:801–811
5. Wang Z, Gerstein M, Snyder M (2009) RNA-Seq: a revolutionary tool for transcriptomics. Nat Rev Genet 10:57–63
6. Horibata S, Guib G, Lack J et al (2019) Heterogeneity in refractory acute myeloid leukemia. Proc Natl Acad Sci U S A 116:10494–10503
7. Patel RP, Kuhn S, Yin D et al (2021) Cross-resistance of cisplatin selected cells to anti-

microtubule agents: role of general survival mechanisms. Transl Oncol 14:100917

8. Tang F, Barbacioru C, Wang Y et al (2009) mRNA-Seq whole-transcriptome analysis of a single cell. Nat Methods 6:377–382
9. Stahl PL, Salmen F, Vickovic S et al (2016) Visualization and analysis of gene expression in tissue sections by spatial transcriptomics. Science 353:78–82
10. Ding S, Chen X, Shen K (2020) Single-cell RNA sequencing in breast cancer: understanding tumor heterogeneity and paving roads to individualized therapy. Cancer Commun (Lond) 40:329–344
11. Cui P, Lin Q, Ding F et al (2010) A comparison between ribo-minus RNA-sequencing and polyA-selected RNA-sequencing. Genomics 96:259–265
12. Zhao S, Zhang Y, Gamini R, Zhang B, von Schack D (2018) Evaluation of two main RNA-seq approaches for gene quantification in clinical RNA sequencing: polyA+ selection versus rRNA depletion. Sci Rep 8:4781
13. Wesolowski S, Birtwistle MR, Rempala GA (2013) A comparison of methods for RNA-Seq differential expression analysis and a new empirical bayes approach. Biosensors (Basel) 3:238–258
14. Seyednasrollah F, Laiho A, Elo LL (2015) Comparison of software packages for detecting differential expression in RNA-seq studies. Brief Bioinform 16:59–70
15. Ji F, Sadreyev RI (2018) RNA-seq: basic bioinformatics analysis. Curr Protoc Mol Biol 124: e68
16. Dobin A, Davis CA, Schlesinger F et al (2013) STAR: ultrafast universal RNA-seq aligner. Bioinformatics 29(1):15–21
17. Kim D, Pertea G, Trapnell C et al (2013) TopHat2: accurate alignment of transcriptomes in the presence of insertions, deletions and gene fusions. Genome Biol 14(4):R36
18. Li B, Dewey CN (2011) RSEM: accurate transcript quantification from RNA-Seq data with or without a reference genome. BMC Bioinform 12:323
19. Patro R, Duggal G, Love MI, Irizarry RA, Kingsford C (2017) Salmon provides fast and bias-aware quantification of transcript expression. Nat Methods 14:417–419
20. Bray NL, Pimentel H, Melsted P, Pachter L (2016) Near-optimal probabilistic RNA-seq quantification. Nat Biotechnol 34:525–527
21. Love MI, Huber W, Anders S (2014) Moderated estimation of fold change and dispersion for RNA-seq data with DESeq2. Genome Biol 15:550
22. Robinson MD, McCarthy DJ, Smyth GK (2010) edgeR: a bioconductor package for differential expression analysis of digital gene expression data. Bioinformatics 26:139–140
23. Trapnell C, Hendrickson DG, Sauvageau M et al (2013) Differential analysis of gene regulation at transcript resolution with RNA-seq. Nat Biotechnol 31(1):46–53
24. Ritchie ME, Phipson B, Wu D et al (2015) Limma powers differential expression analyses for RNA-sequencing and microarray studies. Nucleic Acids Res 43(7):e47
25. Subramanian A, Tamayo P, Mootha VK et al (2005) Gene set enrichment analysis: a knowledge-based approach for interpreting genome-wide expression profiles. Proc Natl Acad Sci U S A 102(43):15545–15550

# Chapter 4

# Generating Mitochondrial-Nuclear Exchange (MNX) Mice to Identify Mitochondrial Determinants of Cancer Metastasis

Danny R. Welch, Melissa A. Larson, Carolyn J. Vivian, and Jay L. Vivian

## Abstract

Understanding the contributions of mitochondrial genetics to disease pathogenesis is facilitated by a new and unique model—the mitochondrial-nuclear exchange mouse. Here we report the rationale for their development, the methods used to create them, and a brief summary of how MNX mice have been used to understand the contributions of mitochondrial DNA in multiple diseases, focusing on cancer metastasis. Polymorphisms in mtDNA which distinguish mouse strains exert intrinsic and extrinsic effects on metastasis efficiency by altering epigenetic marks in the nuclear genome, changing production of reactive oxygen species, altering the microbiota, and influencing immune responses to cancer cells. Although the focus of this report is cancer metastasis, MNX mice have proven to be valuable in studying mitochondrial contributions to other diseases as well.

**Key words** Mitochondria, Genetics, Mitochondria-nuclear exchange mice, Animal model

## 1 Introduction

Many diseases, like neoplasia, diabetes, neurodegenerative diseases, Alzheimer's disease, Parkinson's disease, and cardiovascular diseases, have heritable underpinnings. Genetic contributions to these complex diseases have focused primarily on mutations or polymorphisms in the nuclear genome, but increasing evidence supports contributions of the mitochondrial genome as well. Unfortunately, the contributions of the mitochondrial genome have proven challenging to study for a variety of reasons summarized below [1].

### *1.1 Challenges to Studying Mitochondrial Genetics*

As organelles, mitochondria have many roles. The function most associated with them is energy production, but controlling chemical oxidation and reduction as well as providing substrates for macromolecular biogenesis are also well recognized. Bioenergetic

Usha N. Kasid and Robert Clarke (eds.), *Cancer Systems and Integrative Biology*, Methods in Molecular Biology, vol. 2660, https://doi.org/10.1007/978-1-0716-3163-8_4,

functions are documented for their contributions to many diseases. Yet, bioenergetics may not be the sole means by which the mitochondrial genome contributes to cellular functions [2–6].

Mammalian mitochondrial genomes are circular DNA consisting of ~16.5 kb encoding 13 protein subunits of the electron transport chain, 2 rRNA (16S and 12S), and 22 tRNA. Mitochondrial DNA (mtDNA) duplexes are distinguished by base composition, and there can be 10's to 1000's of copies per cell. Since an Alphaproteobacterium was first engulfed, it has lost most of its genome to the nucleus. Indeed, the nuclear genome encodes >90% of the molecules responsible for mitochondrial structure, function, bioenergetics, replication, and repair. In recent years, some of the genes remaining encoded in the mtDNA have been shown to have identical copies in the nuclear DNA [7, 8]. Look-alikes within the nuclear genome require caution when interpreting sequencing-based associations. Furthermore, mitochondria are dynamic, changing shape and function depending on tissue type and physiological needs. Mitochondria can even change subcellular localization, depending upon cellular conditions [9–12].

Mitochondrial diseases are traditionally characterized as being caused by mutations in genes involved in oxidative phosphorylation and ATP synthesis [1]. The current prevalence of mitochondrial diseases is ~1:5000 people, but newer whole-genome analyses appear to be increasing the measured rates. Clinical diagnoses and definitive cause–effect relationships have been challenging for multiple reasons since penetrance and disease severity vary greatly, depending upon copy number variants per cell and thresholds of heteroplasmy leading to a detectable phenotype [13]. That is, if the mutational burden is below a certain level, cells may be phenotypically indistinguishable from fully wild-type, homoplasmic cells. However, if the percentage of mutant mtDNA within a cell exceeds a threshold, functions are impaired. Making matters more complicated, cellular levels of heteroplasmy can change with age and depending upon stress [14–16], revealing age-associated or environmentally stimulated mitochondrial disease. Additionally, cross talk between mitochondria, the nucleus, and other cells influences epigenetic marks or the extracellular microenvironment which impact disease penetrance and/or severity [17–19]. Further, despite matrilineal inheritance, mitochondrial diseases can appear autosomal recessive or dominant, maternal, or X-linked. Within the mitochondrial genome, there are single-nucleotide variants that represent polymorphisms that confer disease susceptibility [4, 20, 21]. Acquired somatic mutations, deletions, and/or mtDNA depletion can be driven by nuclear gene defects as well as comparatively high mutation rates due to high concentrations of reactive oxygen species in the mitochondrial milieu coupled with poor or incomplete repair of mtDNA. Despite all of these technical and biological issues related to difficulty in studying mitochondrial genetics, mutations and polymorphisms in mtDNA are increasingly becoming recognized as important contributors to complex diseases.

### 1.2 Models for Studying Mitochondrial Genetics

Previous studies of the contributions of mitochondria in normal physiology and pathologies have utilized generation of "rho-null" cells which contain no mtDNA and replacing the mitochondrial genome with the mtDNA of choice [22–24]. The newly constructed cells are termed cybrids. While powerful, the generation of rho-null cells involves prolonged exposure of cells to low-grade mutagens, like ethidium bromide. The possibility of non-mtDNA mutations is high, and subsequent interpretation of experimental results could be challenged. Other approaches to study mitochondrial genetics involve repeated (>8–10×) backcrossing to create congenic mice or xenomitochondrial animals which involved a rho-null intermediate. Bussard and Siracusa reviewed the relative strengths and weaknesses of these historical approaches [25]. We desired an experimental model in which no further genotype complexity was introduced by nuclear crossover or recombination in filial generations or exposure of cells to mutagens.

These considerations led us to develop mitochondrial-nuclear exchange (MNX, pronounced minx) mice. MNX mice contain nuclear DNA from one strain of mouse and mitochondrial DNA from a different strain of mouse. Details regarding the methods used to generate MNX mice are provided below. However, the salient point is that micropipettes were used to physically extract and replace pronuclei of one mouse strain as one-cell embryos and replacing them with pronuclei from another strain. This methodology, which eliminates mutagen exposure and backcrossing, can be applied, in theory, to any mouse strains. To date, MNX mouse strains that have been generated from different strain combinations, are stable, and are successfully bred through many generations for more than a decade.

The methodology we developed allows for the efficient introduction of transferred pronuclei into the enucleated zygote using the membrane fusion machinery of the Sendai virus. This technique is distinguished from other nuclear transfer methods, which use direct microinjection of pronuclei for transfer into enucleated zygotes. Although the two methodologies have not been rigorously compared, the Sendai virus fusion method described here may be gentler to increase survival and may also reduce the likelihood of transfer of perinuclear mitochondria from the karyoplast donor strain. With appropriate technical skill and robust assessment of homoplasmy, mice derived from these reconstituted embryos can be identified in which the associated cytoplasmically localized mitochondria from the first strain constitute the sole mitochondrial population.

### 1.3 Mitochondrial Genetics in Cancer Metastasis

Our focus has been to study the role of mtDNA in cancer, specifically cancer metastasis using MNX mice [26–31]. However, MNX mice have already proven useful for studying the role(s) of the mitochondrial genome in other diseases [32–35], development [36], and aging [29] as well.

The rationale for exploring the contributions of the mitochondrial genome in cancer metastasis stemmed from an elegant study by Hunter et al. who discovered metastatic susceptibility loci in a genetic screen crossing MMTV-PyMT on the FVB/HeN genetic background with different mouse strains resulting in $F_1$ progeny with different primary tumor latencies and metastatic potentials [37]. His lab subsequently identified nDNA-encoded metastasis modifier genes and, using comparative genomic analyses, showed inheritance of metastasis susceptibility [38–45]. However, the experimental design—male transgenic mice crossed with females of other strains—left open the possibility that maternal mtDNA inheritance was contributing to the observed changes in metastatic behavior.

To address the possibility that differences in metastatic efficiency were due to differences in mtDNA, we mated MNX mice with PyMT mice, keeping the nuclear genome constant (FVB/HeN) but crossing hemizygous PyMT sires with MNX females with wild-type (FVB/HeN), C57BL/6J, or BALB/cJ mtDNA. Because of maternal inheritance, the only experimental variable using this design is the mtDNA. The results obtained using MNX mice were nearly superimposable for primary tumor latency and metastasis [26].

The experiment was extended by testing the impact of mtDNA on tumor development and metastasis with a different oncogenic driver (HER2) [27]. Similar changes in metastasis were observed, but the results showed that mtDNA is a quantitative trait locus that alters efficiency of tumorigenicity and metastasis in a driver-dependent manner. Further, simply aging MNX mice for more than 2 years did not develop spontaneous tumors at different rates compared to parental strains [29]; but metastatic potentials with C57BL/6J mtDNA consistently developed fewer, less aggressive tumors, regardless of nDNA composition [31]. Critically, male MNX crossed with MMTV–Her2 (mtDNA is rarely paternally transmitted) $F_1$ progeny did not cause a change in tumorigenicity or metastasis [27].

To provide further confidence that the pronuclear manipulations were not responsible for the phenotypic changes observed, we recently replicated many of the non-cell autonomous studies utilizing conplastic backcrosses (>10 generations) and showed identical patterns of metastasis. No significant differences in mitochondrial mass, development of heteroplasmy, and transmission of mtDNA were observed (C.J. Vivian and D.R. Welch, unpublished observations).

Recognizing that genetic crosses of transgenic with MNX mice also changes mtDNA in host (stromal) cells as well, we asked whether wild-type tumor cells behaved differently when injected into nDNA-matched (i.e., histocompatible) MNX mice. Syngeneic mammary and melanoma tumor cells were injected into wild-type or age-matched MNX mice, and lung colonization was measured. Interestingly, whenever host cells contained C57BL/6J mtDNA, metastasis was inhibited; whenever stromal cells had C3H/HeN mtDNA, metastasis was promoted [31]. These findings showed that mitochondrial signals from the stroma could directly or indirectly alter tumor cell behavior.

We assessed several possible mechanisms for mtDNA exerting effects on the phenotypes above. C3H/HeN mtDNA consistently produced higher reactive oxygen species than C57BL/6J mtDNA. Metastasis increased with higher ROS levels, and scavenging ROS with MitoTEMPO resulted in fewer metastases [27]. Recognizing that ROS is commonly used by immune cells for killing, we also asked whether baseline immune profiles differ in MNX mice compared to their nDNA-matched wild-type counterparts and whether there were differences in immune cell infiltrates into lung metastases [30]. Although still under investigation, in healthy MNX mice, the immune compositions indeed differed, and the number and polarization states of myeloid populations infiltrating metastases were significantly different [30].

Because mitochondria are central to metabolism, we reasoned that mtDNA SNP might change metabolomic profiles. Using the Metabolon platform, we examined >5000 metabolites from mammary gland and lung compared in wild-type and MNX mice. As expected, MNX mice clustered differently from wild-type. However, specific metabolites corresponding to changing metastatic potential have not yet been identified, consistent with previous reports from our lab [27, 32] and other laboratories [46–49].

Recognizing that metastasis requires coordinated expression of multiple genes, we asked whether mtDNA changed nDNA expression patterns by altering DNA methylation and/or histone marks [28]. Using age- and sex-matched samples, we showed that there were selective changes in the transcriptomes of MNX mice [28] that corresponded with sites of cytosine methylation in a whole-genome analysis as well as histone methylation and histone acetylation/methylation using ChIP-Seq.

Lastly, since mitochondria are thought to have evolved from ancient bacteria [50] and since bacterial ecosystems form when microbes communicate with each other [51–53], we reasoned that mitochondria could retain the ability to communicate with some bacteria. If so, we reasoned that selective communication would result in promotion/inhibition of some bacterial species, resulting in a change in the microbiome. Deep sequencing of gut microbiota identified selective changes (8–17 bacterial species) in the gut microbiome associated with mtDNA content and clear differences in microbiota depicted by principal component analysis (manuscript in preparation). Fostering pups with mothers with different mtDNA content did alter the microbiomes, but the underlying genetic changes were largely maintained. These data are presented as additional evidence of the specificity of mtDNA signals.

Collectively, these data show that the MNX mouse is a powerful, relatively new tool to study the contributions of mitochondrial DNA to physiological homeostasis and pathogenesis of disease.

## 2 Materials

1. Five prepubertal embryo donor female mice of the first desired strain (e.g., C57BL/6J) (*see* **Notes 1** and **2**).
2. Five prepubertal embryo donor female mice of the second desired strain (e.g., FVB/HeN).
3. Five sexually mature stud male mice of the first desired strain (C57BL/6J), individually housed.
4. Five sexually mature stud male mice of the second desired strain (FVB/HeN), individually housed.
5. Pregnant mares' serum gonadotropin (PMSG), reconstituted to 50 IU/mL in sterile PBS. Aliquot and store at −20 °C. Avoid repeated freezing and thawing.
6. Human chorionic gonadotropin (hCG), reconstituted to 50 IU/mL in sterile PBS. Aliquot and store at −20 °C until use. Avoid repeated freezing and thawing.
7. 1 cc insulin syringes with attached 28-gauge needles.
8. M2 HEPES-buffered embryo handling medium, available commercially.
9. KSOM bicarbonate-buffered embryo culture medium, available commercially (*see* **Note 3**).
10. Light mineral oil or liquid paraffin.
11. Hyaluronidase, reconstituted to 10 mg/mL in sterile PBS. Make 100 μL aliquots and store at −20 °C until use.
12. Sterile plastic petri dishes, 35 mm and 60 mm.
13. Surgical instruments, including watchmakers' forceps (4 pair), 4″ straight scissors, micro-scissors.
14. Glass transfer capillary and pipetting apparatus.
15. Stereomicroscope with transmitted light, 8–80× magnification.
16. Micromanipulation assembly, including an inverted microscope with 40–400× magnification, a pair of manipulators, a

micrometer (CellTram Air or equivalent), a SAS air syringe or equivalent, and an XYClone laser objective or equivalent (e.g., Piezo drill).

17. Fire-polished glass-holding pipettes, pulled on a needle puller and polished on a microforge. Alternatively, holding pipettes are available commercially.
18. TransferTips (ES) or other embryonic stem cell injection needle.
19. Demecolcine added to M2 medium, final concentration 0.1 μg/mL.
20. Cytochalasin B added to M2 with demecolcine, final concentration 5 μg/mL.
21. Sendai fusion virus (HVJ-E), resuspended to working concentration at manufacturer's suggestion, 50 μL aliquots stored at −80° (*see* **Note 4**).
22. Incubator at 37 °C, 6% CO2, and 5% O2.

## 3 Methods

### *3.1 Collection of Presumptive Zygotes*

1. Thaw a 1 mL aliquot of PMSG at RT or by rolling between your hands (do not heat in a water bath). Administer 5 IU (0.1 cc) PMSG by intraperitoneal injection to donor females 3 days prior to scheduled procedure between 12:00 and 2: 00 pm.
2. Forty-six to forty eight hours later, administer 5 IU (0.1 cc) human chorionic gonadotropin (hCG) and mate females to individually housed stud males of their respective strain.
3. Check copulation plugs the following morning.
4. Prepare two 35 mm petri dishes with 3 mL M2, two dishes with 3 mL M2 plus 100 μL aliquot of 10 mg/mL hyaluronidase (final concentration ~ 300 μg/mL), and two dishes with three 100 μL drops of M2 overlaid with oil.
5. Euthanize females according to your IACUC-approved Animal Care and Use Protocol.
6. Lay females on their back and spray abdomen with 70% ETOH.
7. Lift the skin over the abdomen and make a transverse incision with the straight scissors. Pull the skin apart and lift the skin toward the head of each mouse.
8. With clean instruments, lift the body wall of one mouse, and cut the tissue open to expose the intestines.
9. Locate the reproductive tract lying along the back wall. The ovaries will lie beneath the kidneys.

10. Grasp one of the uterine horns with forceps just below the oviduct, and cut away the connective tissue with the spring micro-scissors. Pull gently to separate the oviduct from the ovary and cut. With the forceps still at the bottom of the oviduct, cut beneath the forceps to separate the oviduct from the uterine horn. Place the dissected oviduct in the dish of M2.
11. Repeat the procedure for the contralateral oviduct in the same female, then repeat **steps 8–10** for the remaining four donor C57BL/6J females until all oviducts have been collected for that strain. Repeat the procedure, collecting the ten oviducts from the FVB donor females, and place in a second dish.
12. Working with one strain at a time, pick up an oviduct with clean forceps, and transfer to a dish of M2 with hyaluronidase.
13. With a pair of watchmakers' forceps, pin the oviduct to the bottom of the dish. With a second pair of forceps, tear open the ampulla and release the cumulus mass containing the presumptive embryos (*see* **Note 5**). Repeat the procedure for the remaining oviducts from that strain.
14. Allow embryos to drop out of the cumulus cells. Pick up clean embryos with a transfer pipette and your pipetting apparatus (mouth pipette), and wash through two drops of M2 (without hyaluronidase). Hold in a drop of M2 overlaid with oil, or place in a drop of KSOM and hold in the incubator at 37 °C, 6% $CO_2$, and 5% $O_2$.
15. Repeat **steps 12–14** for the oviducts from the second strain (*see* **Note 6**).

### 3.2 Zona Drilling

1. Place a dish of embryos in M2 overlaid with oil on the inverted microscope, and remove the lid of the dish. (If embryos were held in KSOM in the incubator, move back to a drop of M2 with oil.)
2. Locate the embryos on low magnification and swirl the dish to collect the embryos in the center of the drop.
3. Rotate the XYClone objective (either 20× or 40×) into position, turn on the computer, and open the laser software.
4. Focus the laser on the zona of one embryo, at a point where the gap between the plasma membrane of the embryo and the zona is the greatest.
5. Activate the laser to cut a hole in the zona.
6. Move around the dish slowly, lasering one hole in the zona of each embryo (*see* **Note 7**).
7. Repeat the entire process for the dish containing embryos of the second strain.

### 3.3 Pronuclear Exchange

1. Prepare a manipulation dish by placing two 100 μL drops of M2 with demecolcine and cytochalasin B in the middle of a low-profile 60 mm petri dish in a north/south orientation. Thaw an aliquot of Sendai fusion virus, and place a 50 μL drop in close proximity but east of the drops of M2. Overlay with oil (Fig. 1).
2. Load the lasered C57BL/6J zygotes into the center of the southern drop of M2.
3. Load the lasered FVB zygotes into the center of the northern drop of M2.
4. Move the dish to the stage of the inverted microscope. Lower the holding pipette attached to the CellTram Air into the northern drop (FVB) of M2 from the left side. Adjust the height with the micromanipulators until the holding pipette is almost touching the bottom, but it does not scrape the bottom.
5. Insert a TransferTip (ES) into the capillary holder attached to the SAS air syringe on the right side of the microscope. Lower the needle into the northern drop of M2 while observing with the 4× objective through the binoculars of the microscope. Stop when the injection needle is in focus with the holding pipette. Both capillaries should be near the group of FVB embryos.

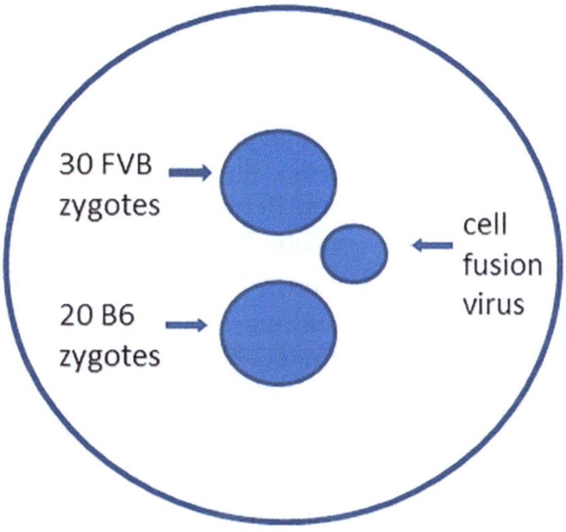


**Fig. 1** Construction of MNX embryos: C57BL/6J karyoplast to FVB cytoplast. Two 100 μL drops of M2 medium with demecolcine (0.1 μg/mL) and cytochalasin B (5 μg/mL) are placed in the center of a 60 mm dish in a north/south orientation. A 50 μL drop of Sendai cell fusion virus (HVJ-E) is placed just east of the drops of M2. Approximately 30 FVB zygotes are placed in the northern drop; approximately 20 C57BL/6J zygotes are placed in the southern drop (both groups possess a lasered opening in the zona). An ES cell injection needle is used to enucleate, then the B6 karyoplast is washed through Sendai cell fusion virus before transfer under the zona of the FVB cytoplast

6. Increase magnification to 200× (or 400× if more comfortable).
7. Pick up an FVB embryo while applying gentle suction with the holding pipette. Bring the two pronuclei of the one-cell embryo into focus. Rotate the embryo until the lasered hole in the zona is at approximately 2 o'clock position.
8. Focus on one pronucleus until the pronuclear membrane is sharp and the interior appears rather clear. Adjust the height of the needle until the tip of the needle is focused on the same plane as the pronuclear membrane.
9. Propel the needle forward until it passes through the lasered hole. This will give you some indication as to whether you have adjusted the needle to the middle of the embryo in the z plane. If you are too high or too low, you will likely roll the embryo.
10. After the needle is within the perivitelline space, focus again on the pronucleus and adjust the height of the needle. Move the needle forward until it is in contact with the plasma membrane near the pronucleus. When you are very close to the pronucleus, provide very gentle suction via the SAS air syringe to draw the pronucleus into the needle. Minimize the amount of cytoplasm that you withdraw. Move to the next pronucleus and repeat the procedure. When both pronuclei have been aspirated, withdraw the needle.
11. Move the enucleated cytoplast to the bottom of the drop. Expel the pronuclei in the needle at the top of the drop. Likewise, move any lysed embryos or cytoplasts to the top of the M2 drop, return to the center, pick up another embryo, and repeat the enucleation process.
12. When all FVB embryos have been enucleated, raise the capillaries slightly, and move the stage of the microscope to the southern drop of C57BL/6 J zygotes at low magnification (*see* **Note 7**).
13. Lower both capillaries into the drop, increase magnification, and immobilize an embryo on the holding pipette. Enucleate the C57BL/6J embryo in the same manner, being especially careful to minimize the amount of cytoplasm withdrawn with the pronuclei (*see* **Note 8**).
14. With the C57BL/6J pronuclei still in the needle, move the stage so that the capillaries move to the drop of virus. Expel the pronuclei into the drop of fusion virus, then draw the pronuclei back into the needle.
15. Move the stage again so that the capillaries move into the northern drop of FVB cytoplasts. Secure an enucleated FVB cytoplast onto the holding pipette and rotate until the lasered hole is in the 2 o'clock position. Insert the needle into the lasered hole, and deposit the C57BL/6J pronuclei in the

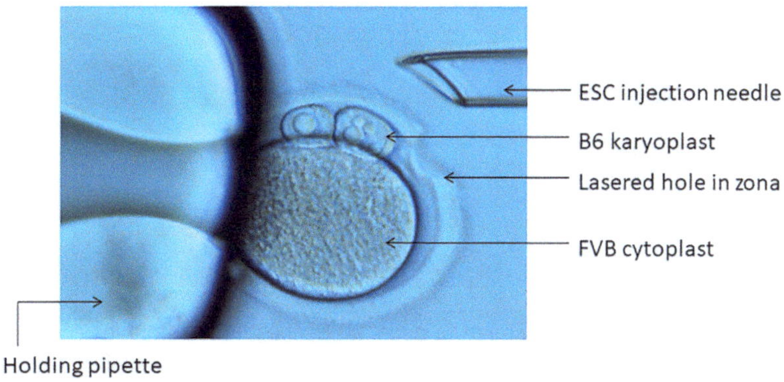


**Fig. 2** MNX embryo after reconstruction. A reconstructed MNX embryo is pictured. The tip of the holding pipette can be seen to the left, and the tip of the ESC needle is evident on the right. The FVB cytoplast contains no pronuclei, and the two C57BL/6J pronuclei have been transferred to the perivitelline space through the hole lasered in the zona of the FVB cytoplast

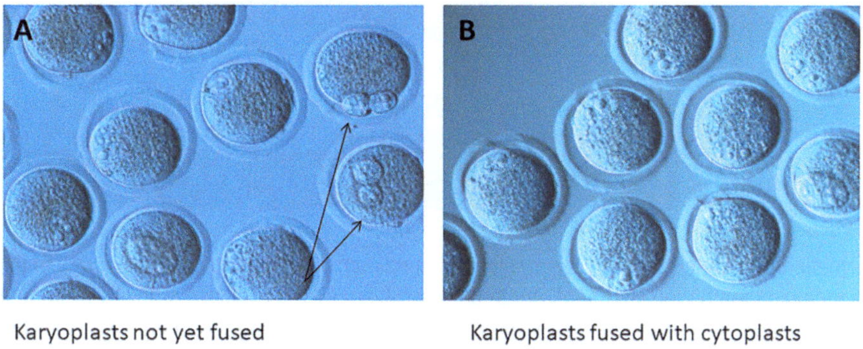


**Fig. 3** Reconstructed embryos. (**a**) Some of the transferred karyoplasts have not yet fused with the cytoplasts. (**b**) The C57BL/6J karyoplasts have fused with the FVB cytoplasts in all but one of the reconstructed embryos

perivitelline space of the FVB cytoplast (Fig. 2). Move the reconstituted embryo to the side of the drop.

16. Move back to the southern drop of C57BL/6J embryos and repeat **steps 13–15**. When all C57BL/6J karyoplasts have been transferred to the FVB cytoplasts, remove the reconstituted embryos, and transfer to a drop of KSOM with oil and incubate overnight. It is already possible at this point to see that the karyoplasts and cytoplasts have fused (Fig. 3).
17. Score MNX embryos for development to the two-cell stage (Fig. 4). Transfer two-cell embryos to 0.5 dpc pseudopregnant recipient females, 10 to 15 per recipient (*see* **Note 9**).

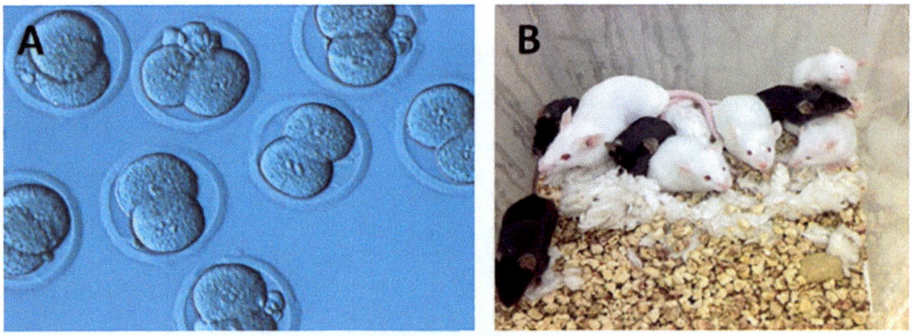


**Fig. 4** MNX embryo development to pups. (**a**) Reconstructed embryos cultured O/N in KSOM have developed normally to the two-cell stage. MNX embryos are transferred to recipient females at this point with FVB control embryos. (**b**) A CD-1 recipient female is shown with her litter of pups that resulted from transfer of MNX embryos in the presence of carrier FVB control embryos. The MNX pups are black, as they contain C57BL/6J nuclear DNA; the FVB control pups are albino

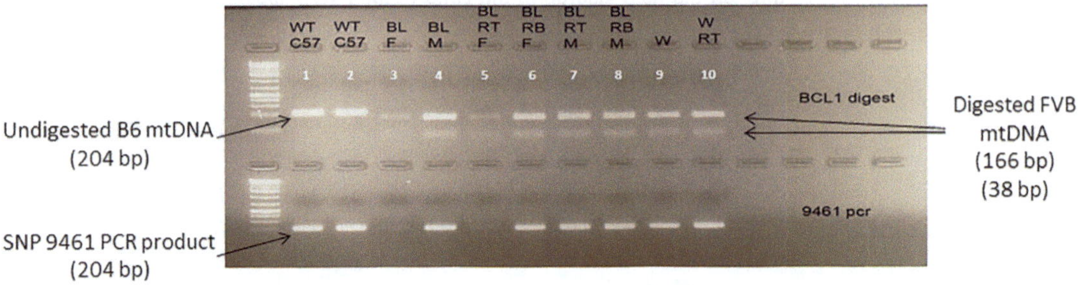


**Fig. 5** Confirmation of homoplasmy by SNP analysis. Homoplasmic mitochondrial background of the MNX pups were confirmed using restriction fragment length polymorphism (RLFP) analysis of the PCR products using the 9461 single-nucleotide polymorphism (SNP) that is distinguishable between the FVB/NJ and C57BL/6J mouse strains. PCR is performed followed by restriction digest with BclI. FVB and B6 mtDNA are distinguishable by a C to T mutation at position 9461. MNX containing B6 mtDNA will not have a PCR product digested with BclI, resulting in a 204 bp product. MNX containing the FVB mtDNA will have BclI digested product resulting in 166 bp and 38 bp products. Top row BclI digest, bottom row 9461 PCR. Lanes 1 and 2 are wild-type B6, lanes 9 and 10 are FVB controls, lanes 3–8 are MNX pups

### 3.4 Genotyping

1. Mitochondrial DNA homoplasmy was confirmed by RFLP (restriction fragment length polymorphism) analysis of PCR products using SNP (single-nucleotide polymorphism) that are distinguishable between mouse strains (Fig. 5). Mouse DNA extraction was performed using REDExtract-N-Amp Tissue PCR kit. Mouse DNA was subjected to 35–40 cycles of PCR using primers that span the SNP of interest. PCR products were then digested with a restriction enzyme. Both PCR product and restriction digested PCR product were run on agarose gel.

## 4 Notes

1. The strains of choice for creating an MNX mouse are completely dependent upon the research needs of the investigator. This protocol depicts transferring C57BL/6J karyoplasts to FVB/HeN cytoplasts, but other strain combinations are certainly possible. It is also important to note that the similarly named mouse strains from different vendors (Jax, Charles River, Envigo, Taconic, etc.) are not genetically identical.
2. Mouse strains differ in their response to superovulatory hormones according to their age and weight [54]. C57BL/6J females superovulate well between 3 and 4 weeks of age, whereas FVB females superovulate best between 8 and 10 weeks.
3. Rather than using two different media (one HEPES buffered for handling embryos in air and one bicarbonate buffered for culture in a $CO_2$ incubator), there are now commercially available media can be used for both environments.
4. Alternatively, fusion between the transferred karyoplast and recipient cytoplast can be accomplished by electrical pulse, if the equipment is available to accomplish this.
5. Scoring of the oviduct near the ampulla (the swollen and clear area of the oviduct) is an efficient means of isolating the cumulus mass. The cumulus mass will be visible in the ampulla, and you need only to nick the oviduct to create an opening; the cumulus mass will ooze out on its own.
6. Collecting presumptive zygotes from two different strains is more complicated than collecting from one strain, and embryos derived from different strains are virtually indistinguishable via microscopic examination. Above all else, be certain to maintain identity of the embryos with which you are working. It is possible to sacrifice the females from both strains and collect oviducts sequentially, but one should not try to process both groups through hyaluronidase at the same time. Begin by processing one group of oviducts from one strain through hyaluronidase, collect the denuded embryos, wash through M2 without enzyme and hold in M2 or KSOM under oil, then return and process the oviducts from the second strain.
7. The procedure of creating an opening in the zona by laser may require practice—both for focusing the laser and for scaling the intensity of the pulse. This procedure may be practiced on denuded presumptive zygotes that do not display pronuclei and will only be discarded. Alternatively, an opening in the zona can be created with a Piezo drill.

8. Mitochondria tend to cluster around pronuclei; therefore, by minimizing the amount of cytoplasm withdrawn with each pronucleus, the number of mitochondria transferred to the recipient cytoplast can be minimized, thus avoiding artificial heteroplasmy, unless this is an intended experimental outcome. It is possible to actually "brush off" extra cytoplasm by holding the enucleated pronucleus at the edge of the needle and brushing against the opening of the holding pipette. However, this will require some practice, as the pronuclei will tend to lyse if performed too aggressively. Critically, all mice generated from this procedure must be tested for homoplasmy for the desired mitochondrial genotype. If heteroplasmy is detected using PCR, then the mouse should be discarded if homoplasmy is desired.
9. Due to the technically challenging aspects of this procedure, it may not result in a large number of surviving MNX embryos, which may not be a sufficient number of embryos for transfer to pseudopregnant females to maintain pregnancy. It is suggested that carrier embryos also be transferred to pseudopregnant female hosts to assist in establishing the pregnancy. Embryos that have been lasered but not enucleated are suitable carriers for this process. If reconstructing FVB/HeN cytoplasts with C57BL/6J karyoplasts, then transfer five to 10 FVB embryos with the MNX embryos (Fig. 4). After confirmation of mitochondrial homoplasmy in the MNX pups, the line must only be propagated and maintained by mating MNX females to wild-type males of the nuclear strain. Similarly, the line should be cryopreserved by freezing embryos collected from MNX females mated to wild-type males.

## Acknowledgments

This work done in the authors' labs has been generously funded by Susan G. Komen for the Cure (SAC110037); the National Foundation for Cancer Research, METAvivor Research and Services Inc.; and Theresa's Research Foundation. Additional funding support was provided by US Army Medical Research Defense Command, W81XWH-18-1-0450 and National Cancer Institute P30-CA168524. D.R.W. is a co-inventor of MNX mice and holds US Patent US 9,040,771 B2 with Scott W. Ballinger at the University of Alabama at Birmingham. D.R.W., M.A.L., C.J.V, and J.L.V. each wrote and edited this manuscript. The authors declare that they have no known competing financial interests or personal relationships that could have appeared to influence the work reported in this paper.

## References


1. Schon KR, Ratnaike T, van den Ameele J et al (2020) Mitochondrial diseases: A diagnostic revolution. Trends Genet 36:702–717. https://doi.org/10.1016/j.tig.2020.06.009
2. Wallace DC (2018) Mitochondrial genetic medicine. Nat Genet 50:1642–1649. https://doi.org/10.1038/s41588-018-0264-z
3. Beadnell TC, Scheid AD, Vivian CJ et al (2018) Roles of the mitochondrial genetics in cancer metastasis: not to be ignored any longer. Cancer Metastasis Rev 37:615–632. https://doi.org/10.1007/s10555-018-9772-7
4. Scheid AD, Beadnell TC, Welch DR (2019) The second genome: effects of the mitochondrial genome on cancer progression. Adv Cancer Res 142:63–105. https://doi.org/10.1016/bs.acr.2019.01.001
5. Scheid AD, Beadnell TC, Welch DR (2021) Roles of mitochondria in the hallmarks of metastasis. Br J Cancer 124:124–135. https://doi.org/10.1038/s41416-020-01125-8
6. Welch DR, Foster C, Rigoutsos I (2022) Roles of mitochondrial genetics in cancer metastasis. Trends Cancer 8:1002. https://doi.org/10.1016/j.trecan.2022.07.004
7. Telonis AG, Loher P, Kirino Y et al (2014) Nuclear and mitochondrial tRNA-lookalikes in the human genome. Front Genet 5:344. https://doi.org/10.3389/fgene.2014.00344
8. Telonis AG, Kirino Y, Rigoutsos I (2015) Mitochondrial tRNA-lookalikes in nuclear chromosomes: could they be functional? RNA Biol 12:375–380. https://doi.org/10.1080/15476286.2015.1017239
9. Furnish M, Caino MC (2020) Altered mitochondrial trafficking as a novel mechanism of cancer metastasis. Cancer Rep (Hoboken) 3:e1157. https://doi.org/10.1002/cnr2.1157
10. Murata D, Arai K, Iijima M et al (2020) Mitochondrial division, fusion and degradation. J Biochem 167:233–241. https://doi.org/10.1093/jb/mvz106
11. Banerjee R, Mukherjee A, Nagotu S (2022) Mitochondrial dynamics and its impact on human health and diseases: inside the DRP1 blackbox. J Mol Med (Berl) 100:1–21. https://doi.org/10.1007/s00109-021-02150-7
12. Boulton DP, Caino MC (2022) Mitochondrial fission and fusion in tumor progression to metastasis. Front Cell Dev Biol 10:849962. https://doi.org/10.3389/fcell.2022.849962
13. Bernardino Gomes TM, Ng YS, Pickett SJ et al (2021) Mitochondrial DNA disorders: from pathogenic variants to preventing transmission. Hum Mol Genet 30:R245–R253. https://doi.org/10.1093/hmg/ddab156
14. Stewart JB, Chinnery PF (2015) The dynamics of mitochondrial DNA heteroplasmy: implications for human health and disease. Nat Rev Genet 16:530–542. https://doi.org/10.1038/nrg3966
15. Szczepanowska K, Trifunovic A (2017) Origins of mtDNA mutations in ageing. Essays Biochem 61:325–337. https://doi.org/10.1042/EBC20160090
16. Hahn A, Zuryn S (2019) The cellular mitochondrial genome landscape in disease. Trends Cell Biol 29:227–240. https://doi.org/10.1016/j.tcb.2018.11.004
17. Xia M, Zhang Y, Jin K et al (2019) Communication between mitochondria and other organelles: a brand-new perspective on mitochondria in cancer. Cell Biosci 9:27. https://doi.org/10.1186/s13578-019-0289-8
18. Meseguer S (2021) MicroRNAs and tRNA-derived small fragments: key messengers in nuclear-mitochondrial communication. Front Mol Biosci 8:643575. https://doi.org/10.3389/fmolb.2021.643575
19. Zhu D, Li X, Tian Y (2022) Mitochondrial-to-nuclear communication in aging: an epigenetic perspective. Trends Biochem Sci 47:645–659. https://doi.org/10.1016/j.tibs.2022.03.008
20. Wallace DC (2005) A mitochondrial paradigm of metabolic and degenerative diseases, aging, and cancer: a dawn for evolutionary medicine. Annu Rev Genet 39:359–407. https://doi.org/10.1146/annurev.genet.39.110304.095751
21. Wallace DC, Chalkia D (2013) Mitochondrial DNA genetics and the heteroplasmy conundrum in evolution and disease. Cold Spring Harb Perspect Biol 5:a021220. https://doi.org/10.1101/cshperspect.a021220
22. Wallace DC, Bunn CL, Eisenstadt JM (1975) Cytoplasmic transfer of chloramphenicol resistance in human tissue culture cells. J Cell Biol 67:174–188. https://doi.org/10.1083/jcb.67.1.174
23. Swerdlow RH, Parks JK, Miller SW et al (1996) Origin and functional consequences of the complex I defect in Parkinson's disease. Ann Neurol 40:663–671. https://doi.org/10.1002/ana.410400417
24. Ishikawa K, Takenaga K, Akimoto M et al (2008) ROS-generating mitochondrial DNA mutations can regulate tumor cell metastasis.

Science 320:661–664. https://doi.org/10.1126/science.1156906

25. Bussard KM, Siracusa LD (2017) Understanding mitochondrial polymorphisms in cancer. Cancer Res 77:6051–6059. https://doi.org/10.1158/0008-5472.CAN-17-1939
26. Feeley KP, Bray AW, Westbrook DG et al (2015) Mitochondrial genetics regulate breast cancer tumorigenicity and metastatic potential. Cancer Res 75:4429–4436. https://doi.org/10.1158/0008-5472.CAN-15-0074
27. Brinker AE, Vivian CJ, Koestler DC et al (2017) Mitochondrial haplotype alters mammary cancer tumorigenicity and metastasis in an oncogenic driver-dependent manner. Cancer Res 77:6941–6949. https://doi.org/10.1158/0008-5472.CAN-17-2194
28. Vivian CJ, Brinker AE, Graw S et al (2017) Mitochondrial genomic backgrounds affect nuclear DNA methylation and gene expression. Cancer Res 77:6202–6214. https://doi.org/10.1158/0008-5472.CAN-17-1473
29. Vivian CJ, Hagedorn TM, Jensen RA et al (2018) Mitochondrial polymorphisms contribute to aging phenotypes in MNX mouse models. Cancer Metastasis Rev 37:633–642. https://doi.org/10.1007/s10555-018-9773-6
30. Beadnell TC, Fain C, Vivian CJ et al (1866) Mitochondrial genetics cooperate with nuclear genetics to selectively alter immune cell development/trafficking. Biochim Biophys Acta Mol basis Dis 2020:165648. https://doi.org/10.1016/j.bbadis.2019.165648
31. Brinker AE, Vivian CJ, Beadnell TC et al (2020) Mitochondrial haplotype of the host stromal microenvironment alters metastasis in a non-cell autonomous manner. Cancer Res 80:1118–1129. https://doi.org/10.1158/0008-5472.CAN-19-2481
32. Fetterman JL, Zelickson BR, Johnson LW et al (2013) Mitochondrial genetic background modulates bioenergetics and susceptibility to acute cardiac volume overload. Biochem J 455:157–167. https://doi.org/10.1042/BJ20130029
33. Betancourt AM, King AL, Fetterman JL et al (2014) Mitochondrial-nuclear genome interactions in non-alcoholic fatty liver disease in mice. Biochem J 461:223–232. https://doi.org/10.1042/BJ20131433
34. Bray AW, Ballinger SW (2017) Mitochondrial DNA mutations and cardiovascular disease. Curr Opin Cardiol 32:267–274. https://doi.org/10.1097/HCO.0000000000000383
35. Dunham-Snary KJ, Sandel MW, Sammy MJ et al (2018) Mitochondrial - nuclear genetic interaction modulates whole body metabolism, adiposity and gene expression in vivo. EBioMedicine 36:316–328. https://doi.org/10.1016/j.ebiom.2018.08.036
36. Kandasamy J, Rezonzew G, Jilling T et al (2019) Mitochondrial DNA variation modulates alveolar development in newborn mice exposed to hyperoxia. Am J Physiol Lung Cell Mol Physiol 317:L740–L747. https://doi.org/10.1152/ajplung.00220.2019
37. Lifsted T, Le Voyer T, Williams M et al (1998) Identification of inbred mouse strains harboring genetic modifiers of mammary tumor age of onset and metastatic progression. Int J Cancer 77:640–644. https://doi.org/10.1002/(sici)1097-0215(19980812)77:4<640::aid-ijc26>3.0.co;2-8
38. Hunter KW, Broman KW, Le Voyer T et al (2001) Predisposition to efficient mammary tumor metastatic progression is linked to the breast cancer metastasis suppressor gene Brms1. Cancer Res 61:8866–8872
39. Park YG, Lukes L, Yang H et al (2002) Comparative sequence analysis in eight inbred strains of the metastasis modifier QTL candidate gene Brms1. Mamm Genome 13:289–292. https://doi.org/10.1007/s00335-001-2151-6
40. Park YG, Zhao X, Lesueur F et al (2005) Sipa1 is a candidate for underlying the metastasis efficiency modifier locus Mtes1. Nat Genet 37:1055–1062. https://doi.org/10.1038/ng1635
41. Lancaster M, Rouse J, Hunter KW (2005) Modifiers of mammary tumor progression and metastasis on mouse chromosomes 7, 9, and 17. Mamm Genome 16:120–126. https://doi.org/10.1007/s00335-004-2432-y
42. Crawford NP, Qian X, Ziogas A et al (2007) Rrp1b, a new candidate susceptibility gene for breast cancer progression and metastasis. PLoS Genet 3:e214. https://doi.org/10.1371/journal.pgen.0030214
43. Crawford NP, Alsarraj J, Lukes L et al (2008) Bromodomain 4 activation predicts breast cancer survival. Proc Natl Acad Sci U S A 105:6380–6385. https://doi.org/10.1073/pnas.0710331105
44. Gaudet MM, Hunter K, Pharoah P et al (2009) Genetic variation in SIPA1 in relation to breast cancer risk and survival after breast cancer diagnosis. Int J Cancer 124:1716–1720. https://doi.org/10.1002/ijc.23919
45. Faraji F, Pang Y, Walker RC et al (2012) Cadm1 is a metastasis susceptibility gene that suppresses metastasis by modifying tumor interaction with the cell-mediated immunity.

PLoS Genet 8:e1002926. https://doi.org/10.1371/journal.pgen.1002926
46. Yu D, Liu C, Guo L (2020) Mitochondrial metabolism and cancer metastasis. Ann Transl Med 8:904. https://doi.org/10.21037/atm.2020.03.42
47. Tiedemann K, Hussein O, Komarova SV (2020) Role of altered metabolic microenvironment in osteolytic metastasis. Front Cell Dev Biol 8:435. https://doi.org/10.3389/fcell.2020.00435
48. Ly T, Harihar S, Welch DR (2020) KISS1 in metastatic cancer research and treatment: potential and paradoxes. Cancer Metastasis Rev 39:739–754. https://doi.org/10.1007/s10555-020-09868-9
49. Tasdogan A, Faubert B, Ramesh V et al (2020) Metabolic heterogeneity confers differences in melanoma metastatic potential. Nature 577:115–120. https://doi.org/10.1038/s41586-019-1847-2
50. Wallace DC (2016) Genetics: mitochondrial DNA in evolution and disease. Nature 535:498–500. https://doi.org/10.1038/nature18902
51. Bassler BL (1999) How bacteria talk to each other: regulation of gene expression by quorum sensing. Curr Opin Microbiol 2:582–587. https://doi.org/10.1016/s1369-5274(99)00025-9
52. Eickhoff MJ, Bassler BL (2018) SnapShot: bacterial quorum sensing. Cell 174:1328. https://doi.org/10.1016/j.cell.2018.08.003
53. Irie Y, Parsek MR (2008) Quorum sensing and microbial biofilms. Curr Top Microbiol Immunol 322:67–84. https://doi.org/10.1007/978-3-540-75418-3_4
54. Luo C, Zuniga J, Edison E et al (2011) Superovulation strategies for 6 commonly used mouse strains. J Am Assoc Lab Anim Sci 50:471–478

# Chapter 5

# 3D-Autologous Culture Method for Precision Oncology

Qian Xu, Meiling Yan, and Yao Tang

## Abstract

The 3D-autologous culture method (3D-ACM) for patient-derived cancer samples utilizes a patient's own body fluid or serum to prepare a 3D scaffold and for the culture medium. 3D-ACM enables tumor cells and/or tissues from an individual patient to proliferate in vitro, in a microenvironment that is very similar to their original, in vivo surroundings. The purpose is to maximally preserve in culture the native biological properties of a tumor. This technique has been employed for two models: (1) cells isolated from malignant ascites or pleural effusions (body fluids) and (2) solid tissues from biopsies or surgically removed cancers. Here we describe the detailed procedures for these 3D-ACM models.

**Key words** Autologous culture, Patient-derived cancers, Malignant ascites, Pleural effusion, Solid tumor tissue, Microenvironment

## 1 Introduction

Three-dimensional (3D) in vitro cell/tissue culture models bridge the gap between traditional 2D culture models and native tissue systems and are widely used in routine laboratory research—including organoid studies [1, 2]. The major difference between previous 3D culture techniques and the 3D-autologous culture method (3D-ACM) described here is that to grow patient-derived cancer samples, 3D-ACM utilizes the patient's own body fluid or serum to prepare a 3D scaffold and as the culture medium, rather than a foreign substance like fetal bovine serum (FBS) [3, 4]. Furthermore, it is not necessary to add any commercial reagents (e.g., growth factors, cytokines, chemokines) or specialized media (e.g., special serum-free media or artificially modified culture media) to 3D-ACM cultures [3].

**Supplementary Information** The online version contains supplementary material available at https://doi.org/10.1007/978-1-0716-3163-8_5.

Usha N. Kasid and Robert Clarke (eds.), *Cancer Systems and Integrative Biology*, Methods in Molecular Biology, vol. 2660, https://doi.org/10.1007/978-1-0716-3163-8_5,

3D-ACM can be employed to culture cancer cells freshly isolated from malignant body fluids (Subheadings 2.1 and 3.1) and small pieces of solid tumor tissues directly removed from cancer patients (Subheadings 2.2 and 3.2). To retain the native biochemical microenvironment, cancer cells/tissues from samples are prepared physically, instead of chemically. Gradient centrifugation is used for body fluid cultures to remove red blood cells while saving all other cells, including tumor cells, mesenchymal cells, and possible white blood cells. Physical dissection methods are employed for biopsies or surgically removed samples, without enzyme digestion. With these methods, almost all cellular components and tissue structures in the original samples are preserved in the in vitro 3D-ACM cultures. In addition, the native growth patterns are protected by replicating the in vivo environment of a tumor; the cells isolated from body fluids are seeded onto 3D gels that resemble the smooth surface of pleural or abdominal cavities, while the solid tumor pieces are embedded into 3D scaffolds. Cells/tissues that we grew in 3D-ACM showed faster proliferation relative to traditional 3D-FBS culture methods and the ability to self-organize to form new tissues. In addition, the original histopathology, immunophenotypes and cytokine production were well-preserved [3, 4].

## 2 Materials

Consent for the use of blood, body fluids, and surgical/biopsy samples is required from the patients and/or their families. Tumor donors must not have had chemotherapy, targeted therapy, or radiation treatment for at least 3 months prior to sample collection. The preparation and use of the experiment materials should be performed at 4 °C, unless indicated otherwise. All equipment (dishes, plates, pipettes, tubes/bottles, etc.) used in these cultures needs to be cooled overnight at −20 °C. Biological waste disposal regulations should be closely followed.

### *2.1 Liquid Samples (Ascites or Pleural Effusions)*

1. Clinical sample: Fresh ascites or pleural effusions (≤5 h old) from a patient with malignant cancer.
2. Autologous medium (AM):
   (a) Centrifuge the fresh ascites or pleural effusion at 2000 rpm for 10 min to precipitate cells (these cells are further processed and used for the cell culture; *see* Subheading 3.1 **step 2**).
   (b) Gently remove the supernatant to a new container, then centrifuge at 10,000 rpm for 20 min to remove impurities in the liquid sample (e.g., cell debris).

(c) Transfer the supernatant to the upper chamber of a sterile vacuum bottle with 0.45 μm pole size (we used the one with PES membrane from the Thermo Fisher company), then turn on a laboratory vacuum pump to filter the body fluid (*see* **Note 1**).

(d) Add the selected antibiotic in an appropriate concentration (we used cefoperazone at a final concentration of 20 μg/mL), then store the AM at 4 °C.

3. Autologous Matrigel (AM-Matrigel): Slowly dissolve the Matrigel (stored at −20 °C) by seat it in an ice box overnight in a 4 °C refrigerator, and then mix it well with AM at a 1:1 ratio just before use. Adjust the antibiotic to the right concentration (we used cefoperazone, final concentration of 20 μg/mL).
4. Antibiotics solution: Follow the pharmaceutical company's directions to make a stock solution that is ready for use in the final concentration at different steps (e.g. culture media).

### 2.2 Solid Tissue Samples

1. Clinical sample: put the fresh solid tumor tissue (≥5 $mm^3$ in size), obtained from either surgical operation or clinical biopsy, in a container with cold antibiotics-prepared PBS (we used cefoperazone at 40 μg/mL). Store the sample at 4 °C no longer than 2 h before use.
2. Autologous medium (AM):

(a) To avoid the effect of anesthesia in blood serum, before surgery, withdraw at least 15 mL peripheral venous blood from the cancer donor into an anticoagulant-free tube, and leave it at room temperature until clotting is complete.

(b) Centrifuge the blood sample at 3000 rpm for 10 min, then gently transfer the serum into a new tube.

(c) Centrifuge again at 16,900 g for 5 min to remove any debris in the serum.

(d) Pass the serum through a 0.22 μm filter, then add the selected antibiotic in an appropriate concentration (we used cefoperazone at a final concentration of 20 μg/mL).

(e) Based on the available volume of patient's blood, the AM could be 100% autologous serum, or ≥50% autologous serum complemented with culture medium (we used RPMI-1640).

3. Autologous Matrigel (AM-Matrigel): Slowly dissolve the Matrigel (stored at −20 °C) by seat it in an ice box overnight in a 4 °C refrigerator, and then mix it well with cold autologous serum at a 1:1 ratio. Adjust the antibiotic to an appropriate

concentration (we used cefoperazone at a final concentration of 20 μg/mL).

4. Antibiotics solution: Follow the pharmaceutical company's directions to make a stock solution that is ready to use in the final concentration at different steps.

## 3 Methods

### 3.1 Cell Culture for Liquid Samples

1. Coating culture dishes/plates: The numbers and the sizes of culture dishes/plates depend on the yield of cells obtained (we usually use 100 mm tissue culture dishes).
   - (a) Coat dishes/plates with AM-Matrigel (*see* Subheading 2.1 **item 3**) to evenly cover the bottom (about 1 mm thickness).
   - (b) Let the gel polymerize in a 37 °C incubator for 20–30 min.
2. Cell preparation:
   - (a) After washing cells (from Subheading 2.1 **item 2-a**) with cold PBS once (*see* **Note 2**), resuspend the cells in cold, serum-free, regular culture medium (we used RPMI-1640).
   - (b) Slowly load the cell suspension onto Ficoll solution (*see* **Note 3**). Centrifuge at 400 g with a descent speed of zero for 35 min at 20 °C.
   - (c) Carefully pipette out the layer that contains most of the cells (Fig. 1), and transfer it to a new tube.
   - (d) Wash the cells with cold PBS again; using the Trypan blue method to count the total yield of live cells (Fig. 2). The counting of cells should focus on those ≥10 μm in diameter, to exclude lymphocytes and cell debris from the yield.
3. Cell Culture
   - (a) Warm the AM in a 37 °C water bath.
   - (b) Resuspend the cells at a concentration of $1 \times 10^5$/mL.

**Fig. 1** Cells (except red blood cells) isolated by Ficoll solution (yellow arrow)

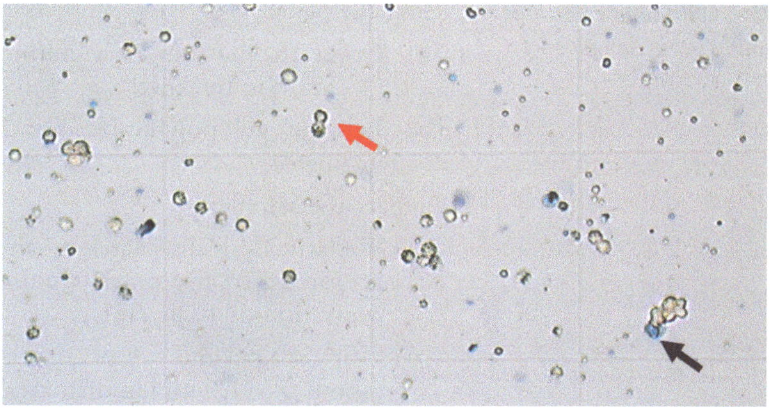

**Fig. 2** Trypan blue stain for cell counting. Red arrow, live cells; black arrow, dead cells

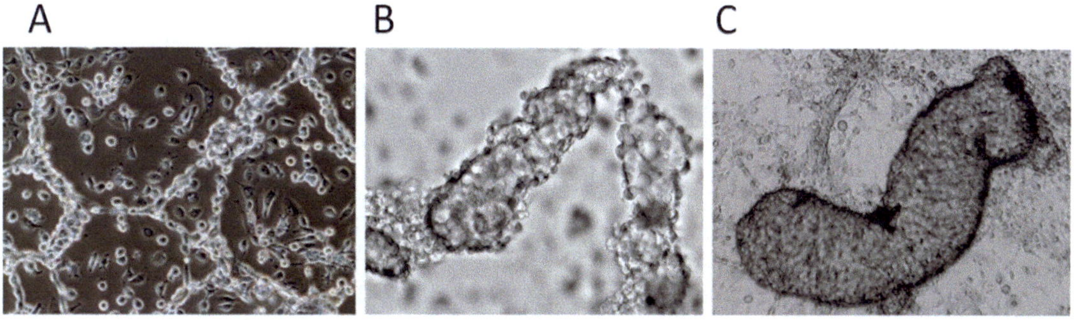


**Fig. 3** Tissue-like structures formed in liquid samples. Microscopy: ACM cultures of single cell suspension isolated from body fluids of cancer patients. (**A**) Gastric cancer (ascites), (**B**) Breast ductal carcinoma (pleural effusion), and (**C**) Lung adenocarcinoma (pleural effusion)

(c) Seed the cells onto the surface of the polymerized AM-Matrigel (we used a 10 mL/dish for $1 \times 10^6$ cells).

(d) Refresh the AM every 3–4 days. Under microscope, cell migration and self-organization may be observed as early as 1–2 days. Cell conflation over 80% usually occurs after 4–5 days in the culture dish, at which tissue-like structures may form in the plate (Fig. 3; Supplementary Video 1) (*see* **Note 4**).

(e) Cell passage can be done based on the experiment's design. When cells conflate over 80–90% in a culture dish/well, gently wash the dish/well with PBS once, then add TrypLE™ Express Enzyme following the company's directions. Use an autologous Matrigel-coated plate (as described in Subheading 3.1 **step 1**) to continue the culture. Drug-sensitivity assay can be performed during the culture period (*see* **Notes 5** and **6**).

### 3.2 Cell Culture for Solid Tissue Samples

1. Coat the plate:
   (a) Cover the bottom of a culture well with AM-Matrigel (1–2 mm in thickness).
   (b) Let the gel polymerize for 20–30 min in a 37 °C incubator.
2. Prepare the sample:
   (a) Wash the fresh clinical tissue samples with cold antibiotics-prepared PBS five times (we used cefoperazone at a concentration of 40 μg/mL).
   (b) Using scalpel and scissors, cut the tissue into small pieces (about 0.5–1.0 mm in diameter), avoiding necrotic areas.
3. Set up the 3D culture
   (a) Place three pieces per well on top of the pre-polymerized AM-Matrigel (tissues prepared as described in Subheading 3.2 **step 2**), positioning them to be 3–5 mm apart (Fig. 4a). Cover these pieces with additional AM-Matrigel (the same used for coating the well).
   (b) Let the gel polymerize in a 37 °C incubator for another 30 min to allow the tissues to become well-embedded in the AM-Matrigel.
   (c) Add pre-warmed AM into the culture well (we used a 24-well culture plate; the AM was 0.3 mL/well).
   (d) Refresh the AM every 3–4 days. Expect new tissues to grow out from the implants and cover the bottom of the culture well after 10–20 days (Fig. 4b). Under microscope or even with the naked eye, migration of tumor and matrix cells from the implanted tumor pieces may be observed as early as 3–5 days; these cells self-organize into tissue-like structures based on the characteristics of the original tumor (Fig. 5; Supplementary Video 2) (*see* **Notes 4** and **6**).

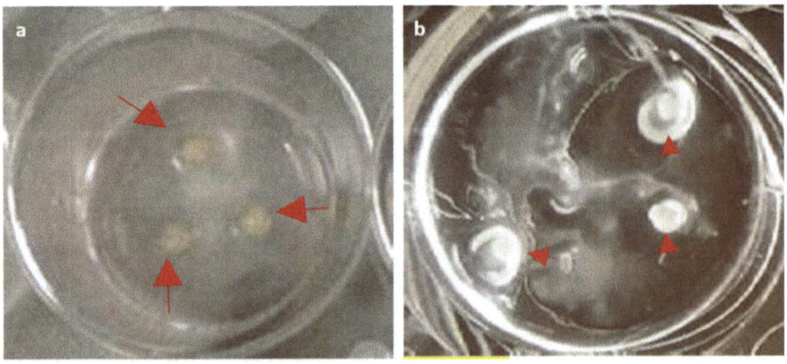


**Fig. 4** Solid tissue in culture well. (**a**) Three pieces of fresh tumor tissue (red arrow) were put on the 3D AM-Matrigel. (**b**) Solid tumor sample in ACM culture for 10 days. Newly grown tissues surrounded the breast cancer implants (red arrowhead) and spread in the culture well

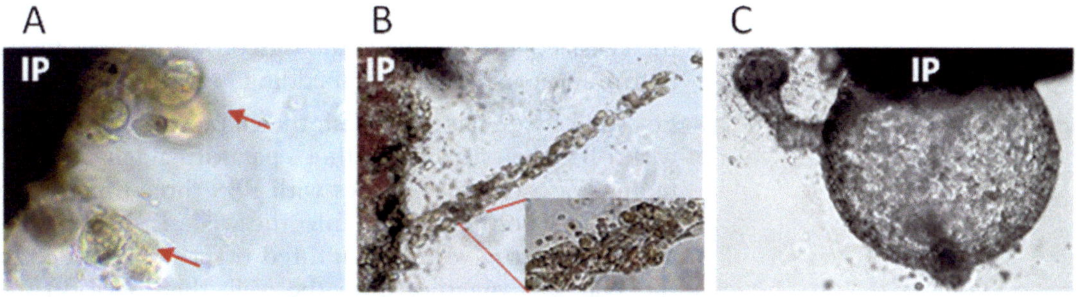


**Fig. 5** Tissue-like growth in solid tumor samples. Microscopy: after 10–20 days in ACM culture, self-organized tissue-like structures formed and grown out of the implanted tumor tissues (IP). (**A**) Breast ductal carcinoma. (**B** and **C**) Lung adenocarcinomas from two different patients

## 4 Notes

1. Passing body fluids through the 0.45 μm filter bottle could cause a blockage; for a large amount of body fluid, new filter units may be needed to complete this step.
2. If there are many red blood cells in the cellular pellet (bloody body fluids), RBC lysis buffer will need to be used following the company's directions.
3. Ficoll graduation solution can be purchased from different companies and should be used following the company's directions.
4. Self-organizations can be observed 1–2 days in body fluid samples (Supplementary Video 1) and 3–4 days in solid tumor cultures (Supplementary Video 2).
5. Drug-sensitivity assay for a body fluid sample can be performed during the culture period (We used CCK-8 kit to compare the cellular viabilities of drug-treated cells and the negative controls) [3]. Drug-sensitivity tests for solid tumors have not been well established; they will require appropriate measurement tools or reagents.
6. All cultured tissues can be harvested for further studies. For immunohistochemistry (IHC), different sample types require different harvest procedures:
   (a) For liquid samples. Wash the cultured tissues gently with PBS, then fix them with 4% paraformaldehyde for 1 h. Scrape the cells from the bottom of the well, and transfer them to a 1.5 mL tube containing PBS buffer, then centrifuge them at 1500 rpm for 5 min. Remove the supernatant; carefully transfer the sediment to a MG tissue paper

(also called Sydney paper), then enfold the paper to cover the cells. Follow routine pathology procedures for tissue dehydration and paraffin embedding.

(b) For solid tissue samples. Wash the cultured tissues gently with PBS, then fix them with 4% paraformaldehyde overnight. Wash the fixed tissues with PBS three times, then pour 4% dissolved agarose into the well. Take the polymerized agarose with the cultured tissues (embedded in the agar gel) together from the well, then perform the routine procedures of dehydration and paraffin embedding.

## References


1. Koledova Z (2017) 3D cell culture: an introduction. Methods Mol Biol 1612:1–11
2. Thermo Fisher Scientific (2018) Cell Culture: 3D cell culture product selection guide for organoids and spheroids. https://www.thermofisher.com/us/en/home/life-science/cell-culture/organoids-spheroids-3d-cell-culture.html?icid=fl-bid-3dmodel
3. Tang Y, Xu Q, Yan M et al (2020) Autologous culture method improves retention of tumors' native properties. Sci Rep 10(1):20455
4. Tang Y (2021) FBSA vs. ACM: how should human cancer cells be cultured? Adv Pharmacoepidemiol Drug Saf 10(4):1–3

# Chapter 6

# High-Plex Spatial Profiling of RNA and Protein Using Digital Spatial Profiler

**Nan Wang, Xia Li, and Zhiyong Ding**

## Abstract

The rapid emergence of spatial multi-omics technologies in recent years has revolutionized biomedical research. Among these, the Digital Spatial Profiler (DSP, commercialized by nanoString) has become one of the dominant technologies in spatial transcriptomics and proteomics and has assisted in deconvoluting complex biological questions. Based on our practical experience in the past 3 years with DSP, we share here a detailed hands-on protocol and key handling notes that will allow the broader community to optimize their work procedure.

**Key words** Spatial biology, Transcriptomics, Proteomics, Formalin-fixed paraffin-embedded (FFPE) specimens, Digital Spatial Profiler (DSP), Whole Transcriptome Atlas (WTA), Cancer Transcriptome Atlas (CTA), Companion diagnostics (CDs), Next-generation sequencing (NGS)-based barcode

## 1 Introduction

Digital Spatial Profiler (DSP) was introduced in 2019 as a breakthrough technology to map the spatial expression of mRNA and proteins designed for formalin-fixed paraffin-embedded (FFPE) specimens [1]. Its usage has extended across broad research fields, with particularly significant applications in oncology [2–5]. User guides are available from the manufacturer delineating DSP's experimental procedures [6, 7]. In addition, we and others have provided experimental designs, including sample selection criteria, morphological selection strategies, region-of-interest (ROI) selection, choices of segmentation, and data analysis methodologies applied under the setting of the cancer medicine field [8–11]. DSP also has the potential as a clinical pathological tool to provide value in companion diagnostics (CDs) [12–14]. DSP now allows a broad spectrum of target profiling from 1,800 genes

Usha N. Kasid and Robert Clarke (eds.), *Cancer Systems and Integrative Biology*, Methods in Molecular Biology, vol. 2660,
https://doi.org/10.1007/978-1-0716-3163-8_6,

(Cancer Transcriptome Atlas, CTA) to more than 18,000 genes (Whole Transcriptome Atlas, WTA). In addition, more than 140 proteins can be profiled simultaneously on a parallel-sectioned slide.

An early protocol describing DSP workflow for low-plex mRNA and protein measurement allows only up to ~90 targets to be quantified in one assay [6]. A recent protocol described RNA profiling in high-plex mode showcasing the principles of the DSP technology but is less detailed [7]. Although there are standard workflows for DSP experimentation, user-experience-based protocols detailing both high-plex RNA and protein profiling, as well as key technical notes, are still lacking. Here, we present a fully optimized step-by-step working procedure for DSP using next-generation sequencing (NGS)-based barcode counting for both high-plex RNA and protein profiling on FFPE tissues.

## 2 Materials

Conducting DSP experiments requires a complicated setup, and the entire workflow can be divided into three parts: (1) sample processing, (2) DSP work procedure, and (3) post-collection procedure. Careful planning is required before experiments, including tissue conditions, choices of antibodies for morphological markers, and analytes of interest, all of which are directly linked to materials and DSP panel selection strategies. Depending on the analytes type (RNA or protein), sample processing methods vary significantly and require different assay setups. A schematic workflow chart is depicted in Fig. 1.

### 2.1 Equipment and Hardware

GeoMx DSP Instrument.

Pressure cooker (BioSB 7008 or equivalent).

Staining jars.

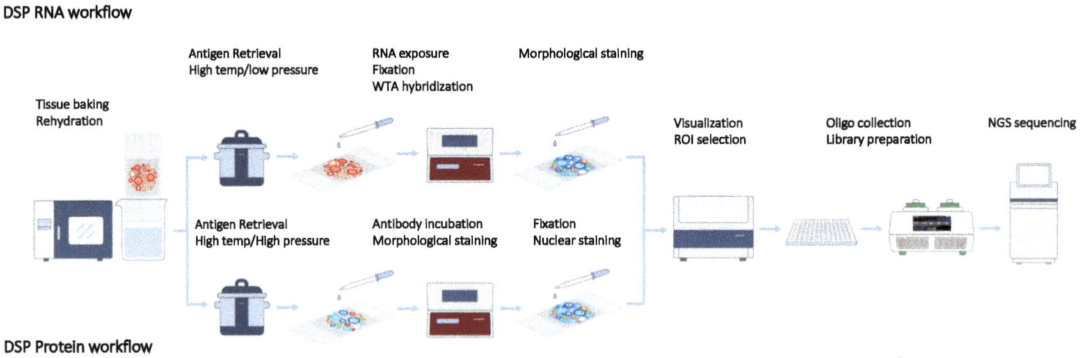


**Fig. 1** Overview of the DSP workflow for protein and RNA profiling using the NGS method

Humidity chamber.

Drying oven.

Centrifuges (Eppendorf 5810R or equivalent, RCF ≥ 2000 g).

Microfuges (compatible with EP microtubes).

Thermocycler (Bio-Rad C1000 Touch or equivalent).

Vortexer.

Icebox.

Heat Sealer (Thermo ALPS 50 V, recommended).

Magnetic stand (Thermo or equivalent).

Water bath.

Hybridization chamber (Boekel 240200-2, recommended).

Thermo Qubit.

Bio-analyzer (Qsep100 or equivalent).

Clean hood.

NGS sequencer (preferably NovaSeq 6000 at a yield of 800 Gigabytes per lane).

### 2.2 Reagents

Xylene.

Ethanol absolute.

DEPC water (PCR grade, nuclease-free).

RNase AWAY (Thermo Scientific 7002).

Proteinase K (Thermo Scientific AM2548).

10× Tris–EDTA (pH 9.0).

10× PBS (pH 7.4, dilute with DEPC water to the working concentration).

10% neutral buffered formalin (NBF).

100% deionized formamide (Thermo Scientific AM9342).

Tris–base.

Glycine.

20× saline sodium citrate (SCC) buffer (Sigma S6639).

10% Tween-20 (Teknova T0710).

GeoMx DSP RNA Slide Prep Kit for FFPE.

GeoMx Morphology Kit – Human RNA compatible (can be replaced with custom-developed antibodies).

GeoMx Whole Transcriptome Atlas – Human RNA for NGS (−80 °C).

DSP collection plates.

GeoMx Seq Code Primer Plates.

Fluoromount-G Mounting Medium (Southern Biotech 0100-01).

Purification beads (Beckman AMPure XP A63881, recommended).

Elution buffer (Qiagen Lot No. 163049988).

Additional or alternative reagents for DSP protein assay.

Tween-20 (Sigma P1379).

10× Citrate buffer (pH 6.0) (Sigma C9999).

10× TBS.

10× TBST.

16% formamide (dilute with PBS to 4%) (Thermo 28908/10).

GeoMx DSP Protein Slide Prep Kit for FFPE.

GeoMx Morphology Kit for Protein.

GeoMx Core and Module Kits for NGS readout.

### 2.3 Consumables and Accessories

Pipettes for 5–1000 μL.

1.5 mL RNase-free EP tubes.

RNase-free Tips (10 μL, 20 μL, 200 μL).

96-well deep-well PCR plates.

AeraSeal PCR films (Sigma A9224).

Easy pierce heat sealing foil (Thermo Scientific AB-0559).

Aluminum foils.

Dust-free wipes.

Heat protective glove.

Disposable gloves and masks.

Hydrophobic PAP pen.

Coverslips.

Scalpels.

Scissors.

## 3 Methods

### 3.1 Morphological Marker Validation on Tissues

Careful evaluation of morphological markers is pivotal for DSP and has a significant impact on the data per se. For commercially provided morphological markers (CD45, pan-cytokeratin, S100B, amyloid-beta, Iba1 co-stained with SYTO13), it may not be necessary to perform pre-experimental validation on tissues. However, as antibody performance varies across different tissue types and for tissues that are preserved longer than 2 years, optimization of the morphological antibody staining is recommended. Although other antibody information is provided on the nanoString website and in literature, morphological optimization is still indispensable based on our experience. We also suggest performing a parallel hematoxylin–eosin (HE) staining of tissues to allow better pathological evaluation of histological features of interest. Apart from nuclear staining, the DSP instrument has three additional fluorescent channels to allow multiplex staining of up to three antibodies. When starting a DSP experiment, we strongly recommend a morphological marker checklist detailing the antibody names, product details, clone numbers, species, and labeling details (direct conjugation or secondary antibody).

Recommended testing concentration for antibody optimization is typically 1:100 for DSP protein assay and 1:50 for DSP RNA assay. However, if no official dilution is provided, we suggest increasing the concentration to 1:20 for the DSP RNA assay and then optimizing after the first-round evaluation. In most cases, if an antibody underperforms for three rounds, it is not likely to work with the DSP procedure, so one shall source alternative antibodies. If no antibodies are available for the target of interest, researchers may consider RNA-Scope as an alternative method.

### 3.2 Tissue Processing

For both RNA and protein assays, ensure the working environment and all equipment are free of nuclease using RNase AWAY, and prepare all working solutions with DEPC water where necessary.

### 3.3 RNA-Based Profiling (WTA and CTA)

#### 3.3.1 Working Solution Preparation

1. Prepare ethanol with DEPC water to 95% (500 mL), and store at room temperature (RT).
2. Prepare ethanol with DEPC water to 70% (50 mL), and store at RT.
3. Prepare 80% ethanol for NGS library construction (made fresh on need).
4. Prepare 1× PBS with DEPC water (500 mL), and store at RT.
5. Dilute 2.5 μL proteinase K (20 mg/mL) in 50 mL 1× PBS (1: 2000, 1 μg/mL) (**Note:** make fresh on need).
6. Prepare 1 L of 4×/2× SSC from the 20× SSC stock solution, store at RT.
7. 2× SSC-T buffer: Add 10% Tween-20 (stock solution) to 2× SSC, store at RT.
8. NBF stop buffer: dissolve 24.5 g Tris base and 15 g glycine in DEPC water to a final volume of 2 L, store at RT.
9. Prepare 1× Tris–EDTA pH 9.0 from 10× Tris–EDTA pH 9.0 with DEPC water, and store at RT for antigen retrieval.
10. Stringent wash solution: Bring 4× SSC and formamide back to RT and mix 1:1 to a total of 100 mL, store at 37 °C in a water bath (**Note:** prepare on day two before use).

*3.3.2 Tissue Baking and Rehydration*

1. Bake FFPE tissue slides in an oven at 60 °C for 30–60 min (prolonged baking time may apply depending on the tissue conditions).
2. Immerse tissues in xylene for 5 min, and repeat twice with fresh xylene each for 5 min.
3. Rehydrate tissues in ethanol absolute for 5 min, and change fresh again for 5 min.
4. Rehydrate tissues in 95% ethanol for 5 min.
5. Rehydrate tissues in 1× PBS for 5 min, and change fresh again until the next step.

*3.3.3 Antigen Retrieval*

1. Add water into the pressure cooker (BioSB), and also add 1× Tris–EDTA (pH 9.0) into one staining jar and DEPC water into another jar. Bring the cooker up to 100 °C and open the valve. Continue boiling for another 30 min until solutions in both staining jars reach 100 °C.
2. Put tissue slides into the preheated DEPC water for 10 s, and then transfer immediately to the preheated Tris–EDTA (pH 9.0).
3. Close the cooker with the valve open to allow antigen retrieval under low-pressure and high-temperature condition (representative retrieval times for different tissue types are listed below and for other tissue types, use 15 min to start the optimization).

| Tissue type | Target retrieval in Tris–EDTA |
|---|---|
| Breast | 20 min |
| Cell pellets | 5 min |
| Colorectal | 20 min |
| Melanoma | 20 min |
| NSCLC | 20 min |
| Prostate tumor | 20 min |
| Tonsil | 15 min |

4. After antigen retrieval, transfer slides immediately into 1× PBS at RT for 5 min.

*3.3.4 RNA Target Exposure*

Pour proteinase K working solution into a tub, warm up in a 37 °C water bath, and then immerse the slides into the buffer. For different tissue types, refer to the chart below, and for other types of tissues, start with 1 μg/mL for 15 min for optimization.

| Tissue type | Proteinase K digest |
|---|---|
| Breast | 0.1 μg/mL for 15 min |
| Cell pellets | 1 μg/mL for 5 min |
| Colorectal | 1 μg/mL for 15 min |
| Melanoma | 1 μg/mL for 15 min |
| NSCLC | 1 μg/mL for 15 min |
| Prostate tumor | 1 μg/mL for 15 min |
| Tonsil | 1 μg/mL for 15 min |

After digestion, immerse the tissues in 1× PBS for 5 min, and proceed immediately to the next step.

#### 3.3.5 Tissue Fixation and in Situ WTA Probe Hybridization

1. Soak tissues in 10% NBF for 5 min.
2. Transfer to NBF stop buffer for 5 min and repeat one more time.
3. Transfer to 1× PBS for 5 min.
4. Defrost the WTA probes at RT, gentle pipette to remix, and store at 4 °C until use.
5. Bring Buffer R (in the GeoMx DSP RNA Slide Prep Kit) to RT.
6. Wipe the hybridization chamber ((Boekel 240200-2) thoroughly with DEPC water and RNase AWAY to prevent RNA degradation or cross contamination (**Note:** this step is essential, so do it with extra care).
7. Prepare hybridization solutions according to the following chart (*n* represents the number of slides); **Note**: Here, we only consider the situation without adding custom spike-in DSP probes for custom-designed genes.
8. Wet dust-free wipes with 2× SSC (**Note:** remove extra dripping water), and place underneath the slide stand in the hybridization chamber.
9. Remove the residual solution on the tissue slide, and place the slides on the slide holding stand. Apply 200 μL of hybridization solution onto each slide and cover carefully with coverslips. To make sure no bubbles are introduced, touch place the cover from one side and lower down gently (use pipette tips to help).
10. Shut the chamber and incubate at 37 °C overnight (16–24 h).

*3.3.6 Removal of Extra Probes, Blocking, and Morphological Staining*

1. Bring 100% formamide to RT before opening and preheat stringent wash solution to 37 °C.
2. Hand-hold the slides, immerse in 2× SSC, and move up and down gently until the coverslips fall off; if coverslips do not fall off within 5 min, then use 2× SSC-T to help dislodge the coverslips (**Note**: avoid force lift the cover manually).
3. Soak slides in stringent wash solution at 37 °C for 25 min and repeat once.
4. Soak in 2× SSC for 2 min and repeat once.
5. Bring the SYTO13 (nuclear stain) to RT, vortex, and pulse spin for 1 min (**Note**: pipette up and down if necessary).
6. Add 200 μL of buffer W (in the GeoMx DSP RNA Slide Prep Kit), and block 30 min in the humidity chamber at RT (**Note**: to avoid dry-out, use clean tips to spread the blocking reagent to cover the entire tissue area).
7. Prepare staining solution according to the chart below (**Note:** avoid touching the bottom of the SYTO13 stock solution when pipetting, and *n* represents the number of slides):

| SYT013 | Morphological marker 1 | Morphological marker 2 | Other markers | Buffer W | Total volume |
|---|---|---|---|---|---|
| 22 μL × *n* | 5.5 μL × *n* | 5.5 μL × *n* | 5.5 μL × *n* | 187 μL × *n* | **~220** μL × *n* |

**Note:** Here, commercial morphological markers are used for demonstration purposes, and custom antibody dilutions may vary; refer to 3.1 for details

8. Remove buffer W, apply 200 μL staining solution onto the slides, and incubate for 60 min in the humidity chamber at RT.
9. Immerse the slides in 2× SSC and repeat once. Cover the tub with aluminum foils before proceeding to scanning on DSP.

*3.3.7 Post-collection Process and Library Preparation*

1. Mount the slides with Fluoromount-G Mounting Medium for temporary storage at 4 °C in the dark (**Note:** usually, slides can be re-scanned for fluorescence imaging within a month).
2. For immediate processing, use permeable membrane (AeraSeal PCR films) to seal the collection plates containing UV-cleaved oligos. (**Note:** plates can be stored for less than 24 h at 4 °C before processing or up to30 days at −20 °C, or more than 30 days at −80 °C).
3. Heat-dry the sealed plate at 65 °C for 60 min in a thermocycler (Bio-Rad C1000 Touch) until wells are thoroughly dried down.

4. Add 10 μL of PCR grade DEPC water into each well and pipette five times, and then seal the plate with PCR films to prevent contamination. Keep steady at RT for 10 min, and quick spin down to ensure all liquid goes to the bottom.
5. Clean the workstation with RNase AWAY, and conduct all PCR-related work within a clean hood. All procedures, including thawing reagents, must be carried out on the ice.
6. Select the correct GeoMx Seq Code Primer Plate corresponding to lab worksheet generated from each DSP run (**Note:** seq codes can be pooled for sequencing given no identical seq code indices are mixed in one library).
7. Set up PCR reaction cocktails for each well as follows:

| PCR amplification reaction components | Volume (μL) |
|---|---|
| 5× PCR master mix | 2 μL |
| Primer mix (F & R, 2.5 μM each) | 4 μL |
| DSP aspirate | 4 μL |
| Total volume per reaction | 10 μL |

8. In a deep-well PCR plate, add 2 μL of PCR Master Mix first and then 4 μL of primers followed by 4 μL of templates. Mix by pipetting (**Note:** well indices must be fully matched between the collection plate and PCR plate).
9. Heat seal the PCR plate with PCR films (easy pierce heat sealing foils) using the heat sealer, quick spin, and load onto the thermocycler (Bio-Rad C1000 Touch) with the lid heated to 100 °C.
10. Program in the cycling conditions as below:

| Step | Temp | Time | Cycles |
|---|---|---|---|
| UDG incubation | 37 °C | 30 min | 1× |
| UDG deactivation | 50 °C | 10 min | 1× |
| Initial denaturation | 95 °C | 3 min | 1× |
| Denaturation | 95 °C | 15 s | 18× |
| Anneal | 65 °C | 60 s | |
| Extend | 68 °C | 30 s | |
| Final extension | 68 °C | 5 min | 1× |
| Hold | 4 °C | ∞ | 1× |

**Note:** PCR products can be stored at 4 °C for 24 h or −20 °C for up to 72 h

#### *3.3.8 Library Purification and Quality Control (QC)*

1. Perform all purification steps on the bench, and bring the AMPure XP beads to RT for 20 min to maximize their working efficiency.
2. Upon finishing PCR, quick centrifuge the PCR products and transfer 4 μL from individual wells, and mix into a 1.5 mL EP tube.
3. Calculate the total volume within the tube and add 1.2× bead-to-sample ratio of AMPure XP beads, pipette ten times to mix, and quick centrifuge.
4. Incubate for 5 min on the bench, place on the magnetic stand for 5 min to pellet beads, and remove clear supernatant with care (**Note:** avoid direct disturbing the beads).
5. Add 200 μL fresh made 80% ethanol to wash beads and incubate 30 s on the stand.
6. Remove supernatant and repeat once (**Note:** remove residue ethanol thoroughly when possible).
7. Air-dry the pellet beads on the stand for 5 min at maximum.
8. Remove the tube from the stand, and add 54 μL of elution buffer and pipette gently to resuspend.
9. Place the tube back to the magnetic stand for 5 min to pellet the beads.
10. Extract 50 μL of supernatant to a fresh tube and add 60 μL of AMPure XP beads, pipette ten times to mix, and quick centrifuge.
11. Incubate for 5 min on the bench, place on the magnetic stand for 5 min, and remove clear supernatant with care (**Note:** avoid direct touch with the beads).
12. Add 200 μL fresh made 80% ethanol to wash beads and incubate 30 s on the stand.
13. Remove supernatant and repeat once (**Note:** remove residue ethanol thoroughly when possible).
14. Air-dry on the stand for 5 min at maximum.
15. Remove the tube from the stand and add elution buffer according to the reference below, and pipette gently to resuspend:

| Number of reactions | Elution buffer volume (μL) |
|---|---|
| 96 | 48 μL |
| 48 | 24 μL |
| 24 | 16 μL |
| 12 | 12 μL |

16. Incubate at RT for 5 min, and transfer to the magnetic stand for 5 min to completely separate the beads.

17. Carefully transfer supernatants to a new 1.5 mL EP tube, and this is ready for library QC.
18. Use the bioanalyzer (Qsep) for library QC, and make sure to perform separate QC for No Template Control (NTC) well. An ideal library is shown in Fig. 2.

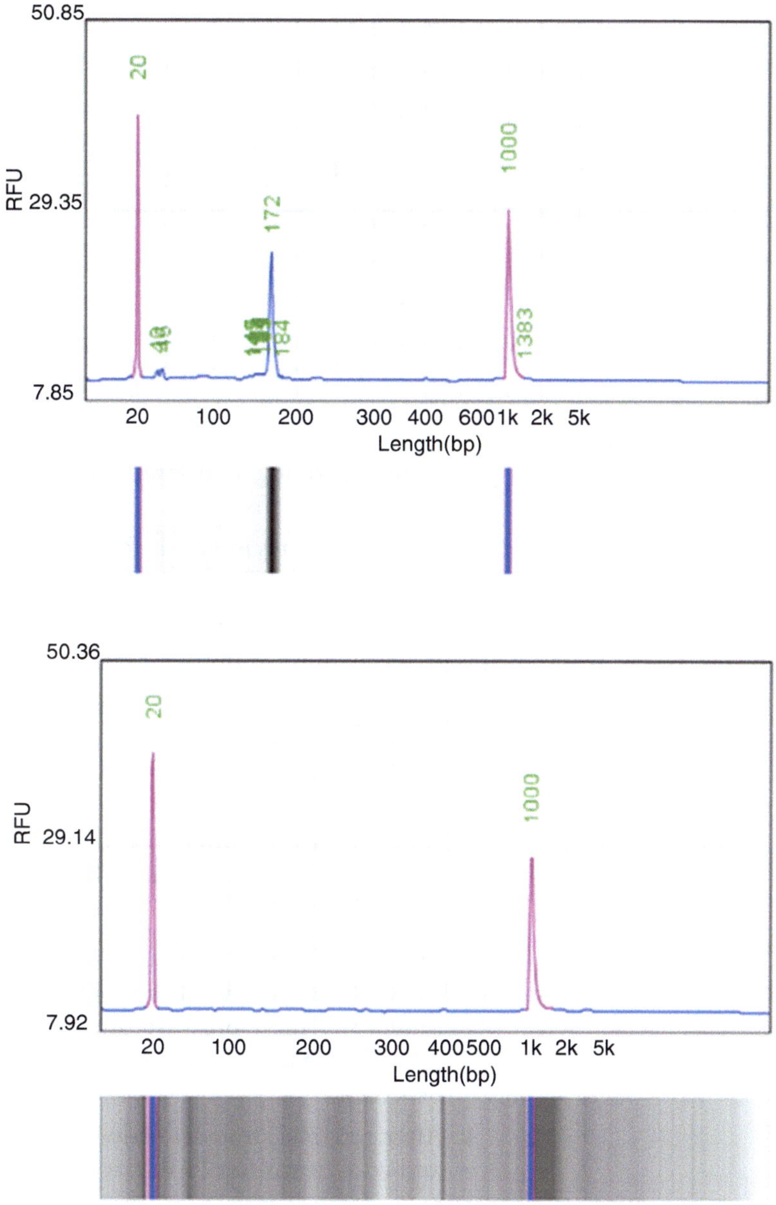


**Fig. 2** Representative plots of the capillary electrophoresis using Qsep for library quality control. The upper plot shows a successfully constructed library with a single peak at around 172 bp. The lower plot is the No Template Control (NTC) negative control without templates

### *3.4 Protein-Based Profiling*

Here we only describe steps specific to the DSP–protein NGS workflow, and the rest of the steps are the same as the DSP–RNA workflow.

#### *3.4.1 Working Solution Preparation*

1. Prepare ethanol with DEPC water to 95% (500 mL), and store at RT.
2. Prepare ethanol with DEPC water to 70% (50 mL), and store at RT.
3. Prepare 80% ethanol for NGS library construction (made fresh on need).
4. Prepare 1× citrate buffer (pH 6) from 10× citrate buffer (pH 6) with DEPC water (**Note**: make fresh on need).
5. Prepare 1× TBS with DEPC water (1000 mL), and store at RT.
6. Prepare 1× TBST by mixing 500 mL 1× TBS with 500 μL Tween-20, store at RT.
7. Prepare 50 mL of 4% formamide by diluting 16% formamide with PBS, and store at RT.

#### *3.4.2 Tissue Baking and Rehydration*

1. Bake FFPE tissue slides in the oven at 60 °C for 30–60 min (**Note:** prolonged baking time may apply depending on the tissue conditions).
2. Immerse tissues in xylene for 5 min, and repeat twice with fresh xylene each for 5 min.
3. Rehydrate tissues in ethanol absolute for 5 min, and change fresh again for 5 min.
4. Rehydrate tissues in 95% ethanol for 5 min and one more time.
5. Rehydrate tissues in dd $H_2O$ for 5 min, and change fresh again until the next step.

#### *3.4.3 Antigen Retrieval*

1. Add 1× citrate buffer into one staining jar and water to the pressure cooker. Make sure the water level does not submerge or float the staining jar.
2. Put tissue slides into the staining jar.
3. Close the cooker with the valve shut to allow antigen retrieval under the high pressure and temperature condition (**Note:** refer to DSP–RNA protocols above for tissue optimization process).
4. After retrieval, transfer slides onto the bench, remove the lid, and cool down for 25 min to RT.
5. Immerse the slides in 1× TBST for 5 min.
6. Remove residue liquid, mark with a PAP pen on the tissues, and ensure the distance to tissue areas is around 5 mm.

*3.4.4 Tissue Blocking and Antibody Incubation*

1. Place the slides in the humidity chamber, add 200 μL buffer W (in the GeoMx DSP Protein Slide Prep Kit), and incubate at RT for 1 h with the lid on.
2. Defrost GeoMx Core and Module Kits for NGS readout on ice, and quick spin down.
3. Aliquot the above antibody reagent into 2/4 slides per assay run depending on the situation (**Note:** reagents can only be freeze-thaw twice before use). Standard morphological panels are stored at 4 °C.
4. Make the protein and morphological antibody mix with the following chart, and apply it onto the slides. Top up to 200 μL with buffer W (n represents the number of slides).

| Core panel (1:25) | Module 1 (1:25) | Module 2 (1:25) | Additional modules | Morphological marker 1 | Morphological marker 2 | Other markers | Buffer W | Total volume |
|---|---|---|---|---|---|---|---|---|
| 8 μL × $n$ | 8 μL × $n$ | 8 μL × $n$ | ... | 5 μL × $n$ | 5 μL × $n$ | ... | Up to 200 μL | 200 μL × $n$ |

5. After blocking, remove all solutions, add 200 μL of antibody mix on each slide, and incubate in the humidity chamber at 4 °C overnight in the dark.

*3.4.5 Tissue Fixation and Nuclear Staining*

1. Immerse the slides in 1× TBST for 10 min, and repeat with fresh TBST twice.
2. Wipe off residue liquid, place the slides in the humidity chamber, and add 200 μL of 4% formamide to fix the tissues for 30 min at RT.
3. Immerse in 1× TBST for 5 min and repeat.
4. Thaw SYTO13 to RT and quick centrifuge for 1 min.
5. Dilute SYTO13 1:10 to a working concentration of 500 nM, apply 200 μL to each slide, and incubate for 15 min in a humidity chamber.
6. Wash twice with 1× TBST, scrape off the PAP pen marks using scalpels, and leave in the TBST with aluminum foil before scanning to DSP (**Note:** do not let the slides and immerse with TBST intermittently when needed).

*3.4.6 Post-collection Processes*

All post-collection processes remain identical to the DSP–RNA assay, and the only difference is to resuspend the dry-down oligos with 80 μL of DEPC water instead of 10 μL due to the higher concentration of oligos in the collection plates (**Note:** refer to Subheading 3.3.7 **step 3**).

## 4 Notes from Our Experience

1. Keep all types of DSP assays (WTA/CTA/protein) in separate spaces as possible to avoid cross-contamination.
2. A thorough clean is always needed when performing a DSP experiment, and use RNase AWAY preferably.
3. Once start processing the RNA probe mix, change gloves after completion of the first step to reduce potential contamination.
4. In case of viscosity, warm the RNA probe mix at RT, briefly vortex, and pulse centrifuge to improve the pipetting accuracy. Unused probes can be stored at 4 °C for up to 6 months with up to five circles of freeze-and-thaw.
5. For Seq-Code plates, used plates can be stored at 4 °C for up to 3 months with one circle of freeze-and-thaw.
6. In case of DSP instrument errors, contact the official remote support team to open the front lid and remove the slides. Store RNA slides in 2× SSC at 4 °C for up to 48 h and protein slides in 1× TBST at 4 °C for up to 72 h. Once fixed, restart the scanning to complete the rest of the experiment.
7. Make sure to prepare 80% ethanol with DEPC water for library construction freshly before use.
8. Due to the significant variation of RNA profiles inter-tissue and intra-tissue, the library concentration may vary significantly. We can simply evaluate the quality by comparing the PCR product (a clear band) with NTC (blank) side by side.
9. The most cost-efficient NGS method so far for DSP is the S4 flowcell on Illumina NovaSeq 6000 with a maximum data output (around 850 GB) on a single lane, which is the preferred way in our laboratory.

## Acknowledgments

We thank the DSP team at Fynn Biotechnologies Ltd for their detailed protocols and thoughtful technical notes based on their experience. We thank the nanoString team for their support.

### References

1. Merritt CR, Ong GT, Church SE, Barker K, Danaher P, Geiss G et al (2020) Multiplex digital spatial profiling of proteins and RNA in fixed tissue. Nat Biotechnol 38(5):586–599
2. Toki MI, Merritt CR, Wong PF, Smithy JW, Kluger HM, Syrigos KN et al (2019) High-Plex predictive marker discovery for melanoma immunotherapy-treated patients using digital spatial profiling. Clin Cancer Res 25(18): 5503–5512
3. Cabrita R, Lauss M, Sanna A, Donia M, Skaarup Larsen M, Mitra S et al (2020) Tertiary lymphoid structures improve immunotherapy and survival in melanoma. Nature 577(7791):561–565

4. Keam SP, Halse H, Nguyen T, Wang M, Van Kooten LN, Mitchell C et al (2020) High dose-rate brachytherapy of localized prostate cancer converts tumors from cold to hot. J Immunother Cancer 8(1)
5. Zugazagoitia J, Gupta S, Liu Y, Fuhrman K, Gettinger S, Herbst RS et al (2020) Biomarkers associated with beneficial PD-1 checkpoint blockade in non-small cell lung cancer (NSCLC) identified using high-plex digital spatial profiling. Clin Cancer Res 26(16): 4360–4368
6. Zollinger DR, Lingle SE, Sorg K, Beechem JM, Merritt CR (2020) GeoMx RNA assay: high multiplex, digital, spatial analysis of RNA in FFPE tissue. Methods Mol Biol 2148:331–345
7. Beechem JM (2020) High-Plex spatially resolved RNA and protein detection using digital spatial profiling: a technology designed for immuno-oncology biomarker discovery and translational research. Methods Mol Biol 2055:563–583
8. Bergholtz H, Carter JM, Cesano A, Cheang MCU, Church SE, Divakar P et al (2021) Best practices for spatial profiling for breast cancer research with the GeoMx((R)) digital spatial profiler. Cancers (Basel) 13(17):4456
9. McCart Reed AE, Bennett J, Kutasovic JR, Kalaw E, Ferguson K, Yeong J et al (2020) Digital spatial profiling application in breast cancer: a user's perspective. Virchows Arch 477(6):885–890
10. Wang N, Wang R, Zhang X, Li X, Liang Y, Ding Z (2021) Spatially-resolved proteomics and transcriptomics: an emerging digital spatial profiling approach for tumor microenvironment. Visualized Cancer Med 2:1. vcm.edpsciences.org
11. Wang N, Wang R, Li X, Song Z, Xia L, Wang J et al (2021) Tumor microenvironment profiles reveal distinct therapy-oriented proteogenomic characteristics in colorectal cancer. Front Bioeng Biotechnol 9:757378
12. Gupta S, Zugazagoitia J, Martinez-Morilla S, Fuhrman K, Rimm DL (2020) Digital quantitative assessment of PD-L1 using digital spatial profiling. Lab Investig 100(10):1311–1317
13. Badve SS, Gokmen-Polar Y (2022) Protein profiling of breast cancer for treatment decision-making. Am Soc Clin Oncol Educ Book 42:1–9
14. Van Herck Y, Antoranz A, Andhari MD, Milli G, Bechter O, De Smet F et al (2021) Multiplexed immunohistochemistry and digital pathology as the foundation for next-generation pathology in melanoma: methodological comparison and future clinical applications. Front Oncol 11:636681

# Chapter 7

# Integration of Single-Cell RNA-Sequencing and Network Analysis to Investigate Mechanisms of Drug Resistance

Stephanie The, Patricia M. Schnepp, Greg Shelley, Jill M. Keller, Arvind Rao, and Evan T. Keller

## Abstract

Innate resistance and therapeutic-driven development of resistance to anticancer drugs is a common complication of cancer therapy. Understanding mechanisms of drug resistance can lead to development of alternative therapies. One strategy is to subject drug-sensitive and drug-resistant variants to single-cell RNA-seq (scRNA-seq) and to subject the scRNA-seq data to network analysis to identify pathways associated with drug resistance. This protocol describes a computational analysis pipeline to study drug resistance by subjecting scRNA-seq expression data to Passing Attributes between Networks for Data Assimilation (PANDA), an integrative network analysis tool that incorporates protein–protein interactions (PPI) and transcription factor (TF)-binding motifs.

**Key words** Single-cell RNA-sequencing, Drug resistance network, Data integration, Protein–protein interactions, Transcription factor-binding motifs, Passing Attributes between Networks for Data Assimilation, Gene set enrichment analysis, Connectivity map analysis

## 1 Introduction

Drug resistance is a frequent therapeutic challenge when attempting to treat cancer patients. Many cancer studies have attempted to identify other drugs and therapies to overcome drug resistance and increase the chance of survival in various cancer patients [1]. The majority of these studies tend to focus on a few genes and examine their change in expression due to drug resistance [1]. Furthermore, they tend to only look at one data type or multiple data types separately. This limits the ability to fully understand the mechanisms underlying drug resistance. To increase the possibility to determine how drug resistance develops and subsequently identify alternative therapies, gene regulatory network modeling, which can integrate multiple data types, should be considered. Passing Attributes between Networks for Data Assimilation (PANDA) is an

Usha N. Kasid and Robert Clarke (eds.), *Cancer Systems and Integrative Biology*, Methods in Molecular Biology, vol. 2660,
https://doi.org/10.1007/978-1-0716-3163-8_7,

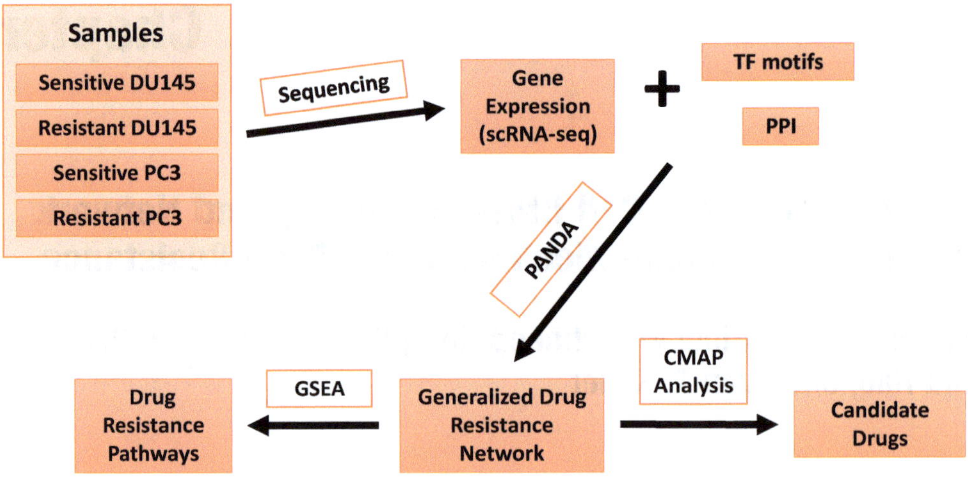


**Fig. 1** General network analysis pipeline. Sensitive and resistant samples are subjected to scRNA-seq, followed by subjecting the scRNA-seq data in combination with transcription factor (TF) motifs and protein–protein interactions (PPI), derived from established databases to PANDA analysis. This will lead to identifications of a generalized drug resistance network which can then be subjected to gene set enrichment analysis (GSEA) and connectivity map analysis (CMAP) to identify drug resistance pathways and candidate drugs, respectively

integrative network analysis tool, which uses a message-passing model to iterate over multiple data types to predict regulatory relationships [2, 3]. Specifically, PANDA integrates gene expression, protein–protein interactions (PPI), gene co-regulation, and transcription factor (TF)-binding motif data. Single-cell RNA-seq (scRNA-seq) is a powerful technique that allows us to study the transcriptome of variable and heterogenous cell populations on a single-cell level, which cannot be examined with traditional bulk sequencing [1]. Incorporating scRNA-seq data into PANDA may lead to an enhanced understanding of mechanisms of drug resistance and identify new, alternative therapies to overcome this resistance. The described protocol will provide a general analysis pipeline to identify regulatory networks and pathways associated with these regulatory networks and discover alternative drugs that could potentially overcome drug resistance. This general pipeline (Fig. 1) was composed and used with previous studies [1, 4].

## 2 Materials

### *2.1 Software (see Note 1)*

1. R.
   - (a) R is an open-source programming coding language mainly used for statistical analysis and graphics [5].
   - (b) Information on and download R here: https://www.r-project.org/

(c) Additional packages needed:

(i) Seurat (https://satijalab.org/seurat/).

1. Seurat is an R package used to analyze quality control (QC) and exploration of scRNA-seq data [6].

(ii) wordcloud2 (https://github.com/lchiffon/wordcloud2),

1. wordcloud2 is an R package used to create word clouds for data visualizations [7].

2. Passing Attributes between Networks for Data Assimilation (PANDA) [2].

(a) Download the bash version here: http://sourceforge.net/projects/panda-net/files/Version2/Version2.tgz/download.

(b) Read the "README.txt" to compile the C++ scripts, and run the commands PANDA and AnaPANDA. Also, read this file to find more information on how the input files need to be structured.

3. Find Individual Motif Occurrences (FIMO).

(a) FIMO is a program used to search and extract sequences that match independent motifs provided [8].

(b) Download the MEME suite, which contains FIMO, here: https://meme-suite.org/meme/doc/download.html.

(c) Install the MEME suite. You can find installation instructions here: https://meme-suite.org/meme/doc/install.html?man_type=web.

(d) To run FIMO, you can find the manual here: https://meme-suite.org/meme/doc/fimo.html.

4. Gene set enrichment analysis (GSEA).

(a) GSEA is a computation method that is commonly used for pathway analysis. It determines if a set of genes that are different between two biological states is statistically significant for a set of pathways [9, 10].

(b) Download and install GSEA here: https://www.gsea-msigdb.org/gsea/doc/GSEAUserGuideTEXT.htm#_Starting_GSEA.

(c) The link above also contains the manual to run GSEA both on command line and GUI program.

### 2.2 Datasets

1. Gene expression dataset.

(a) General structure:

(i) Gene expression, done with scRNA-seq, from at least two cell lines is required. Each cell line requires at least two conditions (e.g., sensitive vs. resistant). In total, there should be at least four samples (two cell lines, each with a sensitive and resistant variant).

(ii) The counts in the gene expression dataset from every sample should be normalized.

(iii) Since the dataset will have multiple single cells per sample, the normalized counts should be combined into "pseudo-bulk" counts for each gene for each sample for the dataset to work in PANDA. Some methods to combine the normalized single-cell counts per condition are by averaging, summation, scoring, etc. These "pseudo-bulk" counts will then be used in PANDA.

(iv) This new matrix must be structured as genes (as rows) by samples (as columns) and exported as tab delimited *.txt* file.

(b) Example:

(i) Raw counts can be found here: https://www.ncbi.nlm.nih.gov/geo/query/acc.cgi?acc=GSE140440.

(ii) The example expression gene dataset used for this protocol is based on another previous study we published [11]. Briefly, this dataset contains gene expression of single cells from two prostate cancer (PCa) cell lines, Du145 (DU145), and PC-3 (PC3). There was a parental (docetaxel-sensitive) and docetaxel-resistant variant for each cell line. In the end, there was four conditions: sensitive DU145, resistant DU145, sensitive PC3, and resistant PC3. For more information about the experiment and sequencing, please refer to this study's paper [11] (*see* **Note 2**).

(iii) The single-cell counts from all conditions were normalized together using the log normalization method (*NormalizeData*) in Seurat. The normalized counts were then aggregated into "pseudo-bulk" counts per condition. In this example, the matrix had genes as rows and conditions as columns, which there were four columns.

2. TF motif prior dataset.

(a) General structure:

(i) A TF motif dataset needs to be downloaded that matches the species of the gene expression dataset (*see* **Note 3**). For example, if the gene expression dataset is from *Mus musculus*, then you will need a TF motif dataset specific to *Mus musculus*. This dataset will contain all the motifs found in this species.

(ii) The full TF motif dataset needs to be filtered for TFs found in the gene expression dataset. FIMO can be used to get these motifs.

1. A sequence file is needed (in *fasta* format) for the dataset's species to run FIMO.
2. To run FIMO, refer to the link in Subheading 2.1 **item 3d**.

(b) Example:

(i) The *Homo sapiens* TF motif dataset was downloaded from the. This dataset contains all the TF motifs found in *Homo sapiens* (*see* **Note 4**).

(ii) A sequence file (in *.fasta* format) was also downloaded. For this study, the human genome assembly dataset GRCh37 (also called hg19) was used (https://genome.ucsc.edu/cgi-bin/hgTables).

(iii) The TF motifs position weight matrices were mapped to using FIMO.

(iv) A TF motif prior specific to the expression dataset was created by only using the TFs found in the expression dataset. There was 240 TFs in this example TF motif prior dataset.

3. PPI files.

(a) General structure:

(i) A PPI dataset needs to be downloaded that contains interaction scores.

(ii) The interaction scores need to be between 0 and 1. The scores will need to be transformed if they are not already in this format. Any self-interactions need to be set equal to one.

(b) Example:

(i) The example PPI dataset that contains interaction scores was downloaded from StringDb v10.5 (*see* **Note 4**). This dataset contains interactions from all data sources in StringDb.

(ii) The interaction scores were then divided by 1000 for the scores to be between 0 and 1. The self-interactions were set equal to one.

## 3 Methods

Subheadings 3.1, 3.2, 3.3, 3.4, 3.5, and 3.6 are delineated in Fig. 2.

### 3.1 Constructing PANDA Regulatory Networks

1. Create a TF regulatory network for each condition using the TF motif prior, PPI, and the expression dataset from Material section. A code example is:

```
./Version2/PANDA -e express_data/DU145_sen.txt -m motif_data/
TF_motif.txt -p PPI_data/PPI.txt -o networks/DU145_sen
```

From this command, you should have four *FinalNetwork.pairs* files.

### 3.2 Compare PANDA Networks

1. Within each cell line, compare networks in the *FinalNetwork.pairs* files from Subheading 3.1 with PANDA with the example code below:

```
./Version2/AnalyzePANDA_v0/AnalyzePANDA -a DU145_resist_Fi-
nalNetwork.pairs -b DU145_sen_FinalNetwork.pairs -P 0.8 -o
DU_resist_vs_sen
```

From this command, there should be two sets of five comparison network files. In this example, there was one comparing resistant and sensitive cells from the DU145 cell line, and another comparing resistant and sensitive cells from the PC3 cell line. Each set should have a *FinalNetwork.pairs*, *RankedList_for_GSEA.rnk*, *Gene_degree_comparison.txt*, *TF_degree_comparison.txt*, and *Subnetworks.pairs* files.

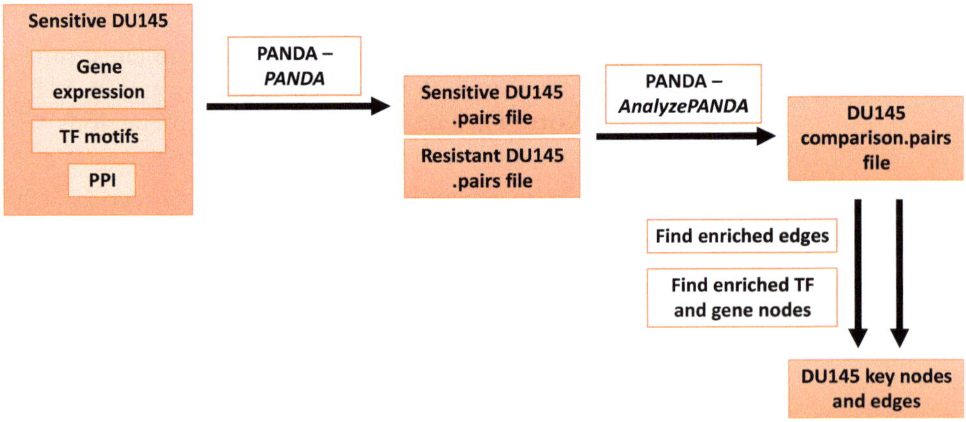


**Fig. 2** Overview of steps for generating a key network for one cell line. This figure describes Subheadings 3.1, 3.2, 3.3, 3.4, 3.5, and 3.6 of the protocol. Details are found within the protocol text. TF, transcription factor; PPI, protein–protein interactions

### 3.3 Finding Enriched Edges

1. Over all the networks (*all*), calculate the median (*med*) and IQR for each edge weight ($w$) between each TF ($t$) and gene ($g$) from the *FinalNetwork.pairs* files from Subheading 3.1.
2. For each edge weight in each network ($n$), calculate a specificity score ($s$). In general, we compare the weight to the median and IQR for each TF-gene found in the last step. The equation to calculate scores is shown below [4]:

$$s_{tg}^{(c)} = \left(w_{tg}^{(c)} - \mathrm{med}\left(w_{tg}^{(\mathrm{all})}\right)\right) / \mathrm{IQR}\left(w_{tg}^{(\mathrm{all})}\right)$$

3. Determine enriched edges. An edge is enriched for a network if $s > N$ where $N$ is a threshold for a specificity score over all networks. $N$ was found through calculating the specificity score for each individual gene and comparing those scores to the median and IQR across all networks. The specificity score can vary between 0 and 1. For this study, we found that $N = 0.4$ where half of the genes were enriched.

### 3.4 Finding Enriched TF Nodes

1. For each network, calculate the in-degree of each TF node, which is the sum of enriched edges connected to a TF node. The enriched edges are from the *TF_degree_comparison.txt* files from Subheading 3.2.
2. Calculate $p$-values to test differences of in-degree values for each TF node between conditions in either cell line using a hypergeometric distribution. In this example, $p$-values were calculated between sensitive and resistant cells in either DU145 or PC3 cell lines.
3. Calculate the edge weight fold change using the in-degree value for each node between the two networks.
4. Determine enriched TF nodes. In this example, these were determined as enriched for a cell line if they had a $p$-value than 0.05 ($p < 0.05$).

### 3.5 Finding Enriched Gene Nodes

1. The process is the same as finding enriched TF nodes in Subheading 3.4, except using the *Gene_degree_comparison.txt* files from Subheading 3.2.

### 3.6 Finding Key TF and Gene Node

1. We followed the following criteria to find key TF and gene nodes:
   (a) The node must have a $p$-value (found in Subheading 3.4 or 3.5) less than 0.05 ($p < 0.05$) for both cell line network comparisons. In this example, a specific node must both have $p < 0.05$ in both DU145 and PC3 cell lines.
   (b) The node must have an edge weight fold change (found in Subheading 3.4 or 3.5) in the same direction for both cell line network comparisons, for instance, the fold change

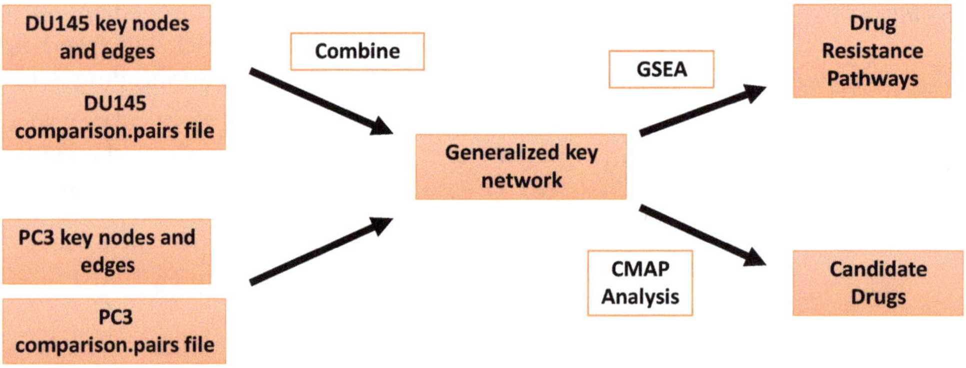


**Fig. 3** Overview of process for generating a generalized key network and further downstream analysis. This figure describes Subheadings 3.7, 3.8, and 3.9 of the protocol. Details are found within the protocol text. GSEA, gene set enrichment analysis; CMAP, connectivity map

must be positive in both comparisons. In this example, a specific node must have an edge weight fold change in the same direction in both DU145 and PC3 cell lines.

Subheadings 3.7, 3.8, and 3.9 are delineated in Fig. 3.

### 3.7 Creating a Generalized Network

1. Combine common key edges and nodes that were identified from Subheading 3.6 to create a generalized network. For this study, this network represents prostate cancer response to docetaxel treatment between DU145 and PC3 cell lines.

### 3.8 GSEA of TF Specific-Targeted Genes

1. Create a pre-ranked gene list using the specificity scores of each TF that was calculated in Subheading 3.4.
2. Run pre-ranked GSEA on GO terms with the pre-ranked gene list. Pathways that were considered as enriched if they had a false discovery rate (FDR) less than 0.05 ($FDR < 0.05$).
3. Cluster the significant pathways using hierarchical clustering.
4. For each cluster, calculate the frequency of each word that appeared in the GO terms.
5. Calculate *p*-values to test word enrichment using a hypergeometric probability. *P*-values were scaled by using $-\log 10$, which the most statistically relevant words would be the largest.
6. Create word clouds for each cluster with the words from the pathway names. The size of the words is determined by the $-\log 10(p)$ values.

### 3.9 Connectivity Map (CMAP) Analysis

1. Go to the CMAP website here: https://portals.broadinstitute.org/cmap (*see* **Note 5** before going to website) [12].
2. Label the gene nodes as a certain condition based on their edge weight fold change found in Subheading 3.6. In this example, the gene nodes were labeled as either sensitive or resistant.

3. Run the gene nodes list from the last step in CMAP to predict response in various drugs. These drugs should either upregulate sensitive gene nodes or downregulate resistant gene nodes. In this example, drugs with positive enrichment would mean that these drugs had the highest potential to reverse docetaxel resistance in PCa.

## 4 Notes

1. The tools used in this protocol can be run on multiple different platforms and coding languages. Alternative versions or coding languages are acceptable to use. Please use the most up-to-date software for all tools mentioned.
2. The scRNA-seq method to create the example expression dataset is very old. However, the analysis pipeline explained in this protocol can be used with current scRNA-seq methods.
3. You can find motif datasets in Catalog of Inferred Sequencing Bind Preferences (CIS-BP), MEME (https://meme-suite.org/meme/db/motifs), etc.
4. The PPI and TF motif datasets used as examples are older versions of what can be found today. Please use the most up-to-date datasets.
5. The CMAP website used for this analysis is not available anymore. The CMAP dataset has moved here: https://clue.io/. For information about the algorithm, please refer to [12].

## Acknowledgments

This work was supported by National Cancer Institute awards P01-CA093900 and P30-CA046592 by the use of the following Cancer Center Shared Resources: the Single Cell Spatial Analysis Shared Resource, the Cancer Data Science Shared Resource, and the Single Cell Spatial Analysis Program.

### References

1. Schnepp PM, Ahmed A, Escara-Wilke J et al (2021) Transcription factor network analysis based on single cell RNA-seq identifies that Trichostatin-a reverses docetaxel resistance in prostate cancer. BMC Cancer 21:1316. https://doi.org/10.1186/s12885-021-09048-0
2. Glass K, Huttenhower C, Quackenbush J, Yuan GC (2013) Passing messages between biological networks to refine predicted interactions. PLoS One 8(5):e64832. https://doi.org/10.1371/journal.pone.0064832
3. Glass K, Quackenbush J, Spentzos D et al (2015) A network model for angiogenesis in ovarian cancer. BMC Bioinform 16:115. https://doi.org/10.1186/s12859-015-0551-y
4. Sonawane AR (2017) Understanding tissue-specific gene regulation. Cell Rep 21(4): 1077–1088. https://doi.org/10.1016/j.celrep.2017.10.001
5. R Core Team (2022) R: a language and environment for statistical computing. R

Foundation for Statistical Computing. https://www.R-project.org/

6. Hao Y, Hao S, Andersen-Nissen E et al (2021) Integrated analysis of multimodal single-cell data. Cell 184(13):3573–3587.e29. https://doi.org/10.1016/j.cell.2021.04.048
7. Dawei L, Guan-tin C (2018) wordcloud2: create word cloud by 'htmlwidget'. R package version 0.2.1. https://CRAN.R-project.org/package=wordcloud2
8. Grant CE, Bailey TL, Noble WS (2011) FIMO: scanning for occurrences of a given motif. Bioinformatics 27(7):1017–1018. https://doi.org/10.1093/bioinformatics/btr064
9. Subramaniana A, Tamayoa P, Mootha VK et al (2005) Gene set enrichment analysis: a knowledge-based approach for interpreting genome-wide expression profiles. Proc Natl Acad Sci USA 102:15545–15550. https://doi.org/10.1073/pnas.0506580102
10. Mootha V, Lindgren C, Erikssonp KF et al (2003) PGC-1α-responsive genes involved in oxidative phosphorylation are coordinately downregulated in human diabetes. Nat Genet 34:267–273. https://doi.org/10.1038/ng1180
11. Schnepp PM, Shelley G, Dai J et al (2020) Single-cell transcriptomics analysis identifies nuclear protein 1 as a regulator of docetaxel resistance in prostate cancer cells. Mol Cancer Res 18(9):1290–1301. https://doi.org/10.1158/1541-7786
12. Lamb J, Crawford ED, Peck D et al (2006) The connectivity map: using gene-expression signatures to connect small molecules, genes, and disease. Science 313(5795):1929–1935. https://doi.org/10.1126/science.1132939

# Chapter 8

# Combination of Tissue Microarray Profiling and Multiplexed IHC Approaches to Investigate Transport Mechanism of Nucleoside Analog Drug Resistance

**Debasis Nayak, Brenna Weadick, and Rajgopal Govindarajan**

## Abstract

Nucleoside analogs (NAs) are an established class of anticancer agents being used clinically for the treatment of diverse cancers, either as monotherapy or in combination with other established anticancer or pharmacological agents. To date, nearly a dozen anticancer NAs are approved by the FDA, and several novel NAs are being tested in preclinical and clinical trials for future applications. However, improper delivery of NAs into tumor cells because of alterations in expression of one or more drug carrier proteins (e.g., solute carrier (SLC) transporters) within tumor cells or cells surrounding the tumor microenvironment stands as one of the primary reasons for therapeutic drug resistance. The combination of tissue microarray (TMA) and multiplexed immunohistochemistry (IHC) is an advanced, high-throughput approach over conventional IHC that enables researchers to effectively investigate alterations to numerous such chemosensitivity determinants simultaneously in hundreds of tumor tissues derived from patients. In this chapter, taking an example of a TMA from pancreatic cancer patients treated with gemcitabine (a NA chemotherapeutic agent), we describe the step-by-step procedure of performing multiplexed IHC, imaging of TMA slides, and quantification of expression of some relevant markers in these tissue sections as optimized in our laboratory and discuss considerations while designing and carrying out this experiment.

**Key words** Nucleoside analogs, Tissue microarray, Immunohistochemistry, Multiplexing, Solute carrier (SLC) transporters, Nucleoside transporter (NT), Equilibrative nucleoside transporter (ENT), Concentrative nucleoside transporters (CNT), Heat-induced antigen retrieval (HIER), Pancreatic cancer

## 1 Introduction

Nucleoside analogs (NAs) are the cornerstones of cancer chemotherapy in many solid tumors including pancreatic and lung adenocarcinoma as well as in hematologic malignancies such as acute myeloid leukemia (AML) and chronic lymphocytic leukemia (CML) [1, 2]. They include purine analogs (fludarabine, clofarabine, nelarabine, 5-mercaptopurine, 6-thioguanine), pyrimidine

Usha N. Kasid and Robert Clarke (eds.), *Cancer Systems and Integrative Biology*, Methods in Molecular Biology, vol. 2660, https://doi.org/10.1007/978-1-0716-3163-8_8,

analogs (cytarabine, troxacitabine), azidopyrimidines (azacitidine, decitabine) and fluoropyrimidines (gemcitabine, capecitabine, fluorouracil, tegafur/uracil). A majority of these agents are established chemotherapeutic drugs being used clinically either as monotherapy or in combination with other therapeutic agents for the treatment of diverse cancers, whereas a fraction of them including troxacitabine and guadecitabine are now in clinical trials being tested in patients with AML and CML [3–5]. Mechanistically, NA anticancer agents act as antimetabolites, which when translocated by the nucleoside transporters into the cytoplasm get phosphorylated by the intracellular enzymes (e.g., deoxycytidine kinase, dCK) into active metabolites and incorporated into newly synthesized DNA (s-phase of cell cycle) leading to inhibition of DNA synthesis and loss of cell proliferation [6, 7]. Additionally, the azidopyrimidine class of anticancer drugs are also called DNA hypomethylating agents since they affect DNA methylation by covalently interacting and inhibiting the enzyme DNA methyl transferase (DNMT) after incorporation into the newly synthesized DNA, thus reducing tumor cell proliferation. The two DNMT inhibitors (azacytidine and decitabine) achieved significant success in clinic for the treatment of myelodysplastic syndrome and AML [5, 8].

Despite of the proven anticancer activities of NAs in vitro, in preclinical murine models, and in patients with primary grade tumors to locally advanced cancers, their clinical utilities are limited in advanced stage cancers because of development of resistance to these drugs by cancer cells [9–11]. One of the primary mechanisms of drug resistance to NAs is the lack of entry of these drugs into cancer cells, which is the first rate-limiting step that undermines their treatment efficacy. For instance, loss of expression or function (s) of some specific drug carrier proteins that transport NAs leads to drug resistance. Since NAs are hydrophilic in nature, they need certain carrier proteins called solute carrier (SLC) transporters for their delivery into the cellular compartments. Depending on the NA, a specific SLC or a combination of SLCs are involved in their tumor entry. Two families of SLC transporters that are involved in the transport of NAs across the cell membrane and across intracellular compartments: equilibrative nucleoside transporters (ENTs, also called SLC29A) and concentrative nucleoside transporters (CNTs, also called SLC28A). There are three members from each of these two families ENT1, ENT2, ENT3, CNT1, CNT2, and CNT3, which are substantially studied for their role in the transport of endogenous nucleosides and NAs; depending on the physicochemical characteristics of the cargos, individual ENTs and/or CNTs participate in drug transport [7, 12, 13]. This methodology article focuses on procedures to evaluate acquired drug resistance due to aberrations in nucleoside transporter (NT) expressions and localization. Although there are numerous other mechanisms that can contribute to nucleoside analog drug resistance in cancer (e.g., changes in metabolic enzymes such as deoxycytidine kinase, ribonucleotide reductase, cytidine deaminase, etc.), those are not dealt here, but the described technique can be easily adapted for studying those determinants as well.

### 1.1 Tissue Microarray (TMA) and IHC Multiplexing

Tissue microarray (TMA) is a valuable tool in oncology that allows researchers to study hundreds of patient tissue samples simultaneously under identical experimental conditions. Common applications include immunohistochemistry (IHC), in situ hybridization (ISH), and in situ PCR to evaluate the expression of diagnostic and prognostic markers in cancer. IHC multiplexing is commonly used for simultaneously visualizing multiple target antigens within a single tissue section. The development of multiplexed IHC coupled with multispectral imaging in recent years has enabled researchers to precisely study the complexities of the tumor microenvironment that expresses a wide array of biomarkers including various tumor cell, stromal cell, and immunological markers [14, 15]. Our paper will focus on the classical laboratory technique of IHC (monoplexing and/or multiplexing) to represent a high-throughput method for the simultaneous assessment of protein expression in hundreds of tumor samples of various grades and stages of cancer progression. This powerful technique can be particularly useful for evaluating the expression of NTs which are the first rate-limiting step required for NA drug action. Furthermore, many NTs are aberrantly localized in cancer cells, e.g., translocation of ENT1 and CNT1 from plasma membrane to cytoplasm in different cell types of pancreatic cancer and cholangiocarcinoma and aberrant localization of ENT3 in some intracellular organelles in human fibrosarcoma cells, which resulted in reduced responses to chemotherapy [13, 16–18]. Thus, this technique allows for the visualization of NTs at the plasma membrane and different subcellular compartments, providing valuable information about transporter function, and, prospectively, NA drug efficacy.

### 1.2 ENT1 as a Candidate for Evaluating Drug Resistance in Pancreatic Cancer

ENT1 is a broadly specific NT that contributes to the transport of an array of purine and pyrimidine NAs in diverse cancers. Expression of ENT1 is correlated with improved drug efficacy and increased survival in pancreatic ductal adenocarcinoma (PDAC) and other solid tumors [17, 19–21]. In contrast, ENT1 becomes intracellular in some aggressive PDAC with high epithelial–mesenchymal transition (EMT) characteristics and loses its ability to transport a mainline treatment agent for PDAC called gemcitabine, as demonstrated by our group [16]. Mechanistic studies showed ENT1 cell surface localization may require different cell adhesion molecules such as EpCAM and the opposing effects of epithelial and neuronal cadherins. Therein the evaluation of ENT1 expression and its associated partner proteins (EpCAM) compositely have diagnostic, prognostic, and therapeutic values since those determine the overall plasma membrane levels and therefore the functionality of ENT1 in cancer cells. Utilizing the example that ENT1 acts as a prototypic transporter for evaluation of drug resistance to NAs, here we present the methodology and laboratory protocol for TMA multiplexing of ENT1 with the cell adhesion molecule EpCAM for evaluation of their expression in tumor and stromal compartments of PDAC.

## 2 Materials

### 2.1 Equipment Needed for IHC Processing of the TMA Slides

(a) Opal slide processing jars (Akoya Biosciences); (b) microwave oven (GE Appliances); (c) orbital shaker (VWR); (d) ImmEdge hydrophobic barrier pen (Vector Laboratories); (e) Falcon tubes, 15 mL; (f) Falcon tubes, 50 mL; (g) microcentrifuge tubes, 1.5 mL; (h) laboratory fume hood, (i) Borosil glass bottles, 1 L; (j) Borosil glass bottles, 250 mL; (k) sterile pipettes, 10 mL; (l) accu-jet pipette controller; (m) adjustable micropipettes (0.5–10 μL, 20–200 μL, and 100–1000 μL); (n) sterile micropipette tips; (o) IHC slide storage boxes; (p) Kimwipes; (q) Vectra automated multispectral imaging system (PerkinElmer) and (r) inForm advanced image analysis software (PerkinElmer).

### 2.2 Chemicals, Reagents, and Antibodies

Chemicals and reagents should be of analytical grade for obtaining better-quality IHC staining and for avoiding issues related to purity of the chemicals. Xylene solvent (cat# 534056), absolute ethanol (cat# E7023), Tris base (cat# T1503), and hydrogen peroxide (cat# 323381) were procured from Sigma-Aldrich (St. Louis, MO). Phosphate-buffered saline (PBS) tablets (cat# 6501), Tween-20 (cat# 9480), and NaCl (cat# 7710) were purchased from Omni-Pur/MilliporeSigma (Burlington, MA). Normal goat serum (cat# PCN5000) and 10% neutral buffered formalin (cat# 5701) were obtained from Thermo Fisher Scientific.

Anti-hENT1 rabbit polyclonal antibody was described earlier [22, 23]. Two other ENT1 antibodies: anti-ENT1 rabbit polyclonal antibody (cat# ANT-051) and anti-ENT1 rabbit polyclonal antibody (cat# ab48607), were obtained from Alomone Labs (Jerusalem, Israel) and Abcam (Cambridge, UK), respectively. Anti-pan-Cytokeratin mouse monoclonal antibody (cat# 914202) was obtained from BioLegend (San Diego, CA). Anti-EpCAM mouse monoclonal antibody (cat# 2929) was purchased from Cell Signaling Technology (Danvers, MA), and anti-N-cadherin mouse monoclonal antibody (cat# 610920) was procured from BD Biosciences (San Jose, CA). Antigen retrieval buffer, pH 6 (AR6) (cat# AR600125ML), was obtained from Akoya Biosciences

(Marlborough, MA). Opal 4 Color Manual Staining Kit (cat# NEL810001KT) was purchased from PerkinElmer (Waltham, MA). This kit includes Opal fluorophores 520 (green), 570 (red), 690 (cyan), and DAPI (blue) that allows staining and detection of multiple target proteins simultaneously in various cellular compartments.

### 2.3 Stock and Working Solutions

Prepare all the solutions using ultrapure deionized water and analytical grade reagents.

*Xylene:* This is a ready-to-use solvent available commercially from Fisher Scientific and other suppliers. Place and use this solvent underneath a fume hood.

*Ethanol gradient:* Prepare 200 mL volume each of 100%, 95%, and 70% ethanol solutions by diluting absolute analytical grade ethanol in ultrapure deionized water.

*PBS:* For making a 1× solution, dissolve 10 PBS tablets in 1 liter of ultrapure deionized water, filter the solution to sterilize.

*Antigen retrieval buffer:* Dilute the antigen retrieval AR6 buffer at 1:10 dilutions in water in a slide processing jar directly prior to the experiment.

*Wash buffer (TBS, 10× stock solution):* Dissolve 24.2 g Tris base and 80 g NaCl in 900 mL of ultrapure water. Then adjust pH to 7.6, and make up the volume to 1 liter by adding ultrapure water.

*Wash buffer (TBST, 1× working solution):* Dilute 100 mL of 10× TBS stock solution in 900 mL of ultrapure water, and then add 0.5 mL of Tween-20 into it. Mix well using magnetic stirrer.

*Blocking solution:* 5% normal goat serum in PBS plus 0.1% detergent (Tween-20). For preparing 5 mL of blocking solution, add 250 μL of normal goat serum in 4.5 mL of PBS, and then add 50 μL of 10% Tween-20 and mix well. This could be used as the antibody diluent as well for preparing primary and secondary antibody solutions.

*1° antibody solution:* Dilute the abovementioned primary antibodies in a ratio of 1:500 for ENT1, 1:100 for each of N-cadherin and EpCAM, and 1:75 for PanCK in the antibody diluent or blocking solution. These dilutions of the mentioned antibodies are optimized based on several pilot IHC experiments conducted in our laboratory on human normal and tumor tissue samples.

*2° antibody solution:* Opal Polymer HRP Ms + Rb detection reagent supplied with the Opal 4 Color Manual Staining Kit was used in this procedure. This is a ready-to-use IHC detection reagent that can react with primary antibodies raised in mouse and rabbit, which provides other advantages including less incubation time (~10 min) and minimal background interference as well as signal amplification.

# 3 Methods

### 3.1 TMA Construction

The TMA used in this described protocol was prepared from tumor samples collected from 114 patients with pancreatic cancer admitted to The Ohio State University Comprehensive Cancer Center, Columbus, OH. These patients were treated with surgical resection with or without postoperative gemcitabine-based chemotherapy. The procedures for the sample collection, ethical approval, and TMA construction and characterization are provided in our published article [16]. Information about the patient demographics, clinicopathological characteristics, and clinical outcomes for each patient was also developed and maintained in a clinically annotated database and described in the abovementioned article by Weadick et al. (2021) [16].

### 3.2 Deparaffinization, Rehydration, and Fixation of the TMA Slides

Deparaffinization of the TMA slides is necessary before staining to remove the paraffin surrounding the tissue sections. Immerse the TMA slides in xylene in slide processing jars twice for a total duration of 15–20 min, and visually confirm removal of paraffin. Repeat this step in fresh xylene if necessary. Next, the tissue sections are rehydrated by immersing the slides in a decreasing ethanol concentration gradient. The deparaffinization/rehydration and fixation steps are summarized below:

1. Bake the paraffin-embedded TMA slides to be stained at 65 °C for at least 1 h or overnight by placing the slides in an oven or BOD incubator. If slides are baked, tilt up to allow for the drainage of paraffin, the movement of this drainage can potentially tear or fold the delicate TMA cores, so baking the slides flat is recommended.
2. Remove slides from the oven and immerse in xylene for 10 min. Use separate slide processing jars for each solvent or solution in which the TMA slides are to be immersed, and label it accordingly (Fig. 1). This step is repeated for a second time in fresh xylene to make sure all traces of paraffin are removed from the slides.

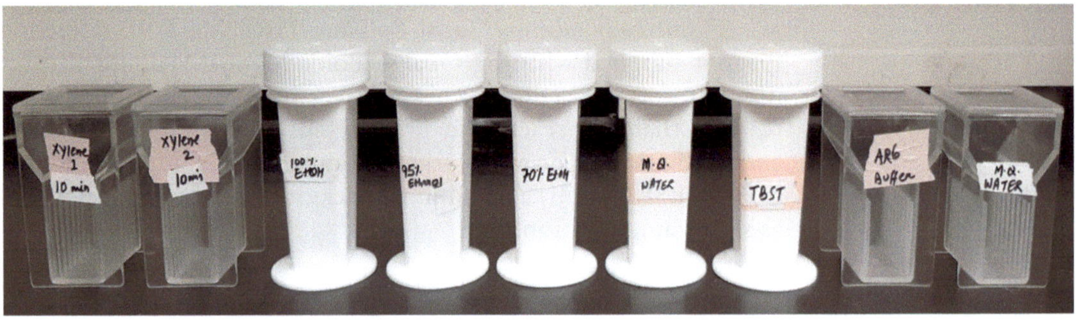


**Fig. 1** Slide processing jars placed in a fume hood are labeled for different solutions required for the deparaffinization and rehydration of the TMA slides

3. Place the slides in 100% ethanol for 10 min.
4. Place the slides in 95% ethanol for 10 min.
5. Place the slides in 70% ethanol for 10 min.
6. Rinse the slides in TBST washing buffer.
7. Rinse the slides in Milli-Q water.
8. Immerse the slides in 10% neutral buffered formalin for 20 min for fixation.
9. Rinse the slides in TBST washing buffer.
10. Rinse the slides in Milli-Q water.

### 3.3 Antigen Retrieval (AR) of the TMA

Frozen tissue sections do not require an AR step; however, most formalin-fixed or paraffin-embedded tissue sections require an AR step prior to the IHC staining to expose the epitope that will bind the primary antibody. This is because formalin fixation forms methylene bridges that cross-link proteins which can mask their epitopes or antigenic sites. AR helps disrupt the methylene bridges, thus exposing the epitopes for the binding of antibodies. Heat-induced antigen retrieval (HIER) is the most used method for AR in IHC. This can be achieved by placing the TMA slides in AR buffer and heating in a microwave or incubating the slides in AR buffer overnight in a water bath raising the temperature to 60 °C. The following steps were optimized for this protocol:

1. In an Opal slide processing jar, dilute the AR buffer 1:10 (each jar holds 100 mL of buffer). Select the buffer with the appropriate pH, usually pH 6 for the first antigen to be detected. However, for some other antibody targets, pH 9 AR buffer gives a better result (Table 1).
2. Take the slides from the last step (rinsing with water), and place them in the appropriate AR buffer required for the target protein.

**Table 1**
**Examples of target proteins and their suitable antigen retrieval (AR) buffers optimized in our laboratory**

| Targets | AR buffer |
|---|---|
| ENT1 | AR6 |
| EpCAM | AR6 |
| N-cadherin | AR6 |
| E-cadherin | AR6 |
| SLC22A15 | AR6 |
| CD20 | AR9 |

3. Microwave for ~45 s on high power until the buffer just boils. Then, reduce the power to 10% and microwave for an additional 15 min. (In the microwave mentioned in the Equipment section above, enter the time, 15:00, and then press power, and 1, which will set the power to 10%). Place the jar toward the outside border of the carousel to ensure even heating.
4. Remove the jar from the microwave and allow to cool at room temperature (RT) for at least 15 min. Optional overnight stopping point: at this step, the slides can be left in AR buffer at RT for overnight.
5. Rinse the slides by immersing in Milli-Q water in a slide processing jar.
6. Rinse the slides in TBST buffer.
7. An optional quenching step is performed here in many IHC experiments to quench the endogenous peroxidases by immersing the slides in 0.3% freshly prepared hydrogen peroxide solution for 10–15 min. However, this step was omitted for the TMA staining protocol described here.

### 3.4 Blocking, Probing with Primary and Secondary Antibodies, and Fluorescent Labeling with OPAL System

In IHC, blocking of the tissue samples with an appropriate blocking solution is an essential step to perform prior to the incubation with primary antibodies. This is to prevent the nonspecific binding of the antibodies to tissues or FC receptors that gives nonspecific or background signals. Normal serum (1–5%, diluted in the washing buffer/TBST) matching to the species of the secondary antibody is recommended to be used as a blocking solution in IHC. Following steps were adopted in this procedure:

1. Partially dry the slides by wiping the perimeter of the tissue section(s) with a Kimwipe, or filter paper cut into small triangles, be careful to avoid disturbing the delicate tissue.
2. Encircle the section with a hydrophobic barrier pen. This forms a contained area that helps retain the blocking solution within the borders during incubation of the tissue with each solution. The barrier may need to be touched up with the pen to maintain its integrity, especially following multiple wash steps.
3. Apply the blocking solution provided in the Opal 4 Color Manual Staining Kit to each section quickly to avoid drying, and incubate the slides for 10 min at RT. Note that 100–150 μL of blocking solution is sufficient to cover up the entire tissue section on one slide. If not using the kit, then prepare the blocking solution fresh on the same day (recipe described above), add to the sections, and incubate for 30 min at RT. This is an optional overnight stopping point; you can also incubate the slides in 4 °C refrigerator or in a cold room by adding the blocking solution.

**Table 2**
**List of primary antibodies and their working dilutions used in this protocol**

| Primary antibodies | Dilutions |
|---|---|
| Anti-ENT1 | 1:500 |
| Anti-EpCAM | 1:100 |
| Anti-N-cadherin | 1:100 |
| Anti-pan-cytokeratin | 1:75 |

4. Dilute the primary antibody in an appropriate antibody diluent solution. In this procedure we diluted the antibodies in the antibody diluent provided with the Opal kit (#ARD1001EA). However, primary antibodies can be diluted in freshly prepared 5% normal serum (blocking solution) described above. The dilutions of various primary antibodies were tested producing the optimal signal intensity in our experimental set up (Table 2).
5. Cut the Whatman filter papers into small triangles. Tilt the slide to allow solution covering the section to collect on one side, and gently absorb with the filter paper.
6. Apply 100–150 μL of primary antibody solution (anti-EpCAM or anti-N-cadherin at the above dilutions) to the tissue sections by adding gently in a dropwise manner. Rapid addition of any of the solutions to the slide could potentially damage the tissue sections, so use care at each step.
7. Incubate the slides for 1 h at RT in a humidified chamber to avoid drying up of the liquid. This can be prepared in a pipette tip box by pouring 10–20 mL of water to the lower chamber of the box and placing the slides in the upper chamber. This is an optional overnight stopping point. If longer incubation time with the primary antibody is required, the slide(s) can be incubated in a refrigerator at 4 °C or inside a cold room for overnight by placing them in the humidified chambers.
8. Rinse the slides gently once with TBST. Wash the slides for three times (2 min each) in TBST with agitation by putting the slides in the white circular slide processing jars (Fig. 1) placed on a benchtop orbital shaker at speed setting 3. These slide processing jars can hold up to five slides at once.
9. Remove excess liquid from the tissue sections by tilting the slides and absorbing the liquid with cut filter papers from one side of the section.
10. Gently add 100–150 μL of Opal HRP polymer supplied with the kit to each section and incubate the slides for 10 min at RT.

11. Rinse the slides once with TBST, and subsequently wash the slides for three times (2 min each) by putting them in the slide processing jars and agitating them as described above.
12. Dilute the appropriate Opal fluorophore in Opal amplification diluent. Do not dilute lower than 1:50, average concentration is 1:100.
13. Remove excess liquid from the tissue sections using filter papers as described above.
14. Incubate the tissue sections with Opal 520 fluorophore supplied in the kit (diluted 1:100) for 10 min at RT.
15. Rinse once with TBST, and then wash with TBST three times for 2 min each.

### 3.5 Staining for Additional Markers and Final Processing of the TMA Slides

After incubating the tissue with Opal fluorophore for the primary antibody applied first, a microwave treatment by immersing the slide in AR buffer strips away any unbound antibodies. Because the fluorophores are covalently bound to the antibodies, they remain largely unaffected by microwave treatment. Then the process is repeated using the next primary antibody and a different fluorophore starting with the blocking step. For each target protein to be detected, a different fluorophore should be used. The following steps describe probing and labeling additional targets:

1. At this point, the process was repeated starting with the microwave treatment in an appropriate AR buffer as described above to strip the unbound antibodies, blocking and the next primary antibody anti-pan-Cytokeratin were applied to the tissue section at 1:75 dilution for 1 h. Subsequently the secondary antibody Opal HRP polymer was applied as described above, and then tissue sections were incubated with Opal 690 fluorophore for 10 min.
2. Microwave treatment was performed again in AR buffer to strip any unbound antibodies, and all the steps up to blocking were repeated. The blocking step was performed to prevent nonspecific antibody binding. Tissue sections were then probed with anti-ENT1 primary antibody at 1:500 dilution for 1 h. Secondary antibody and Opal 570 fluorophore at 1: 100 dilution were applied subsequently for 10 min each.
3. After all the antibodies are applied, the final microwave treatment should be performed with AR 6 buffer.
4. Rinse the slides with water.
5. Rinse once in TBST.
6. Prepare DAPI working solution by diluting spectral DAPI provided in the Opal kit in TBST (1 drop of DAPI in 500 μL of TBST).

7. Remove excess liquids from the tissue sections using filter papers as described above.
8. Incubate the sections with DAPI working solution for 1–2 min.
9. Wash with TBST for 2 min.
10. Wash with water for 2 min.
11. Remove excess liquids from the tissue sections using filter papers as described above, and allow the sections to air dry.
12. Mount with coverslip using fluorescent mounting media without DAPI. Use glass coverslips sufficient to cover all of the TMA cores to avoid seams between multiple coverslips.
13. Store slides at −20 °C for imaging at a late time. The Opal fluorophores are photostable and retain signal for at least 1 year.

### 3.6 Imaging of the Multiplexed TMA Slides

We used Vectra 2.0 multispectral imaging system (Perkin Elmer) to image the multiplexed IHC slides (Fig. 2). This is an automated and multimodal imaging system that not only captures high-quality images of the entire tissue section or within the regions of interest but also allows to study the morphometric and biochemical organization of an intact tissue section both qualitatively and quantitatively. Its multiplexing capability can separate up to eight markers/fluorophores within the field of view in a single tissue section. For more information on how to use the latest version of the microscope, please find the user manual from the manufacturer in this link: https://resources.perkinelmer.com/corporate/content/lst_software_downloads/vectra-user-manual-3-0-3.pdf.

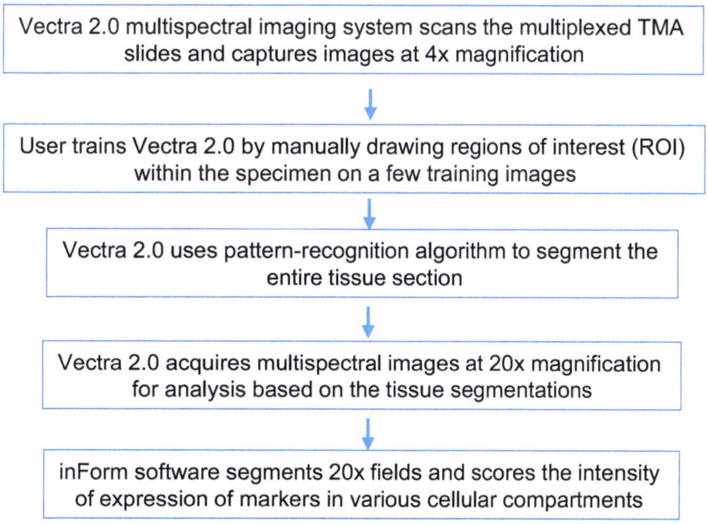


**Fig. 2** Schematic workflow of multiplexed TMA imaging and quantification using Vectra 2.0 multispectral imaging system and inForm analysis software

The following steps represent an overview of using the instrument for imaging of the IHC multiplexed TMAs. Note that these steps are specific for the Vectra 2.0 multispectral imaging system (Perkin Elmer) installed at The Ohio State University Comprehensive Cancer Center:

1. Switch on each of the associated devices sequentially. First, turn on the computer connected to the microscope, and let windows to start. After a minute turn on the microscope module controller box to supply power to the microscope, and then switch on the fluorescence illuminator since the goal is to perform the multispectral imaging. Finally, turn on the Vectra stage controller module to input power to the automated microscope stage.
2. Double-click on the Vectra software icon on the desktop to start. It approximately takes 30 s to initialize the software, and adjust the hardwares including objective lenses. After initialization, the software's home page opens that allows to access through different options.
3. Load a slide on to the stage, and then use the stage control joystick to move the stage in different directions. This is needed to focus the objectives/lenses on the field of interest and to optimize the exposure time for the fluorescent tissue samples. Advanced version of this microscope also allows to load multiple slides in cassettes into a slide loader, and the Vectra microscope can automatically recognize the slide position and image the samples in sequence.
4. Click on file and then load protocol, in case you have a pre-optimized imaging protocol saved earlier in the instrument. You can also create or edit a protocol according to your experiment by setting up the fluorophore wavelengths, name, exposure time, etc.
5. Click setup, then HP imaging, and then 20× (magnification).
6. Set up the order of the Opal fluorophores to be imaged. This is based on the relative brightness of the fluorophore in Vectra 2.0 (e.g., 1, Opal 520; 2, Opal 690; 3, Opal 570).
7. Correct exposure of the saturated fields and press "open shutter." If the field looks saturated, then reduce the exposure or select "auto-exposure," and then press "accept" and "auto-expose."
8. Press "Take one" and then "acquire" to capture an image of the selected field.

If your experimental goal is to detect only one antigen with a single fluorophore (e.g., if you are interested to look at the membranous expression of ENT1 in your samples tagged with the Opal 570 fluorophore) and you would like to capture the images from multiple random fields, then the following steps would apply:

9. Start the Vectra software as described above, and then open through your lab ID or an already saved protocol for the same. Otherwise, click on "WT1" and then "monochrome imaging."
10. Click on "setup" then "settings" then "monochrome imaging" and then "scan area."
11. Adjust the scan area on the sample by moving the specimen through using joystick. Then click on "find specimen" and "select all."
12. Select the "LP imaging option," then "select filed," and then click on "HP imaging."
13. After you chose the fluorophore (Opal 570) and DAPI, adjust the exposure as described above, then click on "acquire" to capture the images from the random fields.
14. Check your data storage location in the computer drive for the captured image files. Vectra usually stores the files in the .im3 format, which can be later imported into the inForm image analysis software for quantification and analysis, and the files can be converted into .tiff images.

### 3.7 Quantification and Analysis of the Acquired Multiplexed IHC Images

We used Vectra multispectral imaging system 2.0 coupled with inForm image analysis software, which helps in automated quantitation of expression of various markers within the tissue sections. inForm is equipped with pattern recognition algorithm and uses machine learning approaches, which enables users to teach the software to identify and segment areas within the tissue section automatically. This capability of the software also helps to accurately measure the protein expression in various cellular and subcellular compartments (e.g., cytoplasmic, nuclear, and membranous) of the TMA sections. Following steps represent an overview of working with inForm image analysis software for quantifying/analyzing multispectral images:

1. Open the inForm software after you switch on the computer in which it is installed. You need the .im3 images saved in this computer, which are already captured using Vectra 2.0. If you are using inForm in the same computer connected to Vectra 2.0, you do not need to turn on the Vectra 2.0 microscope, the microscope module controller box, or the fluorescence illuminator during the image analysis.
2. Click on file, open, and then open the .im3 images from the raw file folder. Then choose the sample format as fluorescence.
3. Click on the spectral library source and inForm, and then select the fluorescence individually from an already saved library (in this case, Opal 520, Opal 570, Opal 690, and DAPI from User's library) or create a library according to the fluorophore used.

4. Click on configure, find features, cell segmentation, and score, and then configure.
5. Click on prepare, spectral library source (inForm), then the saved library (e.g., Brenna's library). Select "all fluorophores" that applies (Opal 520, Opal 570, Opal 690, and DAPI), then "prepare all."
6. Click on advance, segment cells, and select nuclei, cytoplasm, and membrane.
7. Uncheck the edge rules, select the general appearance as "counter stain based," and choose the "nuclear counterstain DAPI."
8. Click on cytoplasm, any component, and then membrane. Select all fluorophores (Opal 520, Opal 570, and Opal 690), then "auto" and then "segment all."
9. After inForm segments the cells and cellular compartments, manual drawing of pattern or tailoring is needed to train this software to score/quantify accordingly. Here you can zoom in and out of the images to inspect the segmentation. The software also shows colored buttons matching to the color of the fluorophores, which allows to turn on and off the staining of one fluorophore and to look at another fluorophore separately.
10. Look at nuclei by unchecking other fluorophores. Shrink the nuclei if necessary and click on segment image.
11. Select the fluorophore for the membrane compartment (Opal 520).
12. Scoring can be done based on the % positivity of the staining intensity (from low to high) into four bins (0+, 1+, 2+, and 3+).
13. Set the threshold by checking the positively stained cells for a particular compartment (membrane), and then press "score all."
14. inForm scores the membrane staining in the entire image based on the set threshold for the positively stained cells and gives the values as histoscores (H-scores).

### 3.8 Statistical Analysis of the Data from Multiplexed TMA

1. TMA slides procured from commercial sources such as biomax. us usually come up with data for patient demographic and clinical characteristics for each core/sample of TMA. These data were analyzed using descriptive statistics and continuous variables expressed as mean (standard deviation) or median (range). Categorical data such as sample size can be presented as count ($n$) and frequency (%). Comparisons can be made using either chi-square test, Student's t test, or one-way ANOVA.

2. Pearson's correlation tests were applied to correlate between the percentages of plasma membrane expression of ENT1-positive and EpCAM-positive cells, plasma membrane N-cadherin-positive and cytoplasmic ENT1-positive, ENT1 plus EpCAM double-positive, and ENT1 plasma membrane positive cells.
3. Kaplan–Meier survival analysis was employed to determine the median overall survival times of patients and survival curves were compared using log–rank test.
4. Univariate and multivariate survival analyses were performed to assess the potential associations between tumor grade, T stage, N stage, adjuvant chemotherapy treatment, H-scores of markers, and median overall survival. Multivariate analysis used proportional hazard regression model of Cox.
5. Two-sided $p$ values of $p < 0.05$ were considered statistically significant in all cases except where analysis was performed on a very limited sample size.
6. Statistical analyses of the TMA were performed using SAS 9.4 (the SAS Institute, Cary, NC), and survival analysis was performed using R3.5.0 (R Foundation for Statistical Computing).

## 4 Notes

1. While transferring the slides from one slide processing jar to another with a different liquid, try to drain the liquid completely from the slides or wipe the slides with Kimwipes to ensure liquids from one bath do not mix with the next bath.
2. The staining process should be continuous, and care should be taken to avoid drying of the tissue sections while transferring the slides from one bath to another or any time during the procedure.
3. Handle the tissue sections and especially the TMA cores with utmost care to minimize peeling or folding of the samples. Forceps can be used to handle the slides by gripping the label to transfer the slides between the washes.
4. The antibodies specified for IHC usually have the method for AR mentioned on the datasheets of those antibodies. However, it is recommended to optimize the best AR buffer and conditions for a specific antibody by doing test experiments. Also, see the humanproteinatlas.org for antibody validations that often describe antigen retrieval used.

5. If appropriate blocking solution is not used or blocking is performed for inadequate period, the detection reagents may bind to numerous nonspecific binding sites and produce background signals, which often mask up the specific antigen–antibody reactivity signals.
6. The primary antibodies should be diluted in the antibody diluent accurately. Commercially available primary antibodies recommended for IHC usually suggest the appropriate antibody dilution to be used. However, it varies depending on the expression of the target protein in the tissue type being stained. Therefore, it is better to optimize the dilution of the primary antibody to be used in your final experiment by running two to three trial experiments with varying antibody dilutions and testing which dilution of the antibody gives the best signal intensity.
7. Temperature during incubation of the tissue section with primary antibody also plays an important role. High temperature promotes binding but decreases specificity. Lower antibody dilutions at low temperatures over a long period of time generally provide the most specific signal.
8. You may want to remove the pen markings at the end of the procedure before adding the coverslip, as it is autofluorescent, but it is not necessary, and there is a risk of disturbing the tissue. To remove the barrier pen markings, use a Kimwipe and ethanol.
9. The best place to find in-depth information about the Opal kit is from Akoya Bioscience's Assay Development guide https://www.akoyabio.com/phenopticstm/resources.

## 5 Loss of ENT1 During EMT Process Opal Panel: A Working Example

The study by Weadick et al. (2021) demonstrates that loss of plasma membrane ENT1 occurs when the epithelial pancreatic cancer cells undergo EMT and attain mesenchymal phenotypes, which also correlate with altered expression of cell–cell adhesion molecules including EpCAM, N-cadherin, and E-cadherin. Loss of ENT1 resulted in reduced efficacy of nucleoside analog gemcitabine in mesenchymal subtypes of pancreatic cancer cells. The following steps describe how the TMA multiplexing experiment was designed and optimized to address the question(s) in this study.

*Question: In human PDAC samples, does loss of ENT1 correlate with EpCAM, or N-cadherin?*

(a) *Panel design*

ENT1: Nucleoside transporter protein with membranous and cytoplasmic expression.

**Table 3**
**List of Opal fluorophores used for different target antigens showing their color and multiplexing order**

| Fluorophore | Relative brightness (Vectra 2.0) | Antigen | Color | Multiplex order |
|---|---|---|---|---|
| Opal 520 | Medium | EpCAM, N-cadherin | Green | 1 |
| Opal 690 | Lowest | Pan-cytokeratin | Cyan | 2 |
| Opal 570 | Highest | ENT1 | Red | 3 |

EpCAM and N-cadherin: Cell–cell adhesion molecules, membranous expression of which alters during EMT (based on the published reports and our preliminary studies).

Pan-Cytokeratin: marker to differentiate epithelial tumors from surrounding stroma.

DAPI: Nuclear counterstain.

Step 1: Assign an Opal fluorophore based on predicted relative expression. For image analysis, it is important that signals are basically balanced in intensity for accurate unmixing. Also, decide which order each target will be detected in the final multiplexed experiment. This may require optimization, as some antigens will become more or less accessible after multiple rounds of microwave treatment (*see* Table 3).

(b) *Make library slides*

1. The key to the Opal multiplexing IHC assay is accurate unmixing of multiple colors. Library slides provide a pure spectrum for each fluorophore to be used for image analysis.
2. Library slides are created in control tissues with a highly expressed marker (we considered CD20 expression in tonsil tissues) and can likely be used in multiple studies, although it may be a good idea to capture new library images as the fluorescent bulb ages.
3. Library slides are made for each fluorophore and processed as usual with no DAPI.
4. One slide is made for DAPI only.
5. One slide of representative sample tissue is processed and left without DAPI or Opal fluorophore to collect an autofluorescence signal.
6. Image library slides by setting the capture settings for the same exposure time (i.e., 150 ms) for every channel. Select the option to correct for saturation, before capturing images.

In InForm software, import the .im3 files for each fluorophore, and extract the spectra for each under "build libraries" tab on the left side of the initial page. Save the name and color for each fluorophore to use for future studies. Additionally, capture an autofluorescence image for the tissue of interest in which the final staining will be done. Draw a region in InForm around the autofluorescence image, and apply it to the library slide images.

(c) *Perform IHC monoplex optimization experiments*

If the lab has not previously performed IHC experiments with the antibody of interest, then follow general suggestions below:

1. Establish that the antibody is working in a positive control tissue.

   The antibody manufacturer or the human protein atlas can provide general information in most cases about where to start with determining antibody dilution. If no information is available, try 1:100 dilution of primary antibody. To find an appropriate positive control, consult the literature or human protein atlas tissue expression section. Human tissue slides, formalin-fixed, paraffin-embedded (FFPE) can be purchased from several sources including Biomax.com and novusbio.com. Alternatively, slides can be created in lab or obtain by collaborating with the Veterinary Pathology Department at your University.

2. Determine optimal conditions in target tissue.

   Once you have established that the antibody is working, test it in the tissue to be used in your final experiments. For example, if you validated the antibody in normal pancreas tissue, now you can try in pancreatic cancer samples.

   A. Consider number of microwave steps that will be used in final experiment:

      - These monoplex experiments should incorporate the correct number of microwave treatments for the panel you have designed. Try changing the multiplex order if needed.

   B. Consider changing primary antibody concentration first:

      - Optimize by trying serial dilutions.

   C. Consider changing primary antibody incubation time:

      - Optimize by shortening or lengthening incubation time and temperature. Higher concentration of antibody requires shorter incubation time but can contribute to background.

D. Consider changing Opal fluorophore dilution as a last resort:

- Dilution should never be lower than 1:50 and ideally at 1:100 or above. Lowering the dilution of Opal fluorophore can increase signal strength if necessary.

Multiple trial experiments are needed to optimize the appropriate antibody, dilution, incubation time, and the ideal Opal fluorophore for that target antigen. Figure 3 shows examples of monoplexed IHC images from the trial experiments, which did not produce optimal staining pattern/intensity.

*Question: How do I know if my signal is optimal to move forward to multiplexing?*

- When capturing images with Vectra, aim for autoexposure times between 25 ms and 250 ms.

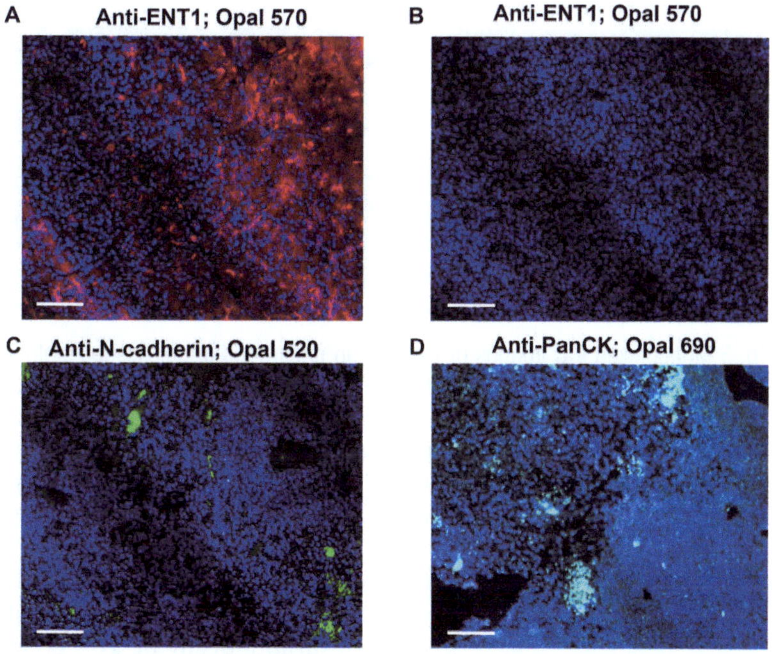


**Fig. 3** Examples of monoplexed IHC staining images that did not work well during our experimental optimization. (**a**, **b**) Opal monoplex images of human pancreatic tumor tissues probed with anti-ENT1 primary antibodies procured from Alomone Labs and Abcam respectively used at 1:2500 dilution each, incubated for overnight at 4 °C and stained with Opal 570 fluorophore (red). (**c**, **d**) Opal monoplexed IHC images of human pancreatic tumor tissues probed with anti-N-cadherin and Anti-pan-Cytokeratin primary antibodies, respectively, used at 1: 2500 dilution each, incubated for 2 h at RT, and stained with Opal 520 (green) and Opal 690 (turquoise), respectively. DAPI staining of nuclei is blue. Images were captured at 20× magnification. Scale bars, 50 μm

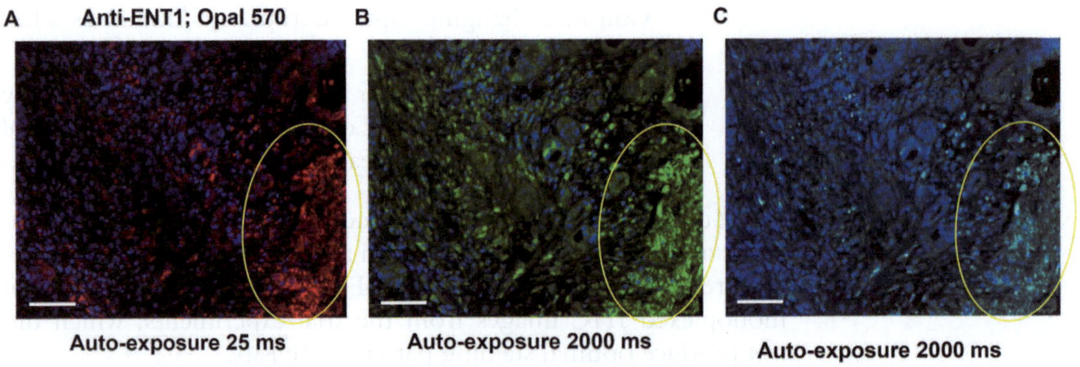


**Fig. 4** Identifying real signal versus autofluorescence. (**a**) Monoplexed IHC image of a human pancreatic tumor tissue probed with the in-house anti-ENT1 primary antibody, stained with Opal 570 fluorophore (red), and counterstained with DAPI; image autoexposure time: 25 ms. (**b**, **c**) Images of the same stained tissue shows autofluorescence in the green and turquoise color filter as well at higher exposure time (2000 ms). Images were captured at 20× magnification. Scale bars, 50 μm

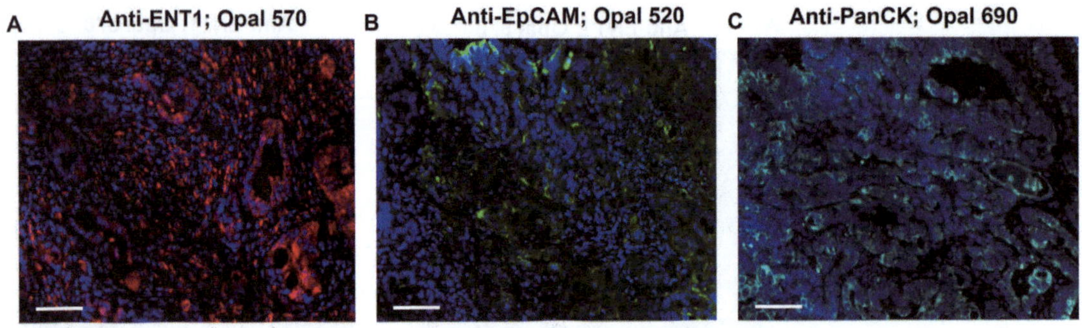


**Fig. 5** Monoplexed IHC images of the human pancreatic tumor tissues stained for various markers. (**a**) Anti-ENT1 primary antibody (1:2500 dilution) incubated overnight at 4 °C, Opal 570 fluorophore (color: red), image autoexposure time 24 ms. (**b**) Anti-EpCAM primary antibody (1:2500 dilution) incubated overnight at 4 °C, Opal 520 fluorophore (color: green), image autoexposure time 82 ms. (**c**) Anti-pan-Cytokeratin primary antibody (1: 2500 dilution) incubated overnight at 4 °C, Opal 690 fluorophore (color: turquoise), image autoexposure time 30 ms. DAPI staining of the nuclei is blue. Images were captured at 20× magnification. Scale bars, 50 μm

- After preparing the images in inForm, determine Opal intensity by hovering over each cell. Acceptable range is between 3 and 30. Signal should ideally be 10× greater than background. This may or may not be achievable considering the relative abundance of the target.
- Within a panel, it is best if intensity levels are within twofold of one another.

Also, it is critical to distinguish between the real signal intensity and autofluorescence (Fig. 4) and to avoid capturing the autofluorescence signals. Figure 5 shows examples of monoplexed IHC images from an optimized experiment in which human pancreatic tumor tissues were stained for the expression of ENT1, EpCAM, and Pan Cytokeratin.

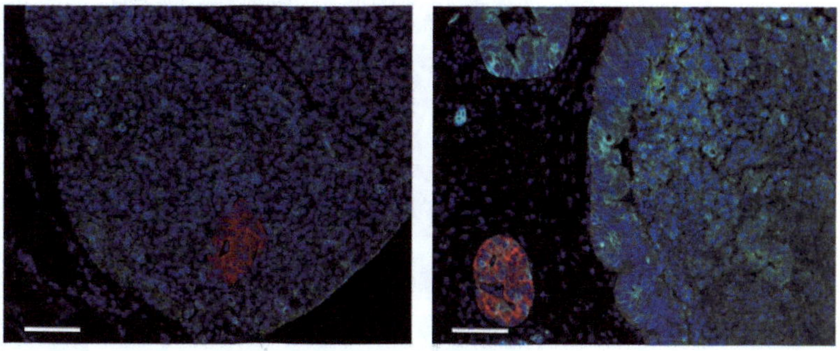


**Fig. 6** Optimization: examples of multiplexed IHC images of a normal human pancreatic tissue (left) and a human pancreatic tumor tissue (right) probed with anti-ENT1 primary antibody (Opal 570 fluorophore, red), anti-EpCAM primary antibody (Opal 520 fluorophore, green), and anti-pan-Cytokeratin primary antibody (Opal 690 fluorophore, turquoise). DAPI staining of the nuclei is blue. Images were captured at 20× magnification. Scale bars, 50 μm

(d) *Multiplex experiment*

Now you are ready to move forward and incorporate all the optimized monoplex conditions. Perform multiplexing experiments in a few normal human tissues and tumor tissues prior to the TMA staining. Figure 6 shows an example of multiplexed IHC staining images of normal human pancreas and pancreatic tumor tissues stained for ENT1, EpCAM, and pan-Cytokeratin expression. Follow the Opal workflow, applying a microwave treatment step before every new primary antibody to strip unbound antibody and fluorophore. Ensure that all signals are balanced at the multiplex level before moving on to the final experiment. Few representative images of the different tumor tissues from a pancreatic cancer TMA stained for the expression of ENT1 (red), EpCAM (green), and Pan Cytokeratin (turquoise) are presented in Fig. 7.

Following points summarize how the expression of various markers were segmented and quantified in the multiplexed TMA described in the study by Weadick et al. [16].

1. Random fields were imaged from each core of the multiplexed TMA slides using Vectra 2.0 multispectral imaging system.
2. inForm analysis software was used to determine the %positive cells for various markers in different subcellular compartments.
3. Cell segmentation was performed based on the DAPI nuclear stain that differentiated nuclear and cytoplasmic compartments.
4. Cell segmentation also distinguished pancreatic tumor cells from stromal cells based on pan-Cytokeratin expression.

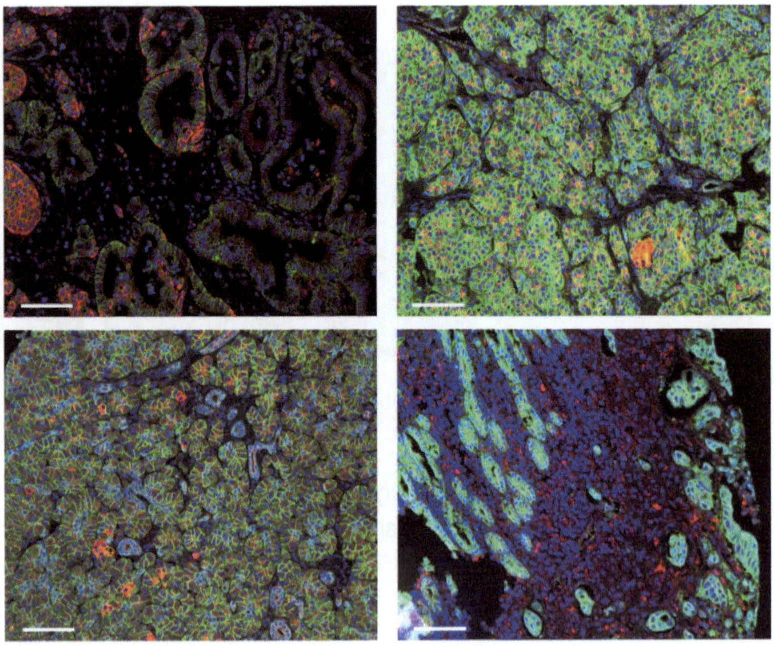

**Fig. 7** Examples of multiplexed IHC images of different human pancreatic tumor tissues from a TMA stained with the in-house anti-ENT1 primary antibody (Opal 570 fluorophore, red), anti-EpCAM primary antibody (Opal 520 fluorophore, green), and anti-pan-Cytokeratin primary antibody (Opal 690 fluorophore, turquoise). The image on the upper left panel is without pan-Cytokeratin stained. Counter staining with DAPI showing nuclei in blue color. Images were captured at 20× magnification. Scale bars, 50 μm

5. Percent positive cells for either EpCAM or N-cadherin and the double-positive cells for both these markers were quantified in the epithelial tumor cells.
6. ENT1 expression on the plasma membrane was quantified using inForm by distributing the fluorescent signal intensity into four bins as mentioned above.
7. H-scores of ENT1 plasma membrane expression for each image were calculated using the formula (1 × (% cells 1+) +2 × (% cells 2+) +3 × (% cells 3+)).

Figure 8 shows example of an IHC image of the human pancreatic TMA stained with anti-EpCAM antibody (green fluorescence) and segmented to study the membranous expression of EpCAM.

EpCAM-membrane positive

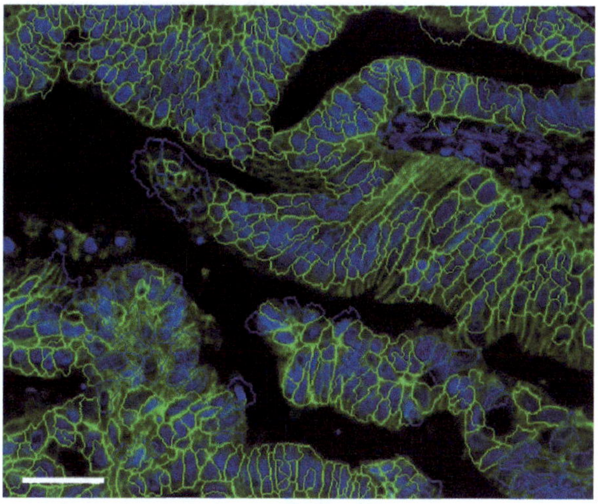

**Fig. 8** Example of an IHC image of a human pancreatic tumor tissue from a TMA stained with anti-EpCAM antibody and DAPI. Cell segmentation was performed using inForm image analysis software showing EpCAM-membrane positive cells in green boundaries. Magnification, 20×. Scale bar, 50 μm.

## 6 Investigating Drug Resistance by Exploiting Multiplexed TMA

The combination of tissue microarray with IHC multiplexing together presents a powerful tool for the next generation oncology and other biomedical research, since it enables researchers for the high-throughput analysis of proteins of interest in hundreds to thousands of tissue samples from patients. The discovery of advanced imaging and quantification technology such as Vectra 2.0 coupled with inForm analysis software has further empowered this technique by allowing scientists for studying expression of multiple markers (up to ~7 or 8) simultaneously in a single tissue section. The alterations in expression of various markers of the tumor microenvironment (oncogenes, tumor suppressor genes, and immune markers) in response to certain chemotherapy and further consequences on drug efficacy can be studied effectively in patient tissue samples using multiplexed TMA. This information can also be correlated with patient demographical characteristics such as age, gender, and race to derive inference about influence of these factors on cellular expression of markers determining drug resistance. Additionally, this technique facilitates comparative analysis of expression of these markers both qualitatively and quantitatively between normal versus cancer tissues and among different grades and stages of tumor progression in patients undergoing neoadjuvant chemotherapy and their impact on survival outcomes including overall survival and disease-free survival [11, 24, 25].

Multiplexed TMA analysis is helping substantially to understand the molecular mechanism(s) of drug resistance, thus providing valuable information about how and when to target such molecular aberrations for overcoming the issues of drug resistance. For instance, by utilizing multiplexed IHC approaches in a TMA comprised of $n = 114$ patient-derived PDAC samples and $n = 25$ adjacent normal pancreatic tissues, our earlier study by Weadick et al. demonstrated that coexpression of EpCAM and ENT1 prevails in epithelial primary tumors from pancreatic cancer patients. In EpCAM-positive cells, ENT1 expression was distinctly visible over the plasma membrane, whereas no plasma membrane expression or cytosolic expression of ENT1 was evident in N-cadherin-positive tumor cells [16]. This was studied both qualitatively and quantitatively by applying cell segmentation feature of Vectra 2.0 and subsequent inForm analysis that allowed distribution of the tumor cells in each image into membrane and cytoplasmic compartments as ENT1-single positive, EpCAM-single positive, N-cadherin-single positive, ENT1 and EpCAM plasma membrane coexpression as double-positive, and N-cadherin-plasma membrane plus ENT1-cytoplasmic expression double-positive, and finally quantification of intensity of expression into H-scores. Additionally, univariate survival analysis considering the H-scores of ENT1 plasma membrane expression demonstrated that high ENT1 expression on the cell surface positively correlated with median overall survival times in gemcitabine-treated patients, whereas multivariate survival analysis suggested that ENT1 membrane H-score, gemcitabine-based chemotherapy, and tumor N stage (nearby lymph nodes having cancer) were independently associated with overall patient survival [16]. These data not only uncover the importance of expression of various cell–cell adhesion molecules that determine ENT1 cell surface expression and gemcitabine/NA drugs accumulation in tumor cells but also provide valuable information such as absence of lymph node metastasis and improved overall survival in patients with ENT1 cell surface dominating pancreatic cancer who are receiving this chemotherapy.

## 7 Future of TMA Multiplexing in Studying Anticancer Drug Resistance

In comparison to the conventional IHC, multiplexing of the TMAs has tremendous potential for use in clinical and translational research, especially in the era of cancer immunotherapy. This multiplexing approach can be successfully applied for the rapid detection and differentiation of rare malignant cells from the infiltrating immune cell, e.g., T- and B-lymphocytes and subsets [15]. In the context of NA chemotherapy, the approach can be utilized to study the immune modulation (T-cell activation, alterations to PD-1 and PD-L1) in response to such therapy and their correlation with chemosensitivity. The technique can be successfully designed and performed to study alterations to markers in the tumor microenvironment (TME) and their consequences on lymph node metastasis in response to NA and other chemotherapeutic drugs. Previous studies demonstrate that expression of nucleoside metabolizing enzymes such as uridine–cytidine kinase (UCK) and deoxycytidine kinase (dCK) greatly influences the efficacy of nucleoside analog chemotherapeutic drugs (azacitidine, decitabine) in hematological malignancies [26]. Alterations to such nucleoside metabolizing enzymes in TMAs from patient samples can be studied effectively with the help of IHC multiplexing to ascertain a more complete picture on drug resistance. Moreover, this tool can be efficiently utilized for visualizing and quantifying the expression of any NA chemosensitivity determinants including the NA-activating enzymes (dCK, deoxyguanosine kinase (dGK)) [27, 28] and NA-deactivating enzymes (5′-nucleotidases and cytidine deaminase) [29, 30] in patient samples, which may be important to overcome the issues of drug resistance in specific cancer subtypes.

## Acknowledgments

This work is supported by a grant award of National Institute of Health (R01GM143217 and R03CA262490) and a Research Scholars Grant (RSG-15-036-01-DDC) from the American Cancer Society awarded to R.G. IHC images presented in this article were generated and analyzed using the instruments and services at the Campus Microscopy and Imaging Facility (CMIF) at The Ohio State University. This facility is partly supported by grant P30 CA016058 from National Cancer Institute, Bethesda, MD. We are grateful to Dr. Terence M. Williams, Department of Radiation Oncology, The Ohio State University Comprehensive Cancer Center and Dr. Wei Chen, Department of Pathology, The Ohio State University Wexner Medical Center for generating the TMA and help in analyzing the IHC images. We acknowledge Pelotonia Fellowship Program of The Ohio State University Comprehensive Cancer Center for supporting D.N.

**Author Contributions** D.N. wrote the original draft and prepared the figs. B.W. optimized the TMA multiplexing, generated the images, and edited the manuscript. R.G. provided the resources, supervised the study, and edited the manuscript.

### References

1. Muggia F, Diaz I, Peters GJ (2012) Nucleoside and nucleobase analogs in cancer treatment: not only sapacitabine, but also gemcitabine, vol 21. Taylor & Francis, pp 403–408
2. Galmarini CM, Mackey JR, Dumontet C (2002) Nucleoside analogues and nucleobases in cancer treatment. Lancet Oncol 3(7): 415–424

3. Pizzorno G, Diasio RB, Cheng Y-C (2003) Pyrimidine analogs. In: Holland-Frei cancer medicine, 6th ed. BC Decker
4. Ravandi F, Kantarjian H, Giles F, Cortes J (2004) New agents in acute myeloid leukemia and other myeloid disorders. Cancer 100(3): 441–454
5. Sato T, Issa J-PJ, Kropf P (2017) DNA hypomethylating drugs in cancer therapy. Cold Spring Harb Perspect Med 7(5):a026948
6. Molina-Arcas M, Pastor-Anglada M (2010) Role of nucleoside transporters in nucleoside-derived drug sensitivity. Nucleosides Nucleotides Nucleic Acids 29(4–6):335–346
7. Pastor-Anglada M, Pérez-Torras S (2015) Nucleoside transporter proteins as biomarkers of drug responsiveness and drug targets. Front Pharmacol 6:13
8. Kordella C, Lamprianidou E, Kotsianidis I (2021) Mechanisms of action of hypomethylating agents: endogenous retroelements at the epicenter. Front Oncol 11:490
9. Qin T, Jelinek J, Si J, Shu J, Issa J-PJ (2009) Mechanisms of resistance to 5-aza-2′-deoxycytidine in human cancer cell lines. Blood 113(3):659–667
10. Gruber E, Franich RL, Shortt J, Johnstone RW, Kats LM (2020) Distinct and overlapping mechanisms of resistance to azacytidine and guadecitabine in acute myeloid leukemia. Leukemia 34(12):3388–3392
11. Jia Y, Gu D, Wan J, Yu B, Zhang X, Chiorean EG, Wang Y, Xie J (2019) The role of GLI-SOX2 signaling axis for gemcitabine resistance in pancreatic cancer. Oncogene 38(10): 1764–1777
12. Carter CJ, Mekkawy AH, Morris DL (2021) Role of human nucleoside transporters in pancreatic cancer and chemoresistance. World J Gastroenterol 27(40):6844
13. Bhutia YD, Hung SW, Patel B, Lovin D, Govindarajan R (2011) CNT1 expression influences proliferation and chemosensitivity in drug-resistant pancreatic cancer cells. Cancer Res 71(5):1825–1835
14. Voduc D, Kenney C, Nielsen TO (2008) Tissue microarrays in clinical oncology. In: Seminars in radiation oncology, Elsevier. pp 89–97
15. Tan WCC, Nerurkar SN, Cai HY, Ng HHM, Wu D, Wee YTF, Lim JCT, Yeong J, Lim TKH (2020) Overview of multiplex immunohistochemistry/immunofluorescence techniques in the era of cancer immunotherapy. Cancer Commun 40(4):135–153
16. Weadick B, Nayak D, Persaud AK, Hung SW, Raj R, Campbell MJ, Chen W, Li J, Williams TM, Govindarajan R (2021) EMT-induced gemcitabine resistance in pancreatic cancer involves the functional loss of equilibrative nucleoside transporter 1. Mol Cancer Ther 20(2):410–422
17. Brandi G, Deserti M, Vasuri F, Farioli A, Degiovanni A, Palloni A, Frega G, Barbera MA, Lorenzo S, Garajova I (2016) Membrane localization of human equilibrative nucleoside transporter 1 in tumor cells may predict response to adjuvant gemcitabine in resected cholangiocarcinoma patients. Oncologist 21(5):600–607
18. Endo Y, Obata T, Murata D, Ito M, Sakamoto K, Fukushima M, Yamasaki Y, Yamada Y, Natsume N, Sasaki T (2007) Cellular localization and functional characterization of the equilibrative nucleoside transporters of antitumor nucleosides. Cancer Sci 98(10): 1633–1637
19. Vos LJ, Yusuf D, Lui A, Abdelaziz Z, Ghosh S, Spratlin JL, Mackey JR (2019) Predictive and prognostic properties of human equilibrative nucleoside transporter 1 expression in gemcitabine-treated pancreatobiliary cancer: a meta-analysis. JCO Precis Oncol 3:1–22
20. Farrell JJ, Elsaleh H, Garcia M, Lai R, Ammar A, Regine WF, Abrams R, Benson AB, Macdonald J, Cass CE (2009) Human equilibrative nucleoside transporter 1 levels predict response to gemcitabine in patients with pancreatic cancer. Gastroenterology 136(1):187–195
21. Vincenzi B, Stacchiotti S, Collini P, Pantano F, Rabitti C, Perrone G, Iuliani M, Baldi A, Badalamenti G, Sanfilippo R (2017) Human equilibrative nucleoside transporter 1 gene expression is associated with gemcitabine efficacy in advanced leiomyosarcoma and angiosarcoma. Br J Cancer 117(3):340–346
22. Jobbagy Z, Ward JL, Toan S-V, Leung GP, Tse C-M (2002) One-step unidirectional cloning of tandem repeats of DNA fragments: an application for fusion protein production. Anal Biochem 303(1):104–107
23. Govindarajan R, Bakken AH, Hudkins KL, Lai Y, Casado FJ, Pastor-Anglada M, Tse C-M, Hayashi J, Unadkat JD (2007) In situ hybridization and immunolocalization of concentrative and equilibrative nucleoside transporters in the human intestine, liver, kidneys, and placenta. Am J Phys Regul Integr Comp Phys 293(5):R1809–R1822
24. Zhu Y-X, Li CH, Li G, Feng H, Xia T, Wong CH, Fung FKC, Tong JH-M, To K-F, Chen R (2020) LLGL1 regulates gemcitabine resistance by modulating the ERK-SP1-OSMR pathway in pancreatic ductal adenocarcinoma.

Cell Mol Gastroenterol Hepatol 10(4): 811–828

25. Nayak D, Weadick B, Persaud AK, Raj R, Shakya R, Li J, Campbell MJ, Govindarajan R (2022) EMT alterations in the solute carrier landscape uncover SLC22A10/A15 imposed vulnerabilities in pancreatic cancer. iScience 25(5):104193
26. Valencia A, Masala E, Rossi A, Martino A, Sanna A, Buchi F, Canzian F, Cilloni D, Gaidano V, Voso MT (2014) Expression of nucleoside-metabolizing enzymes in myelodysplastic syndromes and modulation of response to azacitidine. Leukemia 28(3):621–628
27. Månsson E, Flordal E, Liliemark J, Spasokoukotskaja T, Elford H, Lagercrantz S, Eriksson S, Albertioni F (2003) Down-regulation of deoxycytidine kinase in human leukemic cell lines resistant to cladribine and clofarabine and increased ribonucleotide reductase activity contributes to fludarabine resistance. Biochem Pharmacol 65(2): 237–247
28. Bergman A, Giaccone G, Van Moorsel C, Mauritz R, Noordhuis P, Pinedo H, Peters G (2000) Cross-resistance in the 2′, 2′--difluorodeoxycytidine (gemcitabine)-resistant human ovarian cancer cell line AG6000 to standard and investigational drugs. Eur J Cancer 36(15):1974–1983
29. Eliopoulos N, Cournoyer D, Momparler RL (1998) Drug resistance to 5-aza-2′--deoxycytidine, 2′, 2′-difluorodeoxycytidine, and cytosine arabinoside conferred by retroviral-mediated transfer of human cytidine deaminase cDNA into murine cells. Cancer Chemother Pharmacol 42(5):373–378
30. Suzuki K, Sugawara T, Oyake T, Uchiyama T, Aoki Y, Tsukushi Y, Onodera S, Ito S, Murai K, Ishida Y (2007) Clinical significance of high-Km 5′-nucleotidase (cN-II) mRNA expression in high-risk myelodysplastic syndrome. Leuk Res 31(10):1343–1349

# Chapter 9

# In Situ Proximity Ligation Assay to Visualize Protein–Protein Interactions in Tumor Specimens


## Yasuhiro Miki, Erina Iwabuchi, and Takashi Suzuki



## Abstract

Protein–protein interactions (PPI) are the basis of various biological phenomena, such as intracellular signal transduction, gene transcription, and metabolism. PPI are also considered to be involved in the pathogenesis and development of various diseases, including cancer. PPI phenomenon and their functions have been elucidated by gene transfection and molecular detection technologies. On the other hand, in histopathological analysis, although immunohistochemical analyses provide information pertaining to protein expression and their localization in pathophysiological tissues, it has been difficult to visualize the PPI of these proteins. An in situ proximity ligation assay (PLA) was developed as a microscopic visualization technique for PPI in formalin-fixed, paraffin-embedded (FFPE) tissues as well as in cultured cells and frozen tissues. PLA using histopathological specimens enables cohort studies of PPI, which can clarify the significance of PPI in pathology. We have previously shown the dimerization pattern of estrogen receptors and significance of HER2-binding proteins using breast cancer FFPE tissues. In this chapter, we describe a methodology for the visualization of PPI using PLA in pathological specimens.



**Key words** In situ proximity ligation assay, Protein–protein interaction, Formalin-fixed, paraffin-embedded tissues, Cultured cells, Frozen tissues, Immunohistochemistry, Antibody, Pathology, Histology


## 1 Introduction

Imaging and serodiagnosis are used for the diagnosis of cancer; however, the final definitive diagnosis is usually based on the histological diagnosis in pathology. Lymph node status, tumor size, and histological grade, also known as the Nottingham Grading System, are the main factors that contribute to the prognosis of breast cancer patients [1]. Histological grade is assessed based on the morphological characteristics of invasive breast cancer, that is, the degree of differentiation of the tumor. Pathological specimens are used not only for the diagnosis but also for devising therapeutic strategies for the patients based on immunohistochemical analysis of estrogen receptor (ER) and HER2 [2]. Both, gene expression


Usha N. Kasid and Robert Clarke (eds.), *Cancer Systems and Integrative Biology*, Methods in Molecular Biology, vol. 2660, https://doi.org/10.1007/978-1-0716-3163-8_9,

analysis techniques, such as PCR, and protein expression analysis techniques, such as western blotting, are often selected for several types of cancer research. The histopathological specimen used for the diagnosis is usually a formalin-fixed, paraffin-embedded (FFPE) tissue. Therefore, molecular biological analysis employing histopathological specimens used for diagnosis is well-known to be limited, and therefore, unfixed frozen tissues are required for such analyses [3]. However, when nucleic acids and proteins are extracted by homogenizing the entire frozen tissue, it becomes impossible to determine which cell type constituting the tumor tissue, including normal epithelial, carcinoma, and fibroblastic stromal cells, is predominantly expressed in the cancer microenvironment [4].

One of the mechanisms by which proteins exert their functions involves interaction of a protein with the same or different proteins and is known as protein–protein interactions (PPI). The analysis of PPI has been advanced by in vitro techniques, such as BRET/FRET assays and co-immunoprecipitation; however, PPI have been considered difficult to analyze using FFPE tissues [5, 6]. The advantages of using FFPE are as described above, and visualization of PPI in FFPE tissue is expected to lead to further developments in the fields of cancer research and treatment. In addition, large retrospective cohort studies are conducted using archival FFPE tissues. An in situ proximity ligation assay (PLA) was developed as an immunohistochemical tool that combines the specificity of antigen–antibody reaction with the sensitivity of PCR [7, 8]. PLA can help in visualizing PPI in archival FFPE tissues, revealing the constituent cells in which PPI occur in the cancer microenvironment. Furthermore, by evaluating the amount of PPI signal, the relationship between PPI and various pathological factors, including prognosis, can be statistically clarified. We have visualized the dimeric pattern of ER subtypes [9, 10] and ER-binding proteins [11–13] by PLA and have elucidated their significance in the pathophysiology of breast cancer. Furthermore, we found a novel HER2-binding protein that has been shown to modify the function of HER2 in breast cancer [14]. In this chapter, we will explain in details the PLA procedure using FFPE tissues. We will also describe PLA for breast cancer cell lines.

## 2 Principle of PLA

Figure 1 shows a conceptual scheme of PLA [15, 16]. PLA can detect PPI in immunohistochemistry-ready samples (cell lines, fresh, frozen, formalin-fixed, paraffin-embedded tissue, etc.). First, primary antibodies specific to different targets, viz., "$\alpha$ protein" and "$\beta$ protein," are allowed to react. Secondary antibody probes with characteristic oligonucleotides are then allowed to

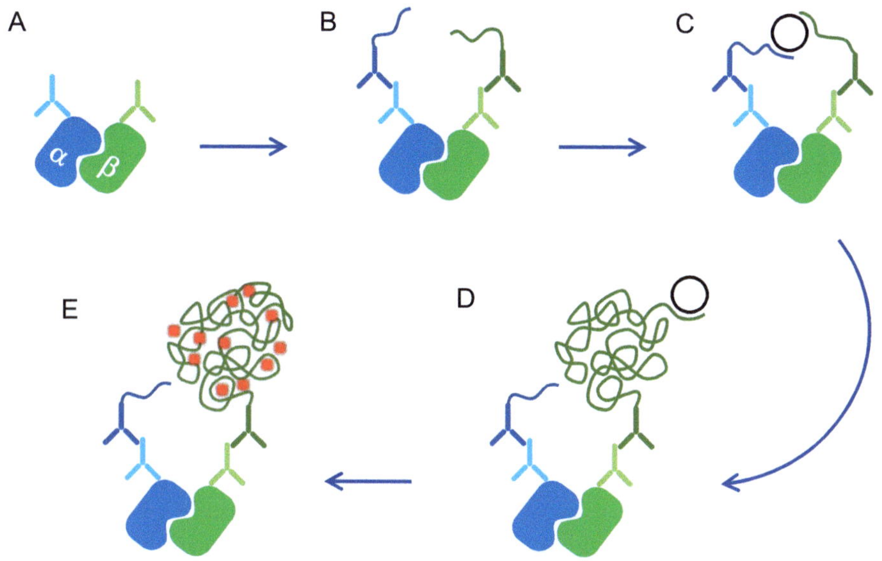


**Fig. 1** Schema of proximity ligation assay. (**A**) Primary antibodies (anti-$\alpha$ and anti-$\beta$ antibodies) specifically bind to two different proteins ($\alpha$ and $\beta$). (**B**) A secondary antibody for each primary antibody binds. These are characteristic secondary antibodies, namely, PLA probes (PLUS and MINUS) with attached oligonucleotides. (**C**) When the $\alpha$ and $\beta$ proteins are in close proximity (<40 nm), they are bridged by hybridization of an oligonucleotide that forms a circular structure between the PLA PLUS and MINUS probes. (**D**) The circular DNA functions as a template for in situ rolling circle amplification. (**E**) Amplified DNA products are detected by complementary binding of the fluorescently labeled oligonucleotide probes

react with the respective primary antibodies. When these two probes are within 40 nm of each other, they are ligated by DNA ligase, which joins the 3′-end of one probe with the 5′-end of the other probe to form a unique target DNA reporter strand. Rolling circle amplification using these reporter strands amplifies the products quantified by real-time PCR. A fluorescence-labeled probe that specifically binds to this amplified product is allowed react, and the intratisular localization of PPI between "$\alpha$" and "$\beta$" proteins is detected under a microscope.

## 3 Materials and Methods

### 3.1 Specimens

#### 3.1.1 Tissue Specimens

In the case of human tissues, FFPE tissue that has been surgically resected and subjected to pathological diagnosis is used for PLA. If the FFPE tissue has been used for immunohistochemical diagnosis, it is considered that there are no issues in the preparation process, such as overfixation. In our laboratory, we were able to obtain good PLA staining using archival FFPE tissues stored at room temperature for over 10 years. In the case of tissues collected for experiments (humans and experimental animals), it is necessary to immerse the tissues in a fixative solution as soon as possible after their excision. Formalin-based fixatives (*see* **Note 1**) are best for PLA, and alcohol-based fixatives solutions may not yield good PLA staining. During the preparation of specimens for PLA, FFPE tissue is sectioned to 3 μm thickness and mounted on slides for immunohistochemical analysis (*see* **Note 2**).

#### 3.1.2 Cultured Cells

Cells should be cultured onto microscopic glassware (e.g., chamber slides or glass bottom dishes) and fixed in formalin-based solution for 10–15 min (*see* **Note 1**). Since antibodies cannot permeate cell membranes, membrane permeabilization with a surfactant is to be performed as necessary (*see* **Note 3**). Samples are washed with phosphate-buffered saline without calcium and magnesium (PBS) and then subjected to PLA.

### 3.2 Primary Antibodies

When considering the primary antibodies to be employed for PLA, it is necessary to set the conditions to obtain specific staining during immunohistochemical analysis. In immunohistochemical conditions, antigen retrieval, and antibody dilution concentration settings are important [17]. Antigen retrieval can be performed using one of the following two methods: epitope retrieval by heating and enzyme digestion [18]. Enzyme treatments and heat treatments along with buffers used for antigen retrieval are summarized in Table 1. In some cases, better staining can be obtained without treatment than with treatment (*see* **Note 4**). It is indeed possible that the analytical conditions for immunohistochemistry for each of these two antibodies might be different. In such cases, stronger antigen retrieval (e.g., using autoclave rather than microwave and pH 9 buffer rather than citrate buffer) should be attempted first. If possible, fluorescent double immunohistochemistry with two antibodies should be attempted.

PLA probes are oligonucleotides linked to secondary antibodies, and secondary antibodies obtained from different animal species with negative and positive strands are used (*details are provided in the next section*). Therefore, to detect PPI between "$\alpha$ protein" and "$\beta$ protein," it is necessary to prepare primary antibodies obtained from different species against each protein. Since anti-mouse, anti-rabbit, anti-goat, and antihuman secondary antibodies are used as PLA probes, it is necessary to select primary antibodies obtained from these species. If the antibodies against "$\alpha$" and "$\beta$" proteins from different species mentioned above are not available, use a PLA kit that ligates oligonucleotides directly to the primary antibodies (*see* **Note 5**).

### 3.3 PLA Protocol

This section describes the PLA detection in pathological tissues using Duolink® In Situ Detection Reagents (Merck, Darmstadt, Germany). The following is the reference URL: https://www.sigmaaldrich.com/JP/en/technical-documents/protocol/protein-biology/protein-and-nucleic-acid-interactions/duolink-fluorescence-user-manual.

**Table 1**
**Enzymes, heat treatments, and buffers for antigen retrieval for immunohistochemistry**

| | |
|---|---|
| Enzymes | |
| 0.1% trypsin | Trypsin |
| | Dissolve in 0.05 M Tris–HCl buffer (pH 7.6) (containing 0.1% $CaCl_2$) |
| | 37 °C, 30 min |
| 0.4% pepsin | Pepsin |
| | Dissolve in 0.01 N HCl |
| | 37 °C, 30 min |
| 0.05% protease | Protease |
| | Dissolve in 0.05 M Tris–HCl buffer (pH 7.6) |
| | Room temperature, 10 min |
| Heating | |
| Autoclave | 121 °C, 5 min |
| Microwave | 500 W, 20 min |
| Buffer for heating | |
| Citrate buffer | 2 mM citric acid, 9 mM trisodium citrate dehydrate (pH 6.0) |
| EDTA buffer | 1 mM EDTA with 0.05% tween 20 (pH 8.0) |
| H buffer | Instant antigen retrieval H (neutral) (LSI Medience, Tokyo, Japan) |
| pH 9 buffer | Antigen Retrieval solution, pH 9 (Nichirei biosciences, Tokyo, Japan) |

*EDTA* ethylenediaminetetraacetic acid

*3.3.1 Specimen Preparation*

Slides with sliced FFPE tissue are deparaffinized and rehydrated in a xylene–ethanol series. Slides are optionally treated for antigen retrieval (*see* Subheading 3.2). Treated slides are kept in PBS until further procedure. Drying of the tissue on the slide leads to nonspecific reactions; therefore, such drying must be avoided in the subsequent operations.

*3.3.2 Blocking*

Initial blocking in immunohistochemistry is carried out to minimize nonspecific reactions resulting in background or false-positive staining. Blocking in PLA is facilitated using a blocking reagent, which is a component of the PLA kit (Duolink® Blocking Solution). If immunohistochemical procedure for primary antibody selection employs skim milk or normal serum as the blocking reagent, they are not used in the PLA procedure. Tap off the PBS by absorbing it with filter paper or wiping paper, and wipe off excess PBS (*see* **Note 6**). The tissue should be loaded with the blocking reagent and incubated at 37 °C for 30 min in a humid chamber (*see* **Note 7**).

### 3.3.3 Reaction Using Primary Antibodies

Dilute the antibodies to the appropriate concentration using an antibody diluent (Duolink® Antibody Diluent). Tap off the blocking reagent. Washing with PBS is not required. Primary antibodies against two targeted proteins are to be simultaneously loaded onto the tissue on the slide. The slides are placed in a humidified chamber and incubated overnight at 4 °C.

### 3.3.4 Reaction Using Secondary Antibodies (PLA Probes)

Dilute PLUS PLA probe (×5) and MINUS PLA probe (×5) to 1:5 dilution using antibody diluent (Duolink® Antibody Diluent). Diluted PLA probes are to be used after further dilution in the following ratio: PLUS probe/MINUS probe/antibody diluent = 1:1:2. Tap off the primary antibodies mixture from the tissue slide, and wash with the wash buffer (1× Wash Buffer A) for 5 min at room temperature; this washing step is to be performed twice. After wiping off the wash buffer, add a mixture of MINUS and PLUS PLA probes to each slide. The slides should then be placed in a humidified chamber and incubated at 37 °C for 60 min.

### 3.3.5 Ligation Reaction

Dilute 5× Duolink® Ligation Buffer to 1:5 dilution with high-purity water. Mix the ligase included in the PLA kit with the diluted Ligation Buffer (×1) in 1:40 ratio. Tap off the PLA probes mixture from the tissue slide, and wash with the wash buffer (1× Wash Buffer A) for 5 min at room temperature; this washing step is to be performed twice. After wiping off the wash buffer, add the ligation mixture onto each slide. The slides should then be placed in a humidified chamber and incubated at 37 °C for 30 min.

### 3.3.6 Amplification

Since light-sensitive reagents are used in the subsequent experiments, these experiments should be performed under light-shielding conditions. The amplification buffer contains fluorescent dyes. Table 2 shows the types of fluorescent dyes and the associated information. The fluorescent dyes to be employed should be selected based on the type of laser available with the fluorescence microscope for detection. Reagents for bright-field detection are also available, allowing observations under a nonfluorescent light microscope (*see* **Note 8**).

Dilute 5× amplification buffer to 1:5 dilution using high-purity water. Tap off the ligation solution from the slides, and wash the slides with 1× wash buffer A for 5 min at room temperature; this washing step is to be performed twice. During the washing step, add the polymerase provided in the PLA kit at 1:80 dilution in 1× amplification buffer. After wiping off the wash buffer, add the amplification mixture onto each slide. The slides should then be placed in a humidified chamber and incubated at 37 °C for 100 min.

**Table 2**
**Fluorescent dyes for detection using in situ proximity ligation assay**

| Dye | Excitation | Excitation[a] | Emission | Filter |
|---|---|---|---|---|
| Red | 594 | 532, 561 | 624 | Texas red |
| Green | 495 | 488 | 527 | FITC |
| Orange | 554 | 532, 561 | 576 | Cy3 |
| Far-red | 644 | 638, 640, 642 | 669 | Cy5 |

*FITC* fluorescein isothiocyanate, *cy* cyanine; [a]wavelength of the laser installed in a general fluorescence microscope; unit, nm

*3.3.7 Nuclear Staining and Mounting*

Tap off the amplification solution from the slides and wash with 1× Wash Buffer B for 10 min at room temperature; this washing step is to be performed twice. Additionally, wash the slides with 0.01× Wash Buffer B for 1 min at room temperature. Nuclei are stained with Duolink® In Situ Mounting Media with DAPI, a water-soluble mounting medium. Tap off the wash buffer from the slides, and add a drop of the mounting medium. Mount the slide with a coverslip, and seal all four sides with clear nail polish. Nail polish sealing can be omitted when using a hardening mounting medium. The slides thus prepared can be stored at 4 °C for about a week and at −20 °C for several months.

## 4 Results

Figure 2 shows PLA of ERα homodimers in a breast cancer tissue and breast cancer cell line. ERα homodimers were detected as red fluorescent dots in breast carcinoma cells (Fig. 2A). ERα homodimers were detected in MCF-7 cell line upon the addition of its ligand estradiol (Fig. 2B). Figure 3 shows the PPI between cell membrane antigens HER2 and CEACAM in a breast cancer cell line. Figure 4 shows PPI of HER2 and CEACAM detected using a bright field microscope; red particles indicating PPI of both proteins were detected in breast cancer cells.

## 5 Conclusion

The key to the success of PLA is antibody specificity in immunohistochemistry. Antibodies employed in pathological diagnosis, such as anti-ERα and anti-HER2 antibodies in breast cancer diagnosis, are well validated and suggest efficiency and success when used in PLA. ERα is well suited for use in PLA due to the use of monoclonal antibodies from different species, including rabbit SP-1 and mouse 6F11, for clinicopathological diagnosis. On the

A

B

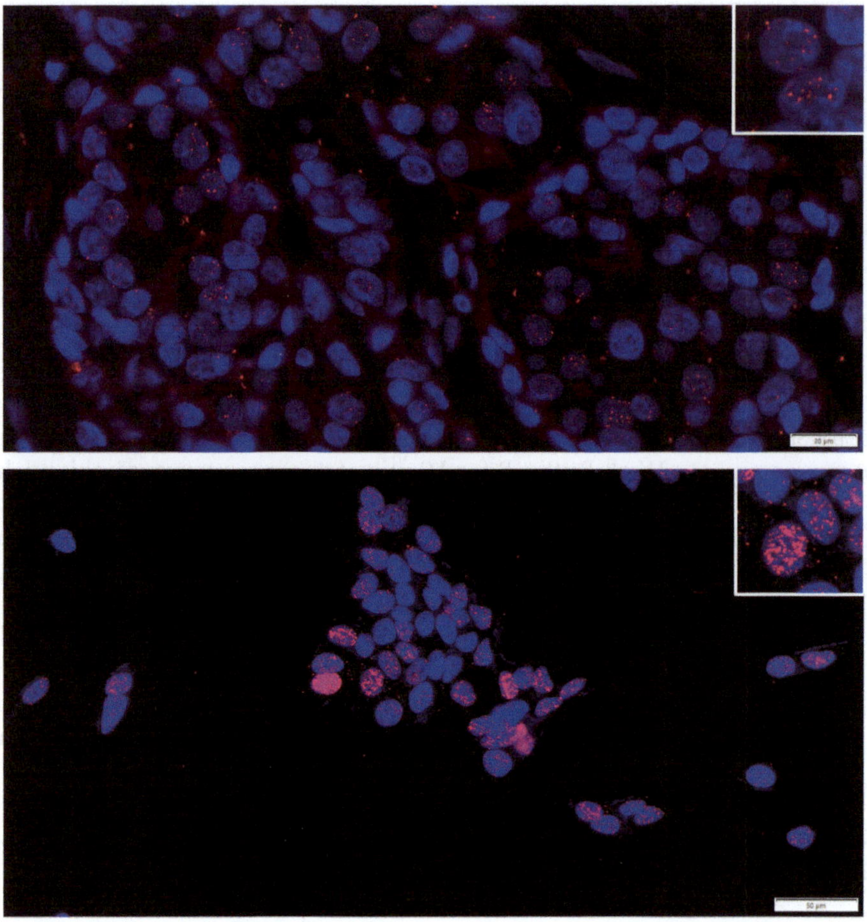

**Fig. 2** Fluorescence PLA of ERα homodimers in breast cancer tissue and breast cancer cell line. The following primary antibodies were used: ERα, mouse monoclonal antibody clone 6F11 (Leica, HE, Germany), and rabbit monoclonal antibody clone SP-1 (Abcam, Cambridge, UK). **(A)** ERα homodimers in breast carcinoma tissue. **(B)** ERα homodimers in MCF-7 breast cancer cell line

other hand, for nondiagnostic proteins, antibodies against various types of proteins have become commercially available; however, it is also true that only a few antibodies have been validated for immunohistochemical analysis. Even for those antibodies advertised for use in immunohistochemistry in the datasheets or in previous reports, pilot studies to standardize experimental conditions are essential in the lab. We spent a considerable amount of time for selecting antibodies and setting conditions for immunohistochemistry to detect PPI of structurally similar proteins, viz., CEACAM6 and CEACAM8 [19]. Even with similar proteins, PLA is successful only when the antigen recognition is accurate.

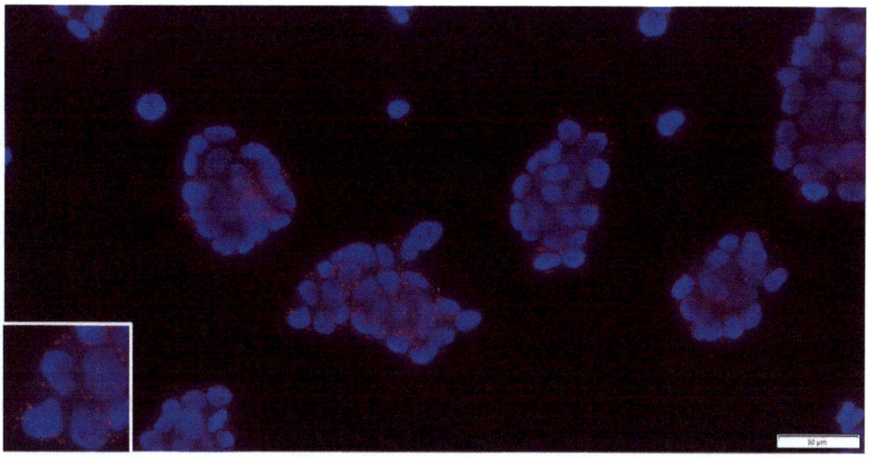

**Fig. 3** Fluorescence PLA of HER2-CEACAM1 interactions in breast cancer tissue. The following primary antibodies were used: HER2, rabbit polyclonal antibody (Dako, Agilent Technologies, Santa Clara, CA, USA) and CEACAM1, mouse monoclonal antibody clone 4D1/C2 (Merck Millipore, Burlington, MA, USA). BT-474 breast cancer cell line was used

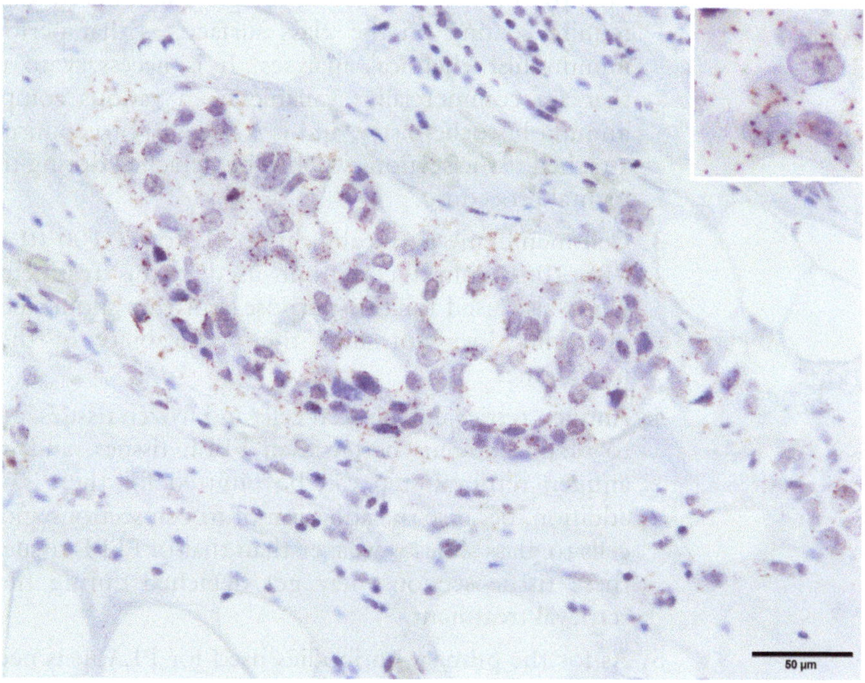

**Fig. 4** Bright-field PLA of HER2-CEACAM6 interactions in breast carcinoma tissue. The following primary antibodies were used: HER2, mouse monoclonal antibody, clone 3B5 (Abcam) and CEACAM6, rabbit polyclonal antibody (Aviva Systems Biology, San Diego, CA, USA)

## 6 Notes

1. Aldehyde-based fixatives: Various aldehyde-based fixatives, including formaldehyde, paraformaldehyde, and glutaraldehyde, are available. Formaldehyde is commercially available as formaldehyde solution (37%; stabilized with methanol). For PLA, 10–20% formaldehyde solution that is commonly used in histochemistry works well. We often use the commercially available 10% formalin neutral buffer solution for our experiments. A 4% paraformaldehyde/phosphate buffer solution is more suitable for fixing tissues and cultured cells. As for alcohol-based fixatives, ethanol or methanol is commonly used for fixation of cultured cells; however, both are not preferable because they lower the stainability during immunostaining.
2. Glass slides: In immunohistochemistry, the adhesion of the tissue sections to glass slides is affected by heat treatment with microwaves or autoclaving as well as by treatment with high pH buffers. Therefore, the glass slides used for PLA also require coating on the glass surface, as that performed for immunohistochemical analyses. It is necessary to use slides that are commercially available from various companies for immunohistochemistry, and it is necessary to confirm that the attached tissue sections will not be detached during the experimental procedure.
3. Cell membrane permeabilization: Triton X-100 (0.1–0.4% in PBS, 10–15 min) is commonly used for this treatment. Digitonin is also used for this purpose, but care should be taken as digitonin does not alter the permeability of the nuclear membrane.
4. Antigen retrieval: Cultured cells and frozen tissues are thought to retain antigens better than FFPE tissues, and, therefore, antigen retrieval step can be omitted for these samples. In addition, because the adhesion of frozen sections and cultured cells to glass slides is weaker than that of FFPE tissue sections, these tissue sections may get detached during the antigen retrieval treatment.
5. As for the primary antibodies used for PLA, it is necessary to prepare monoclonal antibodies against "$\alpha$ protein" and "$\beta$ protein" that are obtained from different animal species. Preparation of "monoclonal antibodies from different species" is one of the main obstacles in performing PLA. In such situations, the PLA probe is conjugated directly with the primary antibodies (Duolink® In Situ Probemaker PLUS and MINUS). Commercially available antibodies often contain salts, such as sodium azide, and, hence, desalting is required for conjugation of the probe with the primary antibody. Because antibody loss

occurs during the desalting process, it is necessary to prepare for that loss.

6. First, place one side of the glass slide on an absorbent paper (e.g., Kim towel) to absorb the solution (Fig. 5). Then, absorb excess solution around the tissue section by wiping the glass using a piece of filter paper of approximately 5 $cm^2$ (Fig. 5). Note that the filter paper or absorbent paper may damage or wipe off the tissue section on the glass slide.
7. A sample humidifier chamber is shown in Fig. 6.
8. PLA bright-field detection: Quenching of endogenous peroxidase is to be performed before "2. Blocking" step. Tap off the PBS in **step 1**, and add hydrogen peroxide provided in the PLA kit. Place the slides in a humidified chamber, and incubate for 5 min at room temperature. Tap off the hydrogen peroxide, and wash with the wash buffer (1× Wash Buffer A) for 5 min at room temperature; this washing step is to be performed twice. Proceed to step "2. Blocking."

After step "6. Amplification," the bright field detection reagent is reacted. Tap off the amplification mixture, and wash with the wash buffer (1× Wash Buffer A) for 2 min at room temperature; this washing step is to be performed twice. During the wash, dilute the 5× bright-field detection stock to 1:5 dilution with high-purity water. Tap off the wash solution and add the detection reagent. Place the slides in a humidified chamber and incubate at 37 °C. Incubation times vary depending on antigen retrieval: 120 min for heat treatment, 60 min for enzymatic treatment.

After the reaction with the detection reagent, the substrate is allowed to be reacted. Tap off the detection reagent and wash with the wash buffer (1× Wash Buffer A) for 2 min at room temperature; this washing step is to be performed twice. During the wash, dilute substrate reagents A (1:70), B (1:100), C (1:100), and D (1:50) with high-purity water to make the substrate solution. The time required for the color development varies depending on the tissue and antigen. Place the slides in a humidified chamber and incubate at room temperature; however, the time required for the color development (a few minutes to 30 min) would vary for different tissues and antigens. While checking the color development with a microscope, if a dot-shaped signal is confirmed, the process is deemed complete.

Finally, counterstaining and mounting are performed. Tap off the substrate solution and wash with the wash buffer (1× Wash Buffer A) for 2 min at room temperature. Add nuclear stain (PLA kit) and incubate for 2 min at room temperature. Wash off the nuclear stain for 10 min using running deionized water. The tissue is dehydrated and permeated by immersing the entire glass slide in ethanol/xylene series. Cover the tissue with a water-insoluble mounting medium, and place the coverslip.

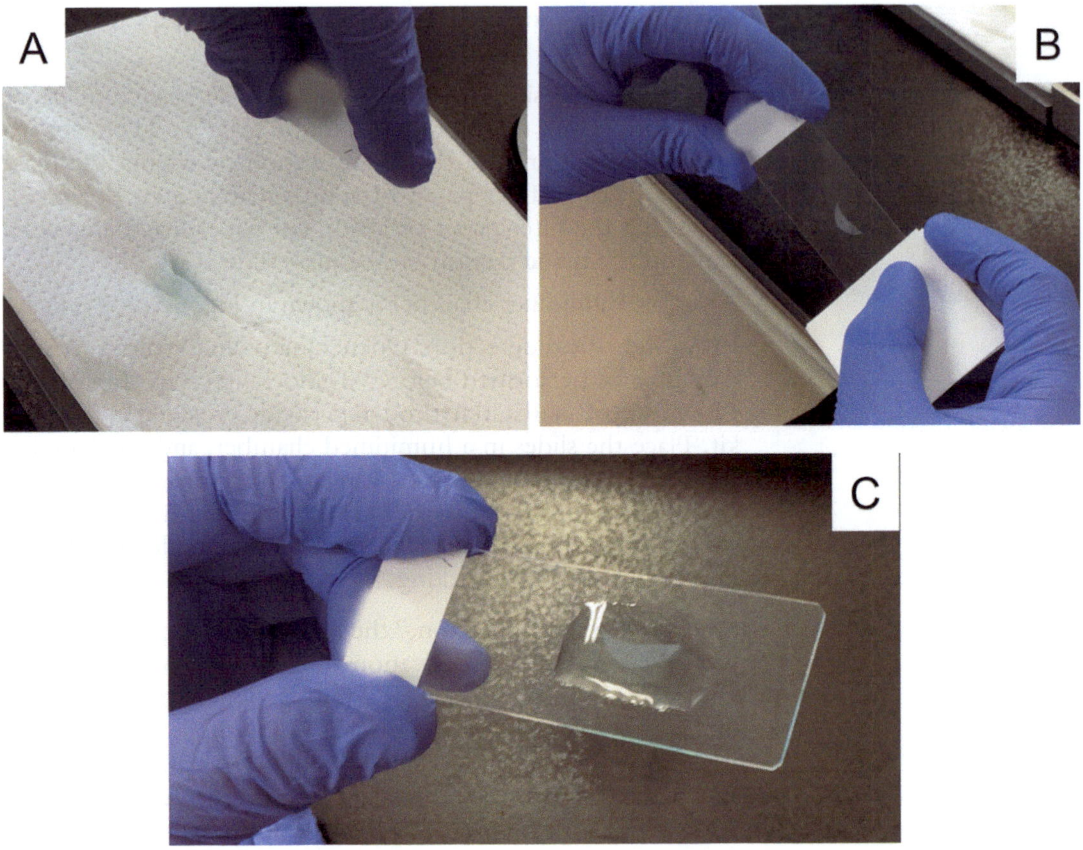


**Fig. 5** Tap off the solution from the glass slide. (**A**) Absorb the solution on the slide glass with an absorbent paper (e.g., Kim towel). (**B**) Wipe off excess solution on the slide with a piece of filter paper (cut into 5 × 5 cm). (**C**) Add reagents. A blue solution has been used in this figure for clarity

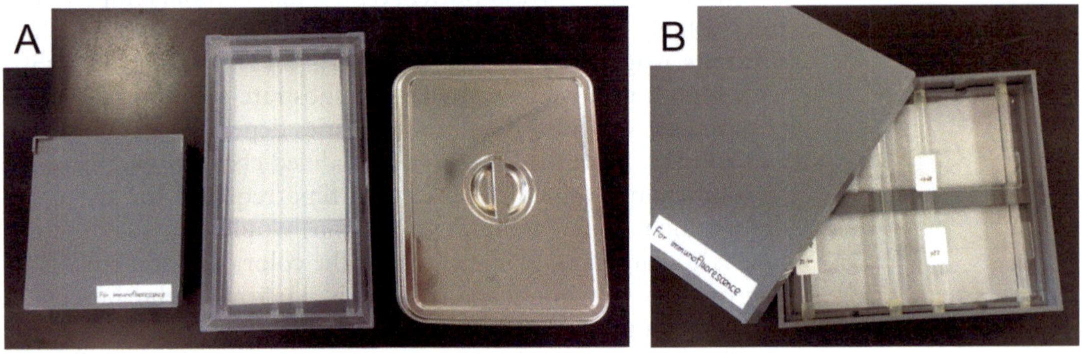


**Fig. 6** Humidified chamber. (**A**) Left, plastic light-shielding chamber (10 slides) (Cosmo Bio, Tokyo, Japan); middle, plastic chamber (not light-shielding, 20 slides) (Cosmo Bio); right, stainless steel light-shielding chamber (12 slides) (Kenis, Osaka, Japan). (**B**) Each chamber is lined with filter paper that are soaked with distilled water

## References


1. Rakha EA, Reis-Filho JS, Baehner F, Dabbs DJ, Decker T, Eusebi V, Fox SB, Ichihara S, Jacquemier J, Lakhani SR, Palacios J, Richardson AL, Schnitt SJ, Schmitt FC, Tan PH, Tse GM, Badve S, Ellis IO (2010) Breast cancer prognostic classification in the molecular era: the role of histological grade. Breast Cancer Res 12:207
2. Gown AM (2008) Current issues in ER and HER2 testing by IHC in breast cancer. Mod Pathol 21:S8–S15
3. Gaffney EF, Riegman PH, Grizzle WE, Watson PH (2018) Factors that drive the increasing use of FFPE tissue in basic and translational cancer research. Biotech Histochem 9:373–386
4. Miki Y (2020) New insights into breast and endometrial cancers. Cancers (Basel) 12:2595
5. Miki Y, Iwabuchi E, Ono K, Sasano H, Ito K (2018) Exploring protein-protein interaction in the study of hormone-dependent cancers. Int J Mol Sci 19:3173
6. Iwabuchi E, Miki Y, Sasano H (2021) the visualization of protein-protein interactions in breast cancer: deployment study in pathological examination. Acta Histochem Cytochem 54:177–183
7. Söderberg O, Gullberg M, Jarvius M, Ridderstråle K, Leuchowius KJ, Jarvius J, Wester K, Hydbring P, Bahram F, Larsson LG, Landegren U (2006) Direct observation of individual endogenous protein complexes in situ by proximity ligation. Nat Methods 3: 995–1000
8. Söderberg O, Leuchowius KJ, Gullberg M, Jarvius M, Weibrecht I, Larsson LG, Landegren U (2008) Characterizing proteins and their interactions in cells and tissues using the in situ proximity ligation assay. Methods 45:227–232
9. Iwabuchi E, Miki Y, Ono K, Onodera Y, Suzuki T, Hirakawa H, Ishida T, Ohuchi N, Sasano H (2017) In situ detection of estrogen receptor dimers in breast carcinoma cells in archival materials using proximity ligation assay (PLA). J Steroid Biochem Mol Biol 165: 159–169
10. Iwabuchi E, Miki Y, Ono K, Onodera Y, Sasano H (2017) In situ evaluation of estrogen receptor dimers in breast carcinoma cells: visualization of protein-protein interactions. Acta Histochem Cytochem 50:85–93
11. Iwabuchi E, Miki Y, Suzuki T, Hirakawa H, Ishida T, Sasano H (2012) Heterogeneous nuclear ribonucleoprotein K is involved in the estrogen-signaling pathway in breast cancer. Int J Mol Sci 22:2581
12. Xu J, Iwabuchi E, Miki Y, Kanai A, Ishida T, Sasano H (2022) FE65 in breast cancer and its clinicopathological significance. Breast Cancer 29:144–155
13. Xu J, Iwabuchi E, Miki Y, Kanai A, Takagi K, Suzuki T, Ishida T, Sasano H (2022) FE65 defines the efficacy of tamoxifen treatment via osteopontin expression in estrogen receptor-positive breast cancer. Pathol Res Pract 234: 153898
14. Iwabuchi E, Miki Y, Kanai A, Miyashita M, Kijima G, Hirakawa H, Suzuki T, Ishida T, Sasano H (2018) The interaction between carcinoembryonic antigen-related cell adhesion molecule 6 and human epidermal growth factor receptor 2 is associated with therapeutic efficacy of trastuzumab in breast cancer. J Pathol 246:379–389
15. Wang S, Yoo S, Kim HY, Wang M, Zheng C, Parkhouse W, Krieger C, Harden N (2015) Detection of in situ protein-protein complexes at the drosophila larval neuromuscular junction using proximity ligation assay. J Vis Exp 95: 52139
16. López-Cano M, Fernández-Dueñas V, Ciruela F (2019) Proximity ligation assay image analysis protocol: addressing receptor-receptor interactions. In: Rebollo E, Bosch M (eds) Computer optimized microscopy, Methods in molecular biology, vol 2040. Humana, New York, pp 41–50
17. MacNeil T, Vathiotis IA, Martinez-Morilla S, Yaghoobi V, Zugazagoitia J, Liu Y, Rimm DL (2020) Antibody validation for protein expression on tissue slides: a protocol for immunohistochemistry. BioTechniques 69:460–468
18. Dunkenberger L, Del Valle L (2022) Antigen retrieval and signal amplification. In: Del Valle L (ed) Immunohistochemistry and immunocytochemistry. Methods in molecular biology, vol 2422. Humana, New York, pp 65–74
19. Iwabuchi E, Miki Y, Onodera Y, Shibahara Y, Takagi K, Suzuki T, Ishida T, Sasano H (2019) Co-expression of carcinoembryonic antigen-related cell adhesion molecule 6 and 8 inhibits proliferation and invasiveness of breast carcinoma cells. Clin Exp Metastasis 36:423–432

# Chapter 10

# Integration of Metabolomic and Proteomic Data to Uncover Actionable Metabolic Pathways

**Christian Heckendorf, Benjamin C. Blum, Weiwei Lin, Matthew L. Lawton, and Andrew Emili**


## Abstract

Mass spectrometry (MS) is an important tool for biological studies because it is capable of interrogating a diversity of biomolecules (proteins, drugs, metabolites) not captured via alternate genomic platforms. Unfortunately, downstream data analysis becomes complicated when attempting to evaluate and integrate measurements of different molecular classes and requires the aggregation of expertise from different relevant disciplines. This complexity represents a significant bottleneck that limits the routine deployment of MS-based multi-omic methods, despite the unmatched biological and functional insight the data can provide. To address this unmet need, our group introduced Omics Notebook as an open-source framework for facilitating exploratory analysis, reporting and integrating MS-based multi-omic data in a way that is automated, reproducible and customizable. By deploying this pipeline, we have devised a framework for researchers to more rapidly identify functional patterns across complex data types and focus on statistically significant and biologically interesting aspects of their multi-omic profiling experiments. This chapter aims to describe a protocol which leverages our publicly accessible tools to analyze and integrate data from high-throughput proteomics and metabolomics experiments and produce reports that will facilitate more impactful research, cross-institutional collaborations, and wider data dissemination.


**Key words** Omics Notebook interface, Omics Notebook R package, Graphical user interface (GUI), Proteomics, Metabolomics, Mass spectrometry, Proteomics search engines, Automated pipeline, Multi-omics data integration, Biological functional modules, Principal component analysis (PCA), Pathway enrichment, Systems biology

## 1 Introduction

The complexity of high-dimensional omics data creates an increasing computational burden in the interrogation of dynamic biological systems compounded by rapid technological advancements in instrumentation. The effect of this bottleneck is to limit the degree to which global profiling is generated for different molecular layers [1]. While next-generation genomics sequencing-based technologies are leading the way in terms of

Usha N. Kasid and Robert Clarke (eds.), *Cancer Systems and Integrative Biology*, Methods in Molecular Biology, vol. 2660,
https://doi.org/10.1007/978-1-0716-3163-8_10,

scaling up sample size and throughput, mass spectrometry (MS)-based global profiling of the proteome and metabolome is also rapidly gaining in popularity due to the unique and complementary functional information they provide [2]. Integrative analysis of the resulting multi-omic data is complicated by the disparate requirements and nuances of each discrete data type, however, which can only be effectively addressed using new computational tools and integrative frameworks to automate analysis and reveal emergent functional insights from combining the underlying complex multi-omic data layers.

While advanced bioinformatic workflows have succeeded in accelerating proteomics and metabolomics research [3, 4], tools leveraging the integration of the two data streams remain relatively uncommon despite the potential for gaining novel biological insight into core cellular systems such as metabolic pathways [5]. Of the publicly available tools created for this approach, few meet the practical logistical requirements of a typical research center needing automated, reproducible, and shareable high-throughput data analysis, while maintaining intuitive usability for researchers with limited computational experience. Existing packages such as ActivePathways [6] and MOMENTA [7] succeed in providing a powerful integrated pathway analysis framework but require knowledge of R to implement them into computational workflows. Packages and RStudio provide a convenient basis for sharing custom analyses to improve reproducibility and extend the utility of existing studies [8, 9], without requiring the recipient to be experienced with programming languages. The use of containers or workflow managers for software development and deployment also solves reproducibility challenges inherent in R packages or other library dependencies [10, 11]. Therefore, a multi-omic data analysis pipeline presented as an RStudio project and R package that makes use of containerization could facilitate code sharing, the integration of different leading R packages, and automation to deliver reproducible exploratory analyses for high volume core labs and specialized research centers. This protocol describes how to perform integrative analysis of MS-based multi-omics data using Omics Notebook [12], an open-source framework with expandable modules for performing exploratory analysis and visualizing MS-based data.

### 1.1 Omics Notebook

The main Omics Notebook interface is formatted as an R script (*see* **Note 1**) with additional code that automates an analysis pipeline incorporating leading bioinformatics R packages. The underlying code is completely customizable, by editing or adding onto the R code or interacting directly with the Omics Notebook R package, but the software comes with many built-in features and a graphical user interface (GUI) for analysis configuration. For example, users can input particular features of interest (see documentation), which are then highlighted in the output plots. The software is primarily designed for analyzing stable isotope (e.g., tandem mass tag)-labeled or label-free protein and peptide/phosphosite data resulting from MaxQuant/Andromeda [13] or other proteomics search engines. Additionally, the software can handle untargeted metabolomic data from tools such as XCMS [14] and automates exploratory analysis and reporting on local computers or shared computing resources. Omics Notebook is distinguished from other similar tools (e.g., Perseus [15] and MSstats [16]) through its automated pipeline and options triggered based on the input data, which is invaluable for MS research centers with many high-throughput projects and collaborations. The output includes exploratory reporting of significantly altered pathways when comparing across sample groups (e.g., case vs control cohorts), multi-omic data integration (e.g., combination of proteomics plus metabolics performed on same sample sets), and biological functional modules (e.g., multi-protein complexes). Additional omics data can be integrated, such as transcriptomics, which are analyzed using standard tabular data formats, as described in the supporting documentation we provide via GitHub (https://github.com/cnsb-boston/Omics_Notebook). This platform creates a seamless and practical laboratory notebook for browsing, interpreting, and visualizing omics data by making use of R markdown to generate concise reports and summary plots for a project and easy sharing of the analysis results among a collaborating team of biologists to accelerate biomedical research.

## 2 Installation

1. Download the Omics Notebook source code:

   *git clone https://github.com/cnsb-boston/Omics_Notebook.git.*

   Or download and extract: https://github.com/cnsb-boston/Omics_Notebook/archive/refs/heads/main.zip.

2. Install R on the host system to run the GUI and notebook (*see* **Note 2**)

   The R installer can be found at: https://cran.r-project.org/.

3. Build a container or manually install the required software dependencies natively. Choose the option that best suits your needs:

   3a. Docker can be used either by building from the Dockerfile bundled with the Omics Notebook source code or using the Docker Hub pre-built image:

   *docker pull cnsbboston/omicsnotebook:latest*

   3b. Singularity, which is commonly used in HPC environments because of its portable containers and ability to run without special user privileges, can also be used (*see* **Note 3**):

*singularity build ON.simg docker://cnsbboston/omics-notebook:latest*

3c. Omics Notebook can be run natively, though perhaps with more difficulty, as follows (read docs/README.md page for current requirements):

Install R 4.1, then in an R session, run the following commands to install the Omics Notebook R library and its dependencies:

*install.packages("remotes")*
*remotes::install_github("cnsb-boston/Omics_Notebook")*

## 3 Prepare Input Files

Example input files can be found at: https://github.com/cnsb-boston/Omics_Notebook_Docs.

### 3.1 Search Result Input Formatting

Proteomics analysis in this method is geared toward analyzing label-free or stable isotope (e.g., tandem mass tag)-labeled protein, peptide, and phosphosite data generated from the MaxQuant/Andromeda search engine. Other proteomics software/search engines (e.g., MSFragger [17]) can be used if the data can be coerced into an appropriate tabular format similar to MaxQuant's proteinGroups.txt or modification Sites.txt (e.g., Phospho (STY)-Sites.txt) format. If wishing to interpret search results other than from MaxQuant, this is the minimal set of columns that will be required for the analysis: protein, gene, and the sample quantification data referenced in the Annotation File (Fig. 1 Subheading 5).

Emulating MaxQuant modification site files can be done using these same columns as a starting point, but the software requires additional columns to filter the results and run an enrichment tool, such as kinase–substrate enrichment analysis (KSEA) [18]: "Sequence.window," "Amino.acid," "Position," "Diagnostic.peak," and "Localization.prob."

Metabolomics data analysis is done on tabular feature (m/z, RT) lists or compound identification results files in a tab or comma separated table format (i.e., .tsv / .csv) containing the following quantitative columns: identifier, rt., and mz. If "identifier" is omitted, the first column will be used to identify features in the Omics Notebook results. In addition to these, a column for each sample should be present containing the measured abundance, analogous to those in the proteomics files which should be referenced in the Annotation File (Subheading 3.2.1 **step 5**). Since metabolites are commonly analyzed by MS in both positive and negative ionization modes, these results can be input as separate input files using the appropriate file formats (Subheading 3.2.1 **step 4**) and optionally combined automatically during the analysis (Subheading 4 **step 6**).

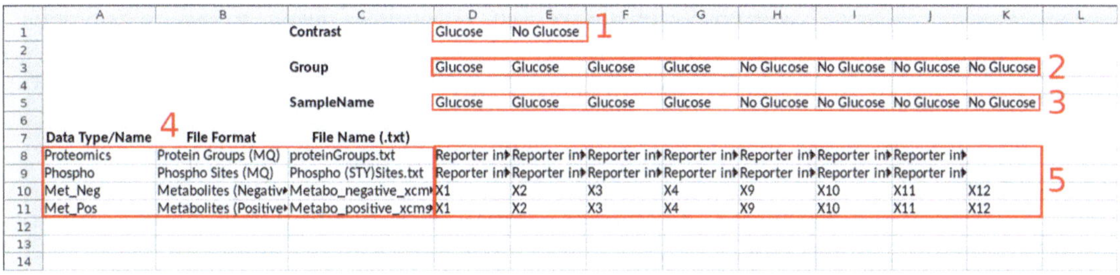

| | A | B | C | D | E | F | G | H | I | J | K | L |
|---|---|---|---|---|---|---|---|---|---|---|---|---|
| 1 | | | Contrast | Glucose | No Glucose | 1 | | | | | | |
| 2 | | | | | | | | | | | | |
| 3 | | | Group | Glucose | Glucose | Glucose | Glucose | No Glucose | No Glucose | No Glucose | No Glucose | 2 |
| 4 | | | | | | | | | | | | |
| 5 | | | SampleName | Glucose | Glucose | Glucose | Glucose | No Glucose | No Glucose | No Glucose | No Glucose | 3 |
| 6 | | 4 | | | | | | | | | | |
| 7 | Data Type/Name | File Format | File Name (.txt) | | | | | | | | | |
| 8 | Proteomics | Protein Groups (MQ) | proteinGroups.txt | Reporter in▸ | Reporter in▸ | Reporter in▸ | Reporter in▸ | Reporter in▸ | Reporter in▸ | Reporter in▸ | | |
| 9 | Phospho | Phospho Sites (MQ) | Phospho (STY)Sites.txt | Reporter in▸ | Reporter in▸ | Reporter in▸ | Reporter in▸ | Reporter in▸ | Reporter in▸ | Reporter in▸ | | 5 |
| 10 | Met_Neg | Metabolites (Negativ▸ | Metabo_negative_xcm▸ | X1 | X2 | X3 | X4 | X9 | X10 | X11 | X12 | |
| 11 | Met_Pos | Metabolites (Positive▸ | Metabo_positive_xcm▸ | X1 | X2 | X3 | X4 | X9 | X10 | X11 | X12 | |
| 12 | | | | | | | | | | | | |
| 13 | | | | | | | | | | | | |
| 14 | | | | | | | | | | | | |

**Fig. 1** The annotation file provides a standard way to read in experimental data. The top section has information about what differential analyses to perform (1), what group each sample belongs to (2), and the sample names for each sample (3). Dataset names, "type" and filename should be specified in the bottom left, with no empty rows (4). Each row in the lower section is a different omics dataset. For a given row, the columns next to SampleName should correspond to the quantitative information related to the associated samples in the data (5). Other columns will be kept as annotation information

## 3.2 *Annotation File*

Parts of Annotation File Main Sheet:

### 3.2.1 *Sheet One*

1. *Contrast* This row is the only row that stands alone and doesn't map to columns in rows below. Include one instance of group names in this row if they should be included in the differential analysis. Include at least one group name; even if differential analysis is not being run, do not leave it empty. The order of the groups impacts the directionality of differential analysis. In the example above, the contrasts will be Group2–Group1, Group3–Group1, and Group3–Group2. If, for example, there is a pool or standard group you don't want to include in the differential analysis, do not include it in this row.
2. *Group* This row labels each sample in the row below with a group name, either from the row above or a new name if that group won't be included in the differential analysis. Samples in the same group should have identical names in this row.
3. *SampleName* Sample names provide short, easy to read names for the samples in the data specified below. In many instances, the columns in the input data may have longer or harder to read headers. The sample names here should be concise since they will be used as labels in many of the plots.
4. *Data File* Information should be provided here about the input data. *Data Type/Name* is a short name that is easy to read to label the dataset in filenames and plots. *File format* is a drop down list that maps to specific data format inputs so the software will know how to handle formatting. *File name* specifies the name of the file in the Analysis Directory and should include the file extension (.txt or .csv).
5. *Sample Quantification* columns specify the name of the column in the accompanied data file in that row for that column's

sample, specified above in Sample Names, which provide quantitative information. Additional rows (Fig. 1, Subheadings 4 and 5) are available for multi-omic data analysis, following the same procedure for additional results files.

#### 3.2.2 Sheet Two

In the second sheet of the annotation file, the sample names are repeated. Additional rows can specify additional annotation information. A couple will control additional functionality.

- Batch: if *Batch* is found, the row will be used to try to perform batch correction. The default is to use ComBat. Care should be taken not to overfit data.
- ColorsHex: if *ColorsHex* is used and the row has functional hexadecimal (hex) color values corresponding to "Group," these colors will be used in figures where possible.
- Group2: if *Group2* is used, this will specify shapes in the PCA plot (*see* **Note 4**).
- TimeSeries: if a numeric value, will run limma as for a time series analysis. Will only work with two groups (*see* **Note 4**). If there is only one group, enter the time point as "Group."
- Pairs: if *Pairs* is found, will try to run differential analysis accounting for paired samples. This list can be extended for custom analysis.

## 4 Running the Omics Notebook

1. Run one of the following commands (*see* **Note 5**) based on your installation method to start the Omics Notebook GUI (Figs. 2 and 3):

   Native: *Rscript Notebook.R*

   Docker: *Rscript Notebook.R Docker*

   Singularity (*see* **Note 3**): *Rscript Notebook.R Singularity -c ON.simg*
2. Provide a custom name for the analysis, if desired. Output will automatically be labeled with a timestamp, so this is useful if running multiple analyses on the same day (e.g., with different parameters, *see* **Note 6**). Additionally, provide the Analysis Directory where the data is and output will be saved (requires write permissions) and may be different from Notebook (software) directory. Finally, specify the annotation file for the analysis (*see* **Note 7**).
3. Option to run the differential analysis (checked is True). If there are many conditions/groups, it may make analysis more manageable to run all groups compared to the first condition

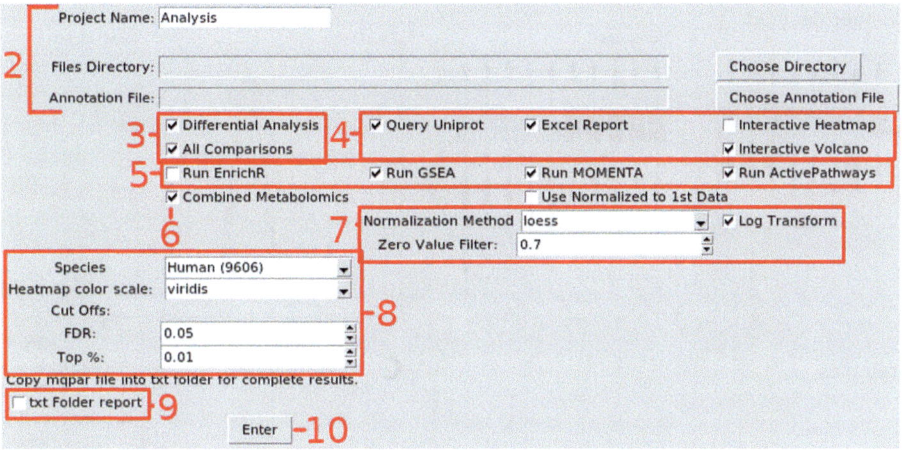


**Fig. 2** The Omics Notebook analysis configuration GUI showing the default set of parameters. Highlighted sections are numbered according to the associated step in the protocol

(i.e., a control group). This should correspond to the first group listed in the contrast row in the annotation file.

4. Options (checked is true) to query UniProt over the Internet for selected feature annotation information (works with UniProt protein IDs, like those provided in MaxQuant searches based on UniProt databases with correct formatting), to generate an excel output file for sharing normalized data and feature level annotation, to generate interactive outputs (heatmaps and volcano).
5. Options to run enrichment analysis based on Enrichr, ActivePathways, and GSEA. Enable MOMENTA (15) for protein–metabolite enrichment and network analysis (*see* **Note 8**).
6. Enabling "Combined Metabolomics" will create an extra set of plots and files by stacking positive and negative mode metabolomics rows into the same output. This may be a useful visualization option for some publications.
7. Normalization parameters specifying method, log transformation, and fraction of nonzero values required to retain as a feature.
8. Additional analysis parameters to specify species (used for some enrichment analysis), the desired heatmap color scale, and cutoffs for significant value plots (*see* **Note 9**).
9. If analyzing MaxQuant data, the entire txt output folder can be provided in the working directory, which will be analyzed using the PTXQC R package for quality control information.
10. When finished with the configuration, press the button labeled *Enter* to begin the analysis.

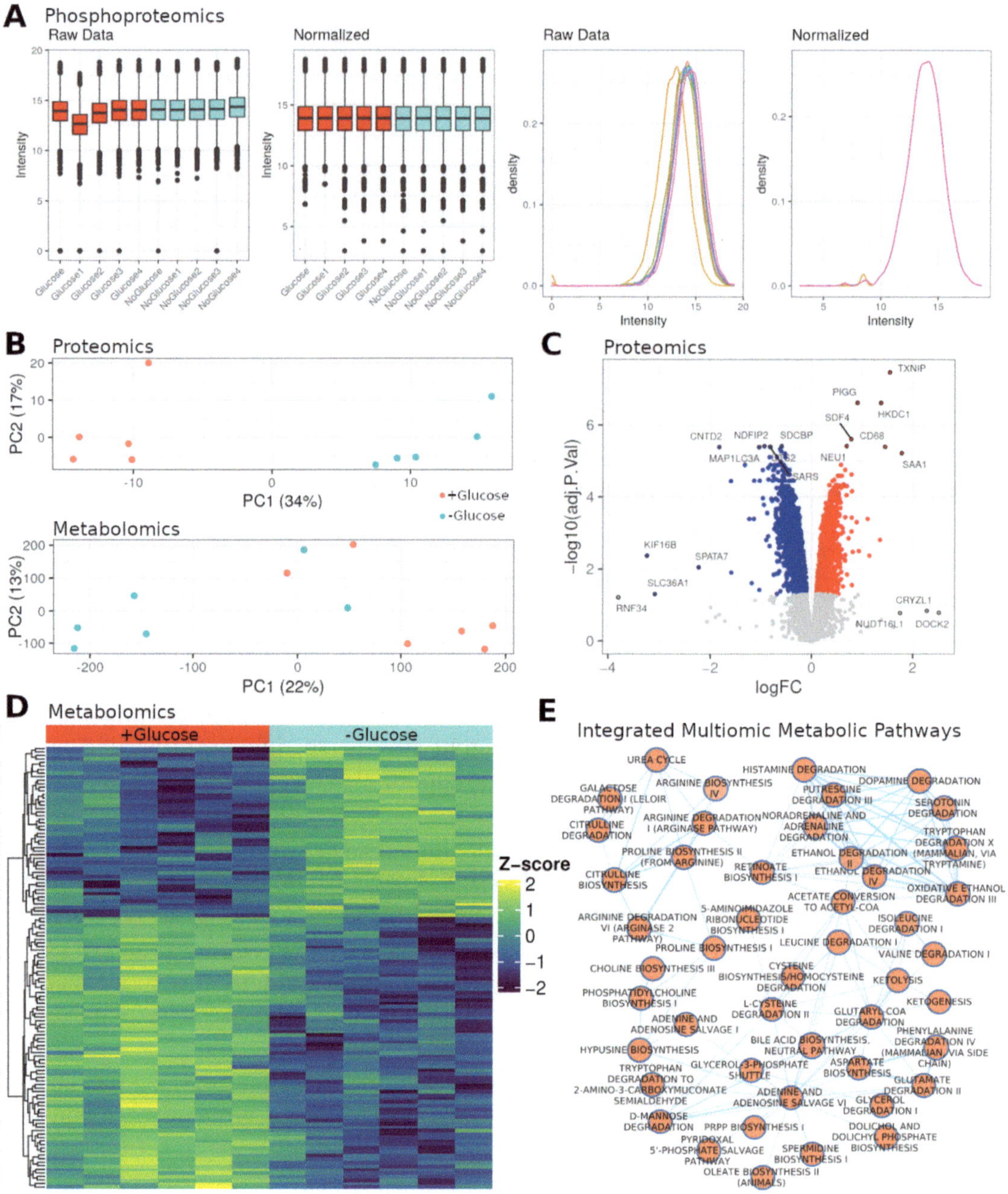


**Fig. 3** A sampling of figures generated by the Omics Notebook using the MB-231 breast cancer cell line dataset published by Blum et al. [7], showing changes in biomolecular activity under glucose starvation. (**a**) Phosphoprotein intensity distributions across samples before and after quantile normalization, shown as boxplots and kernel density estimates. (**b**) PCA plots for the proteomics and metabolomics datasets. (**c**) volcano plot for the proteomics differential analysis. (**d**) heatmap of the top 1% differential features detected in the metabolomics data. (**e**) selection of enriched metabolic pathways from the MOMENTA integrated analysis visualized with the Cytoscape EnrichmentMap module [25]

## 5 Interpreting Results

By creating standardized outputs and figures, Omics Notebook allows researchers to more quickly orient to and understand the analysis of new results. Standard output from the Omics Notebook pipeline includes visualizations for quality control and to aid in the optimization of normalizations. Normalized data and plots are exported in standardized file formats and provide frequently used visualizations, such as principal component analysis (PCA), in order to globally represent and assess data. To assist data visualization, the results from differential analysis, performed using the limma R package [19], are presented as volcano plots and heatmaps, among other standardized annotations. Statistical enrichment analysis is useful for functional interpretation, but different methods may be better suited to different datasets. Therefore, several methods are routinely implemented to provide optimal in-depth functional exploratory analysis. Enrichr and gene set enrichment analysis (GSEA) are two standard yet powerful methods used to provide pathway enrichments for gene-based data (annotated with gene symbols) [20–22]. For phosphoproteomics (phosphosite) data (e.g., MaxQuant Phosphosite files), KSEA is also included, as is metabolomic pathway enrichment analysis using MetaboAnalystR (version 3.2) [23], when metabolomic data is input.

The outputs of Omics Notebook have been designed to shift the bottleneck from analysis to biological interpretation. For example, an excel file summary of all data types is generated with key annotation information and statistical results to easily share data with collaborators and avoid common pitfalls [24]. File outputs with the prefix "Network_" aid in the overlay of experimental data onto network analysis tools, such as Cytoscape. These files also represent examples of how the Notebook can be customized to produce outputs for specific downstream purposes. Pathway enrichments are formatted in standardized tables that also enable subsequent network analysis and clustering of pathways to find functional themes [22].

Integrative analysis occurs across datasets in several ways. First, for studies that include multiple 'omics data types with gene annotation (e.g., proteomics and phosphoproteomics), enrichment is performed with a combined ranked list, where the most significantly altered instance of a gene is retained (noted as "_combined" in enrichment result files). These combined analyses are useful where gene sets may have been annotated from sources where different kinds of changes influence a pathway change, such as protein abundance of one gene and phosphorylation of another. At the pathway level, integrative analysis is performed across data types with gene annotation using ActivePathways, where multiple ranked lists (one for each dataset) are input together and a combined statistical test is used to test for enrichment of a pathway in all datasets at once.

Integration of gene annotated data (such as proteomic data) and metabolomic data is inherently more challenging since the data do not share a common set of IDs. To address this challenge, we use multiple sources of pathway information annotated with both protein and metabolites from MetaboAnalyst [23, 18]. First, for gene-based analysis, we perform enrichment with metabolic pathway subnetworks [7] to identify metabolic pathways that are implicated as being altered through changes in functionally associated proteins. This permits comparison and integration at the pathway level based on enrichment results obtained for metabolomic data since the enrichments are performed on the same underlying pathway model. If metabolomics data is provided together with gene-based proteomic data, a combined enrichment using MOMENTA will also report metabolic pathways using the integrated gene and metabolite feature pathway definitions. By automating multiple integrative analysis approaches, functional patterns appearing across disparate data types can be identified and prioritized more rapidly.

## 6 Notes

1. While the Omics Notebook's main interface is an R script and GUI, options exist for manually running specific functions through RStudio or using the Omics Notebook R package directly.
2. R can be ignored as a requirement in a completely headless automated setup relying on Docker or Singularity. In this case you'll need to have a way of generating an appropriate Parameters.R file (which would otherwise be done by the GUI) and mimic one of the commands in Notebook.R to run the actual Notebook. Alternatively, if the container can be configured to display windows to the host system, the container's version of R can be used to run the notebook GUI.
3. The Singularity image can be stored anywhere you like. Replace the name ON.simg according to your preferences. You may need superuser/administrator permissions to build images, but you can typically run them as any user. If you intend on running the Omics Notebook on a machine where you don't have permission to build, you can simply build the image on a personal computer and copy the image to the destination machine.
4. More functionality may be added by the Omics Notebook developers later, but the code may also be edited by end users for custom analyses.

5. There are a variety of ways of installing and running R commands and scripts. Common methods are to use a GUI-based development environment (IDE) such as RStudio or simply to use the built-in tools found in the default R installation: *Rterm.exe* or *R* on Windows or macOS/Linux, respectively, for typing commands, and *Rscript* (via the command prompt or terminal) for running scripts. Consult the R documentation for your platform for more information.
6. The GUI will always write the parameters as Parameters.R.
7. Make sure the annotation file and all omics results files are present in the same directory selected in **step 2**.
8. Disabling GSEA will also disable MOMENTA, which depends on it.
9. Many features will be bypassed internally when analyzing non-human samples due to lack of organism-specific infrastructure embedded in the notebook.

## References


1. Manzoni C, Kia DA, Vandrovcova J et al (2018) Genome, transcriptome and proteome: the rise of omics data and their integration in biomedical sciences. Brief Bioinform 19:286–302
2. Perez-Riverol Y, Zorin A, Dass G et al (2019) Quantifying the impact of public omics data. Nat Commun 10:3512
3. Kopczynski D, Sickmann A, Ahrends R (2017) Computational proteomics tools for identification and quality control. J Biotechnol 261: 126–130. https://doi.org/10.1016/j.jbiotec.2017.06.1199
4. Schrimpe-Rutledge AC, Codreanu SG, Sherrod SD, McLean JA (2016) Untargeted metabolomics strategies-challenges and emerging directions. J Am Soc Mass Spectrom 27(12): 1897–1905. https://doi.org/10.1007/s13361-016-1469-y
5. Blum BC, Mousavi F, Emili A (2018) Single-platform 'multi-omic' profiling: unified mass spectrometry and computational workflows for integrative proteomics–metabolomics analysis. Mol Omics 14(5):307–319
6. Paczkowska M, Barenboim J, Sintupisut N, Fox NS, Zhu H, Abd-Rabbo D, Mee MW, Boutros PC (2020) Integrative pathway enrichment analysis of multivariate omics data. Nat Commun 11(1):735. https://doi.org/10.1038/s41467-019-13983-9
7. Blum BC, Lin W, Lawton ML et al (2022) Multiomic metabolic enrichment network analysis reveals metabolite–protein physical interaction subnetworks altered in cancer. Mol Cell Proteomics 21(1):100189. https://doi.org/10.1016/j.mcpro.2021.100189
8. Leprevost Fda V, Barbosa VC, Francisco EL et al (2014) On best practices in the development of bioinformatics software. Front Genet 5:199
9. Reznik E, Luna A, Aksoy BA et al (2018) A landscape of metabolic variation across tumor types. Cell Syst 6:301–313.e3
10. Boettiger C, Eddelbuettel D (2017) An introduction to rocker: Docker Containers for R. R J 9:527–536. https://doi.org/10.32614/RJ-2017-065
11. Decan A, Mens T, Claes M, Grosjean P (2016) When GitHub Meets CRAN: an analysis of inter-repository package dependency problems. Paper presented at IEEE 23rd International Conference on Software Analysis, Evolution, and Reengineering (SANER), vol 1, pp 493–504
12. Blum BC, Emili A (2021) Omics notebook: robust, reproducible and flexible automated multiomics exploratory analysis and reporting. Bioinform Adv 1(1):vbab024. https://doi.org/10.1093/bioadv/vbab024
13. Cox J, Mann M (2008) MaxQuant enables high peptide identification rates, individualized p.p.b.-range mass accuracies and proteome-wide protein quantification. Nat Biotechnol 26:1367–1372
14. Smith CA, Want EJ, O'Maille G et al (2006) XCMS: processing mass spectrometry

data for metabolite profiling using nonlinear peak alignment, matching and identification. Anal Chem 78:779–787

15. Tyanova S, Cox J (2018) Perseus: a bioinformatics platform for integrative analysis of proteomics data in cancer research. Methods Mol Biol 1711:133–148. https://doi.org/10.1007/978-1-4939-7493-1_7
16. Choi M, Broudy D, Killeen T et al (2014) MSstats: an R package for statistical analysis of quantitative mass spectrometry-based proteomic experiments. Bioinformatics 30:2524–2526
17. Kong AT, Leprevost FV, Avtonomov DM, Mellacheruvu D, Nesvizhskii AI (2017) MSFragger: ultrafast and comprehensive peptide identification in mass spectrometry–based proteomics. Nat Methods 14(5):513–520
18. Casado P, Rodriguez-Prados JC, Cosulich SC et al (2013) Kinase-substrate enrichment analysis provides insights into the heterogeneity of signaling pathway activation in leukemia cells. Sci Signal 6:rs6
19. Ritchie ME, Phipson B, Wu D et al (2015) Limma powers differential expression analyses for RNA-sequencing and microarray studies. Nucleic Acids Res 43:e47
20. Chen EY, Tan CM, Kou Y et al (2013) Enrichr: interactive and collaborative HTML5 gene list enrichment analysis tool. BMC Bioinform 14:128
21. Korotkevich G, Sukhov V, Budin N et al (2021) Fast gene set enrichment analysis. bioRxiv. https://doi.org/10.1101/060012
22. Reimand J, Isserlin R, Voisin V et al (2019) Pathway enrichment analysis and visualization of omics data using g: profiler, GSEA, Cytoscape and EnrichmentMap. Nat Protoc 14:482–517
23. Chong J, Xia J (2018) MetaboAnalystR: an R package for flexible and reproducible analysis of metabolomics data. Bioinformatics 34:4313–4314
24. Ziemann M, Eren Y, El-Osta A (2016) Gene name errors are widespread in the scientific literature. Genome Biol 17(1):177
25. Merico D, Isserlin R, Stueker O, Emili A, Bader GD (2010) Enrichment map: a network-based method for gene-set enrichment visualization and interpretation. PLoS One 5(11):e13984. https://doi.org/10.1371/journal.pone.0013984

# Chapter 11

# Linking Expression of Cell-Surface Receptors with Transcription Factors by Computational Analysis of Paired Single-Cell Proteomes and Transcriptomes

**April Sagan, Xiaojun Ma, Koushul Ramjattun, and Hatice Ulku Osmanbeyoglu**

## Abstract

Complex signaling and transcriptional programs control the development and physiology of specialized cell types. Genetic perturbations in these programs cause human cancers to arise from a diverse set of specialized cell types and developmental states. Understanding these complex systems and their potential to drive cancer is critical for the development of immunotherapies and druggable targets. Pioneering single-cell multi-omics technologies that analyze transcriptional states have been coupled with the expression of cell-surface receptors. This chapter describes SPaRTAN (Single-cell Proteomic and RNA-based Transcription factor Activity Network), a computational framework, to link transcription factors with cell-surface protein expression. SPaRTAN uses CITE-seq (cellular indexing of transcriptomes and epitopes by sequencing) data and cis-regulatory sites to model the effect of interactions between transcription factors and cell-surface receptors on gene expression. We demonstrate the pipeline for SPaRTAN using CITE-seq data from peripheral blood mononuclear cells.

**Key words** Single-cell Proteomic and RNA-based Transcription factor Activity Network (SPaRTAN), Cellular indexing of transcriptomes and epitopes by sequencing (CITE-seq), Droplet-based scRNA-seq, Antibody-derived tags (ADTs), scADT-seq, Affinity regression (AR), Python package, Jupyter notebook, scVerse and Scanpy ecosystems, pySPaRTAN package, DoRothEA database

## 1 Introduction

The development of specialized cell types and their functions are controlled by external signals that initiate and propagate cell type-specific transcriptional programs. Activation or repression of genes by key combinations of signaling-regulated transcription factors (TFs) drives these transcriptional programs and controls cellular characteristics and functions. For example, ectopic expression of the TF factors Oct4, Sox2, Klf4, and c-Myc reprograms fibroblasts into induced pluripotent stem cells [1]. Conversely, disruption of

Usha N. Kasid and Robert Clarke (eds.), *Cancer Systems and Integrative Biology*, Methods in Molecular Biology, vol. 2660,
https://doi.org/10.1007/978-1-0716-3163-8_11,

TF activity causes a broad range of diseases, including cancer. Hence, identifying cell-specific signaling-regulated TFs is important in understanding human health and treating disease.

Single-cell technologies can now measure RNA, protein, and chromatin states in individual cells [2–5]. However, each technology measures distinct aspects of TF activity with strengths and weaknesses in the identification of cell-specific TFs. Single-cell RNA-sequencing (scRNA-seq) characterizes the transcriptomes of cells only partially. With a bias for highly expressed genes, it misses many TFs with modest levels of transcription. Moreover, cellular processes depend on the expression levels and activities of proteins, notably signaling receptors and TFs, which may differ from mRNA levels. Recent innovations such as CITE-seq (cellular indexing of transcriptomes and epitopes by sequencing) [2] have coupled the relatively sparse scRNA-seq signal with robust detection of highly abundant and well-characterized surface proteins using index sorting [6] and barcoded antibodies [2, 3], providing better cell type discrimination. During CITE-seq, barcoded antibodies—the second set of barcodes—are incubated with the single-cell suspension using a droplet-based scRNA-seq protocol. These barcodes, which have polyA tails, are linked to the barcodes with beads at the same time as mRNAs are linked. Reads containing barcodes associated with each bead are separated by cell. Then, reads that align to transcripts are used to quantify mRNA levels, while those from barcoded antibodies (antibody-derived tags [ADTs]) are used to quantify protein levels.

To date, these single-cell genomics datasets have not been used to link surface proteins to TFs in individual cells. Recently, we developed the first computational framework, SPaRTAN (Single-cell Proteomic and RNA-based Transcription factor Activity Network), to exploit single-cell proteomic (scADT-seq) and corresponding scRNA-seq datasets, both obtained using CITE-seq, to link the expression of surface proteins with inferred TF activities [7] (Fig. 1). SPaRTAN provides a conceptually novel and mechanistically inspired approach for integrating cell-specific transcriptomic and proteomic data with regulatory genomics resources, providing a significant advance in the modeling of cell-specific signaling and gene regulatory programs. The cell-surface phenotype is well-known to immunologists through flow cytometry, but the signaling downstream of cell-surface receptors/co-receptors drives transcriptional and chromatin state changes. Thus, it is important to connect the "cell-surface phenotype" to downstream transcriptional programs and resulting transcriptomic phenotypes. SPaRTAN models this flow of information at single-cell resolution.

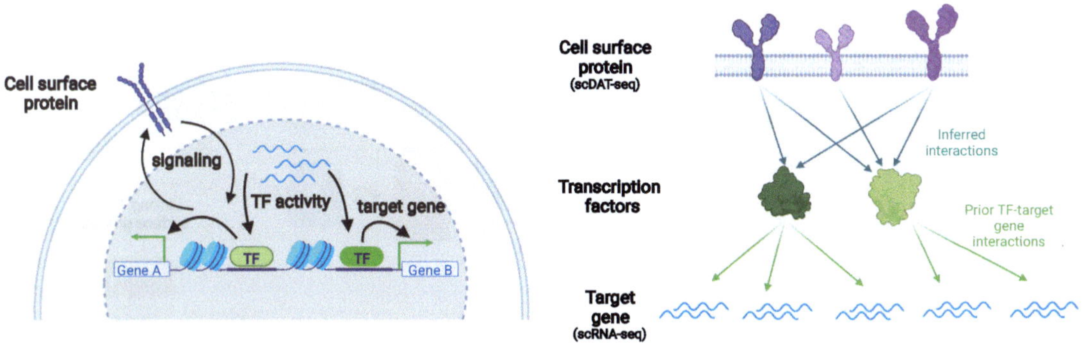


**Fig. 1** Integrative computational model linking cell-surface receptors to TFs. Our integrative model (SPaRTAN, Single-cell Proteomic and RNA-based Transcription factor Activity Network) infers the flow of information from cell-surface receptors to transcription factors (TFs) to target genes by learning the interactions between cell-surface receptors and TFs that best predict target gene expression. (Created with BioRender.com)

Our model uses the expression of surface proteins as a proxy for their activities. Signaling via these proteins converges on particular TFs, whose activities, in turn, regulate the expression of their target genes. Formally, we use a regularized bilinear regression algorithm called affinity regression (AR) [8–10], which provides a statistical framework when the observed data can be explained as interactions between two kinds of inputs. Here, AR establishes an interaction matrix between surface proteins and TFs that predicts target gene expression. We use curated TF-target gene interactions [11] to determine the set of TFs that potentially regulates each gene. The model captures statistical relationships between surface proteins, TFs, and gene expression. We use the trained interaction matrix to evaluate different views of a CITE-seq dataset, e.g., to predict TF activity from a cell's surface protein expression profile or to predict surface protein expression from a cell's gene expression profile. Intuitively, information flows down from observed surface protein levels through the learned interaction matrix to infer TF activities and observed mRNA expression levels or propagates up through the TF-target gene edges and interaction network to infer surface protein expression. We can use inferred TF activities and surface protein expression to gain insights into the transcriptional programs. We applied SPaRTAN to peripheral blood mononuclear cells (PBMC) and malignant mesothelioma CITE-seq datasets to predict the coupling of signaling receptors with context-specific TFs. We validated predictions by prior knowledge, flow cytometry, and immunohistochemical analyses. SPaRTAN greatly enhanced the utility of CITE-seq datasets to reveal TF and cell-surface receptor relationships in diverse cellular states.

In this chapter, we explain how to use SPaRTAN to link cell-surface receptors to TFs using CITE-seq data. A typical SPaRTAN analysis consists of four steps: (1) loading data and preprocessing,

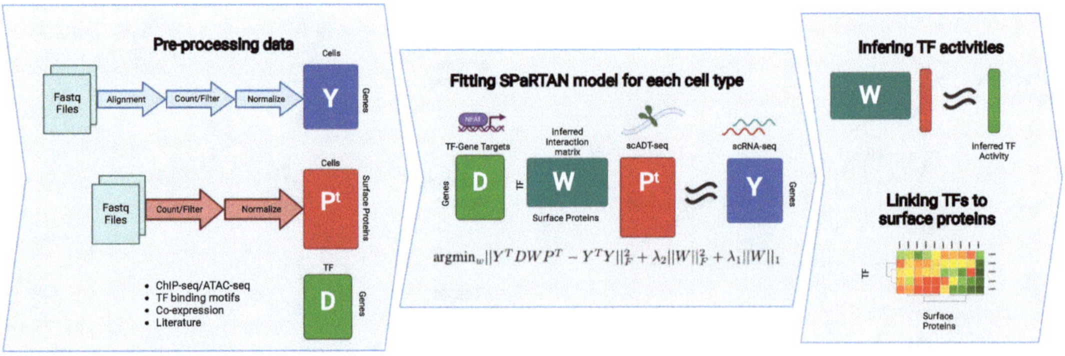


**Fig. 2** General SPaRTAN workflow for CITE-seq, as described in detail in the text, starting from data preprocessing, building cell type-specific SPaRTAN models, inferring TF activities based on the model and downstream analysis (e.g., computing correlation between TF activities and surface protein expression). (Created with BioRender.com)

(2) building cell type-specific TF surface protein models, (3) inferring TF activities, and (4) downstream analysis (e.g., computing correlations between TF activities and surface protein expression) (Fig. 2). Our model pipeline, including preprocessing and downstream analysis, is done in the Python programming language and is compatible with the scVerse and Scanpy ecosystems. We demonstrate the pipeline using publicly available PBMC CITE-seq data. The Python package is open source and available on GitHub, along with a Jupyter notebook containing the demonstration analysis shown in this chapter.

## 2 Materials

### 2.1 Python Packages

Python version 3.7 or higher must be installed, along with the package pySPaRTAN. Our package can be installed using the pip package installer:

```
pip install pySPaRTAN
```

Alternatively, the package can be installed from the source code in the GitHub repository as follows:

```
git clone https://github.com/osmanbeyoglulab/SPaRTAN.git
cd SPaRTAN/
Python setup.py build_ext --inplace
Python setup.py install --user
```

In addition to the pySPaRTAN package, our pipeline depends on the following Python packages:

Scanpy
AnnData
SciPy
NumPy
pandas
statsmodels
Cython

To visualize results, we use seaborn and matplotlib. For reproducibility, we recommend using a Jupyter notebook for the entire pipeline [12]. The example in this chapter is included as a Jupyter notebook on the pySPaRTAN GitHub page (https://github.com/osmanbeyoglulab/SPaRTAN).

### 2.2 Input Files

The input files are a gene expression (mRNA) matrix, cell-surface receptor expression (ADT) matrix, and TF-target gene prior matrix (Fig. 2). Specifically, SPaRTAN requires jointly measured mRNA and surface protein count data from single cells. Raw sequencing data must first be aligned to a reference genome, and the ADT counts are obtained using separate tools, such as CellRanger and CITE-seq-count. In this chapter, we will demonstrate the pipeline for the SPaRTAN model using PBMC CITE-seq data, which are publicly available on the 10×Genomics website. The sample dataset contains 5527 cells with an average of 30,853 mRNA reads per cell and counts for 46 antibodies measuring cell-surface protein expression (Table 1).

Because our model is specific for each cell type, we also require cell-type assignments for each cell in the dataset. For the PBMC data used in our demonstration, we compute a neighborhood graph for the cells using both surface protein and gene expression

**Table 1**
**Links to online resources**

| | |
|---|---|
| PBMC CITE-seq data | https://www.10xgenomics.com/resources/datasets/5-k-peripheral-blood-mononuclear-cells-pbm-cs-from-a-healthy-donor-with-cell-surface-proteins-next-gem-3-1-standard-3-1-0 |
| DoRothEA TF-target gene interaction database | https://github.com/saezlab/dorothea |
| pySPaRTAN GitHub | https://github.com/osmanbeyoglulab/SPaRTAN |
| pySPaRTAN Jupyter notebook | https://github.com/osmanbeyoglulab/SPaRTAN/blob/main/notebooks/pbmc_spartan_tutorial.ipynb |

and cluster the cells using the Leiden algorithm. Cell types are assigned to the clusters based on known marker genes and surface proteins. A full tutorial on the process we used for this dataset can be found in the Scanpy documentation [13].

Several approaches can be used to construct the prior TF-target gene matrix. For this demonstration, we use a gene set resource containing TF-target gene interactions from DoRothEA [11]. Those interactions were collected and curated from several types of evidence, such as curated literature resources, chromatin immunoprecipitation (ChIP)-seq peaks, TF binding site motifs, and interactions inferred directly from gene expression. This prior TF-target gene matrix (**D**) defines a candidate set of associations between TFs and target genes. The DoRothEA database contains 20,244 genes and 1395 TFs, with 486,676 interactions between them. We filtered out TFs that were not expressed across all cell types. The database is available on GitHub at https://github.com/saezlab/dorothea and is available in the pySPaRTAN package (Table 1).

## 3 Methods

Let $n_c$, $n_g$, $n_p$, $n_t$ denote the number of cells, genes, surface proteins, and TFs. Denote the matrix of gene expression with $n_g$ rows and $n_c$ columns as **Y**, the matrix of protein expression with $n_c$ rows and $n_p$ columns as **P**, and the binary matrix for indicating which genes and TFs are known to interact with $n_g$ rows and $n_t$ columns as **D**. The model relates TFs to proteins to predict gene expression using an interaction matrix **W** with $n_t$ rows and $n_p$ columns:

$$\mathbf{Y} \sim \mathbf{DWP}^{\mathrm{T}}$$

We solve for $W$ by the convex optimization problem with L1 and L2 norm regularizers as:

$$\mathrm{argmin}_{\mathrm{W}} \|\mathbf{DWP}^{\mathrm{T}} - \mathbf{Y}\|_F^2 + \lambda_1 \|\mathbf{W}\|_1 + \lambda_2 \|\mathbf{W}\|_F^2$$

The parameters $\lambda_1$ and $\lambda_2$ can be specified, or the optimal choices can be estimated using grid search and cross-fold validation. In practice, we use an approximation of this problem which can be solved much faster (*see* **Note 1**).

The pySPaRTAN package and the pipeline outlined in this chapter are designed around the Scanpy/scVerse ecosystem, which stores data as `AnnData` objects. An AnnData object contains a data matrix with columns (observations) as cells and rows (variables) as genes or ADTs. The object has dataframes `.obs` and `.var.` containing information on each cell and each gene, respectively. Additionally, the fields `.obsm` and `.varm` can store additional dataframes with features describing the cells and genes. For example, we will store the surface protein count matrix in `.obsm`.

### *3.1 Importing and Preprocessing CITE-seq Data*

1. We first import all the packages used in the pipeline:

```
import scanpy as sc
import seaborn as sns
import pandas as pd
from matplotlib import pyplot as plt
import numpy as np

import pySPaRTAN
```

2. We load CITE-seq data into memory using one of Scanpy's i/o functions. In this case, the counts for mRNA and protein ADTs are stored in the same .h5 file, so we use the `sc.read_10x_h5` function. Refer to the Scanpy documentation for other formats, such as h5ad, loom, csv, or excel files. After the data are loaded into an AnnData object, the `var_names_make_unique` function appends a number to the end of gene names that are not unique:

```
path="../data/cite-seq/5k_pbmc_protein_v3_nextgem_filtere
d_feature_bc_matrix.h5"
adata=sc.read_10x_h5(path, gex_only=False)

adata.var_names_make_unique()
```

We separate the gene expression counts from the ADT counts. Variables are annotated as either "Antibody Capture" or "Gene Expression" in the `adata.var["feature_types"]` field. We move the proteomics data ("Antibody Capture") to a separate data frame.

```
protein=adata[:,adata.var["feature_types"]=="Antibody Capt
ure"].copy()
protein=protein[:,[x for x in protein.var_names if "control"
not in x]]
adata.obsm["protein_raw_counts"]=protein.to_df()

adata = adata[:, adata.var["feature_types"] == "Gene
Expression"]
adata.layers["raw_counts"] = adata.X.copy()
```

3. Some cells have a sequencing depth that is too low for our analysis. We recommend filtering out cells with fewer than 1,000 or greater than 5,000 unique gene sequences and cells in which more than 30% of genes are mitochondrial genes. The

function sc.pp.filter_genes removes genes that are not present in at least 3% of cells, and sc.pp.calculate_qc_metrics computes the metrics used to filter cells and stores the results in adata.obs and adata.var:

```
sc.pp.filter_genes(adata, min_cells=0.03*adata.n_obs)
gene_list=adata.var_names
sc.pp.filter_cells(adata, min_genes=1000)
adata.var['mt'] = adata.var_names.str.startswith('MT-')

sc.pp.calculate_qc_metrics(adata,
                           qc_vars=['mt'],
                           percent_top=None,
                           log1p=False,
                           inplace=True)

adata=adata[adata.obs.query(
  "n_genes_by_counts < 5000 and pct_counts_mt<30").index]
```

We also remove all mitochondrial genes from the dataset:

```
adata=adata[:, adata.var['mt']==False]
```

We use the sc.pp.normalize_total function to scale the raw count data to a sum of $10^4$ for each cell. Then, we use sc.pp.log1p to log-transform the data to a form more suitable for a linear model. We. Store the normalized log-transformed counts in adata.layers["log1p"]:

```
adata.X=sc.pp.normalize_total(adata,
                              target_sum=10000,
                              layer="raw_counts",
                              inplace=False)['X']

sc.pp.log1p(adata)

adata.layers["log1p"]= pySPaRTAN.pp.normalize(
     np.array(adata.X.todense()),
     axis=1
)
```

We limit our analysis to the top 5,000 most highly variable genes. Genes that are not highly variable are of less interest as they give less insight into intercellular heterogeneity:

```
sc.pp.highly_variable_genes(
    adata,
    n_top_genes=5000,
    subset=True,
    flavor='seurat_v3')
```

4. For the protein count data, we use the *center log ratio* (CLR) transform using the `clr` function in pySPaRTAN. The CLR transform scales the data by the geometric mean of each cell, and then log-transforms the data:

```
adata.obsm["protein"]=pySPaRTAN.pp.clr(
      adata.obsm["protein_raw_counts"].T).T

adata.obsm["protein"].loc[:]=pySPaRTAN.pp.normalize(
       adata.obsm["protein"],
         axis=1)
```

5. SPaRTAN performs best when it models data from different cell types separately. There are several models for cell-type assignments from CITE-seq data. For this demonstration, we used the Leiden algorithm to cluster the data using both gene and protein expression. Then we manually mapped the clusters to cell types based on the mean expression of the marker genes and proteins for each cluster. This process is omitted here but is available on the SPaRTAN GitHub repository and the Scanpy documentation. We use the marker genes and surface proteins shown in Table 2 to assign cell types to each cluster.

   After cell types are assigned, the results are stored in a csv file with two columns, one for the cell's barcode and another for its cell type. Then we read in the data and store it in the field `adata.obs["cell_types"]`:

```
ct_df=pd.read_csv("cell_types_PBMC.csv",index_col=0)
adata=adata[np.intersect1d(adata.obs_names, ct_df.index)]
adata.obs["cell_types"] = ct_df.loc[adata.obs_names]
```

   It is important to note the number of cells for each cell type, which can be found using the `.value_counts()` function as shown below. The SPaRTAN model will take more time and memory for cell types that have more than approximately 600 cells:

**Table 2**
**Marker genes used to assign cell types to clusters**

| Cell type | Marker genes | Marker proteins |
|---|---|---|
| CD8+ | *CD8A, CD8B, FCER1G* | CD8a, CD4 |
| Naive CD4+ T | *IL7R, CCR7, CD3E* | CD45RA, CD4 |
| Memory CD4+ | *IL7R, S100A4, CD3E* | CD45RO, CD4 |
| NK | *GNLY, NKG7* | CD56 |
| DC | *FCER1A, CST3* | |
| CD14+ Mono | *CD14, LYZ* | CD14 |
| FCGR3A+/CD16+ Mono | *FCGR3A, MS4A7* | CD16 |
| B | *MS4A1* | CD20, CD19 |

```
adata.obs["cell_types"].value_counts()
CD14+MONO  1201
CD4nav      907
CD4mem      702
NK          352
B           332
CD8         318
CD16+MONO   140
DC          100
Name: cell_types, dtype: int64
```

We recommend down-sampling for cell types that have more than 1000 cells when fitting the model becomes intractable:

```
adata=pySPaRTAN.pp.subsample_celltype(
     adata,
     obs_name="cell_types",
     n_cells=600
)

adata.obs["cell_types"].value_counts()
CD14+MONO     600
CD4mem        600
CD4nav        600
NK            352
B             332
CD8           318
CD16+MONO     140
DC            100
Name: cell_types, dtype: int64
```

6. A dataframe that shows interactions between TFs and genes is loaded and stored in the AnnData object as follows. The dataframe has a column for each TF and a row for each gene. An entry is zero if there is no interaction or one if a TF and gene interact. We use an abridged version of the database that filters out TFs and genes with few interactions. Additionally, we filter TFs that are not present in mRNA data, referencing the gene_list variable defined prior to filtering non-highly variable genes:

```
tf_gene=pySPaRTAN.datasets.load_dorthea()
tf_names=np.intersect1d(tf_gene.columns, gene_list)
tf_gene=tf_gene[tf_names]
```

We store the dataframe in our AnnData object, keeping only the genes present in both the TF-gene interaction matrix and the CITE-seq dataset. The dataframe is stored in the AnnData object as `adata.varm["tf_gene"]`:

```
common_genes=np.intersect1d(tf_gene.index,
                            adata.var_names)
tf_gene=tf_gene.loc[common_genes]
adata= adata[:, common_genes]
tf_gene.loc[:]=pySPaRTAN.pp.normalize(tf_gene)
adata.varm["tf_gene"]=tf_gene
```

### 3.2 Fitting SPaRTAN Models for Each Cell Type

To recap, after successfully completing part 1, we have an AnnData object with the following fields:

- `adata.layers["log1p"]`
- `adata.obsm["protein"]`
- `adata.varm["tf_gene"]`
- `adata.obs["cell_type"]`

The prerequisite can be verified by printing the AnnData object and seeing each field present in the output:

```
print(adata)
AnnData object with n_obs × n_vars = 3042 × 2851
    obs: 'n_genes', 'n_genes_by_counts', 'total_counts', 'to-
tal_counts_mt', 'pct_counts_mt', 'cell_types'
      var: 'gene_ids', 'feature_types', 'genome', 'n_counts',
'mt', 'n_cells_by_counts', 'mean_counts', 'pct_dropout_by_
counts', 'total_counts', 'highly_variable', 'highly_variable_
rank', 'means', 'variances', 'variances_norm'
   uns: 'hvg', 'log1p'
   obsm: 'protein_raw_counts', 'protein'
```

```
varm: 'tf_gene'
layers: 'raw_counts', 'log1p'
```

1. To evaluate the model's performance, we train the model using 80% of all cells in the dataset. The remaining 20% can be used for testing the ability of the model to predict gene expression based on surface protein expression:

```
adata.obs["training"]= np.random.rand(adata.n_obs) < 0.8
```

2. Because transcriptional regulation and intercellular signaling are highly dependent on the cell type, we train SPaRTAN separately on each cell type. For example, for CD8+ T cells, we extract two subsets from the entire dataset, one for training the CD8+ T cell model and one for evaluating the model:

```
adata_cd8=adata[(adata.obs["cell_types"] == "CD8")]
adata_cd8_training=adata[(adata.obs["cell_types"] ==
"CD8")
                          & (adata.obs["training"] ==
                          True)]
adata_cd8_testing=adata[(adata.obs["cell_types"] == "CD8")
                          & (adata.obs["training"] ==
                          False)]
```

3. The SPaRTAN class in pySPaRTAN can cross-validate, fit the model, make predictions, infer TF activities, and estimate the correlation between surface proteins and TFs. The algorithm used to fit the model approximates the matrix P by a truncated singular value decomposition, using the number of singular vectors that capture 70% of the full spectrum (*see* **Note 2**).

   The model has two parameters for regularization: `lamda` (the Greek letter "$\lambda$" is intentionally misspelled to avoid conflicting with built-in python syntax) to specify the regularization coefficient and `alpha` as the trade-off between the $l_1$ and $l_2$ regularizer. Setting `alpha=0` will train the model using only the $l_1$ regularizer, `alpha=1` will use only the $l_2$ regularizer, and `alpha=0.5` will weigh the $l_1$ and $l_2$ regularizers equally.

   To use cross-validation and grid search to find the optimal values of `lamda` between 1e-5 and 10 and `alpha` between 0 and 1, we initialize the model specifying a range for `lamda` and `alpha` as follows:

```
model_cd8= pySPaRTAN.SPaRTAN(lamda=[0.0001,10],
                              alpha=[0,1],
                              spectrum=0.7)
```

4. The simplest way to fit the model is by calling the `.fit()` function and providing the AnnData object containing the training data:

```
model_cd8.fit(adata_cd8_training)
```

Alternatively, we can fit the model without using an AnnData structure by directly providing the three matrices needed: the gene expression (**Y**), the protein expression (**P**), and the TF-target gene matrix ($D$):

```
gene_expression=adata_cd8_training.layers["log1p"]
protein=adata_cd8_training.obsm["protein"]
tf_gene_matrix=adata_cd8_training.varm["tf_gene"]

model_cd8.fit(Y=gene_expression,
                          P=protein,
                          D=tf_gene_matrix)
```

5. We evaluate the performance using the `.score` function, which predicts gene expression for each cell in the testing set using the surface protein data and compares this value to the true gene expression based on correlation analysis. Because the model is fit using an approximation of the AR problem, a correction is made during the prediction process (*see* **Note 3**):

```
model_cd8.score(adata_cd8_testing)
0.7606217426058479
```

As with fitting the data, predictions and performance evaluation can be done without using an AnnData object, as follows:

```
protein_test=adata_cd8_testing.obsm["protein"]

gene_expression_test=np.asarray(
     adata_cd8_testing.layers["log1p"])

model_cd8.score(P=protein_test.to_numpy(),Y=gene_express
ion_test)
0.7606217426058479
```

Once we have evaluated the model and verified that it fits the held-out testing data well, we retrain the model with the entire dataset:

```
model_cd8.fit(adata_cd8)
```

6. We repeat the process for each cell type in a loop, storing the models in a dictionary. Note that, as before, the accuracy of the model is verified by fitting a subset of the data and testing with the remaining data before the model is fit using the entire dataset:

```
models=dict()

for ct in np.unique(adata.obs["cell_types"]):
    adata_ct =adata[(adata.obs["cell_types"] == ct)]
    adata_training=adata[(adata.obs["cell_types"] == ct)
                    & (adata.obs["training"] == True)]
    adata_testing =adata[(adata.obs["cell_types"] ==ct)
                    & (adata.obs["training"] ==False)]

    models[ct]= pySPaRTAN.SPaRTAN(lamda=[0.0001, 10],
                    alpha=[0.01,1],
                    spectrum=0.7
                           )
    models[ct].fit(adata_training)

    print("R2 value for "+ct+" cells: \t"
          +str(models[ct].score(adata_testing)))

     models[ct].fit(adata_training)
R2 value for B cells:               0.7752362355217464
R2 value for CD14+MONO cells:       0.6283707877952782
R2 value for CD16+MONO cells:        0.817826164720477
R2 value for CD4mem cells:          0.6175386479974388
R2 value for CD4nav cells:          0.6404709873661867
R2 value for CD8 cells:             0.7604804395513874
R2 value for DC cells:              0.8454503278598255
R2 value for NK cells:              0.7139027245066331
```

### 3.3 Inferring TF Activities

1. From the fitted model, TF activity is inferred from protein expression by calculating $\mathbf{WP^T}$. TF activity for every CD8+ T cell is calculated by the `get_TF_activity` function, using only the protein expression as input:

```
tf = model_cd8.get_TF_activity(adata_cd8.obsm["protein"])
```

2. The significance of the inferred TF activity is assessed using a permutation test. We generate the distribution for an empirical null hypothesis by shuffling the genes for each cell (randomly permuting the rows of **Y**) a specific number of times (typically, 1000), refitting the model, and calculating TF activity. For each TF in each cell, we compare the inferred TF activity to the

empirical null distribution of 1000 inferred TF activities and calculate a *p*-value. The *p*-value is the proportion of values in the null distribution that is higher in absolute value than the true inferred TF activity and has a minimum value inversely proportional to the number of trials (1/1000, typically).

For each cell, *p*-values are corrected using the Bonferroni procedure for multiple hypothesis testing, which is implemented as part of the statsmodel package (`statsmodels.stats.multitest.multipletests`). We use an adjusted *p*-value of 0.15 in this example. The permutation test is done using the `get_TF_activites` function, this time specifying the number of permutations to use for the null distribution:

```
tf, tf_p_val = model_cd8.get_TF_activity(
    adata_cd8.obsm["protein"],
    n_trials=1000,
    verbose=True
)
```

To identify TFs that are important in CD8+ T cells, we look for TFs that have a *p*-value of less than 0.15 in a high proportion of cells:

```
prop_sig=(tf_p_val<0.15).mean()
prop_sig.sort_values(ascending=False).head(20)
```

```
HMBOX1     1.000000
FOSL1      1.000000
SOX6       1.000000
MEIS1      1.000000
MAF        1.000000
MYC        1.000000
ELF1       1.000000
ATF7       1.000000
SSRP1      1.000000
PKNOX1     1.000000
E2F7       1.000000
TP73       1.000000
RFX5       1.000000
ZNF263     0.990566
HOXB13     0.886792
NCOA1      0.883648
TCF25      0.751572
IRF2       0.726415
E2F2       0.566038
FOS        0.553459
dtype: float64
```

The inferred TF activity and corresponding $p$-values are stored in the AnnData object:

```
adata_cd8.obsm["tf_activity"]=tf
adata_cd8.obsm["tf_p_val"]=tf_p_val
```

3. Repeat the process for each cell type, and create a dataframe with the proportion of cells for each cell type where each TF is significant:

```
tf=dict()
tf_p_val=dict()

for ct in np.unique(adata.obs["cell_types"]):
    print("Calculating TF activities for "+ct+" cells")
    tf[ct], tf_p_val[ct] =models[ct].get_TF_activity(
        adata[adata.obs["cell_types"] == ct].obsm["prot
        ein"],
        n_trials=1000,
        verbose=True
   )

adata.obsm["tf"]=pd.concat(tf.values(),
                           axis=0)
adata.obsm["tf_p_val"]=pd.concat(tf_p_val.values(),
                                 axis=0)
adata.obsm['tf_sig']=adata.obsm['tf_p_val']<0.15
```

4. Using the TF activities, we calculate and plot the t-distributed stochastic neighbor embedding (tSNE) for all cells in the dataset, coloring the cells by cell type. The output is shown in Fig. 3a:

```
sc.tl.tsne(adata, use_rep="tf")
sc.pl.tsne(adata, color="cell_types")
```

5. We create a dot plot of the importance of TFs across cell types using seaborn. We plot only the top five TFs that are most important in the transcriptional program for each cell type. The output is shown in Fig. 3b:

```
pySPaRTAN.pl.tf_dotplot(adata,
                        tfs_to_plot=5,
                        group="cell_types"
                        )
```

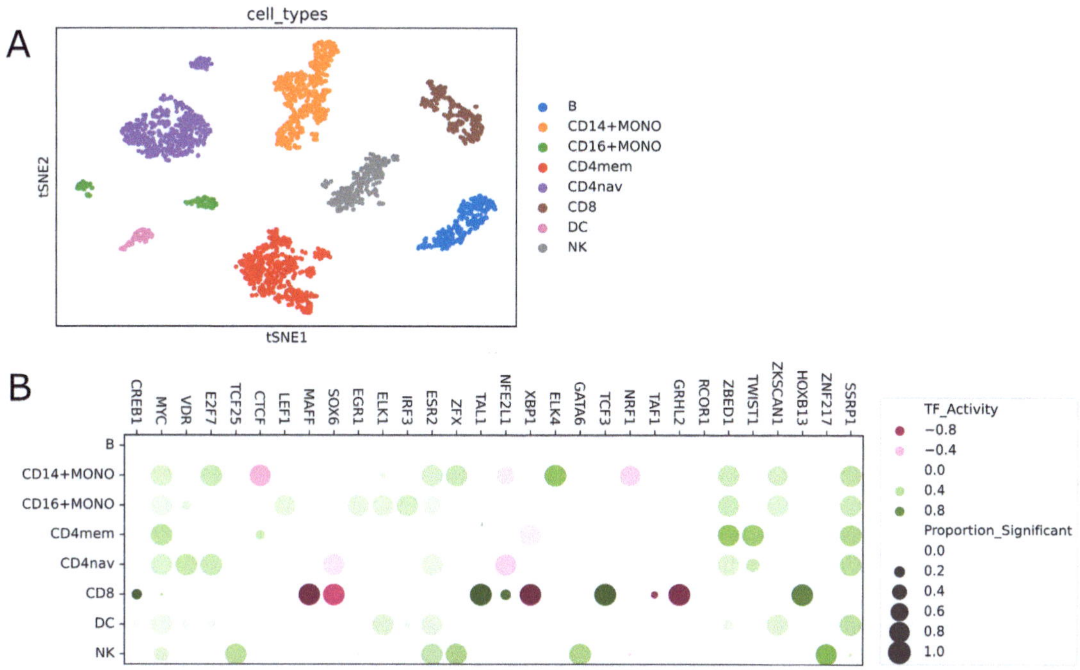


**Fig. 3** Inferred TF activities in PBMCs. (**a**) Plot of tSNE embedding of TF activities for each cell in the PBMC dataset. Each dot corresponds to a cell and is colored by cell type. (**b**) Dot plot showing inferred TF activities for each cell type. Sizes correspond to the proportion of cells in each cell type in which the TF activity is significant, and dots are colored according to the z-scored mean TF activity for each cell type

### *3.4 Identifying TF-Surface Protein Relationships*

1. The matrix **W** in the model can be interpreted as the interactions between TFs and surface proteins. However, because different proteins are expressed on different scales for each cell type, we instead look at the correlations between protein expressions (**P**) and the inferred TF activity ($\mathbf{WP}^{T}$). The `get_tf_protein_cor` is used to compute the correlation matrix using all cells in the training data. The function returns a dataframe with TFs as columns and surface proteins as rows:

```
protein_tf=model_cd8.get_tf_protein_cor()
protein_tf.head()
```

2. We visualize the results using a clustermap, using only the top 15 most significant TFs:

```
tfs_to_plot=(adata_cd8.obsm["tf_p_val"]<0.15).mean().
sort_values(ascending=False).head(15).index.tolist()
```

```
sns.clustermap(
    protein_tf.loc[tfs_to_plot],
    cmap="bwr",vmin=-1, vmax=1
).fig.suptitle('CD8+ T Cells', y=1.05,fontsize=20)
```

We also create a scatter plot of TF-protein correlations with the function `pySPaRTAN.tf_protien_line_plot()`. The clustermap and scatter plots are shown in Fig. 4:

```
pySPaRTAN.pl.tf_protien_line_plot(protein_tf, "CD27",
        title="CD27 - TF correlations in CD8+ T Cells")
pySPaRTAN.pl.tf_protien_line_plot(protein_tf, "TIGIT",
        title="TIGIT - TF correlations in CD8+ T Cells")
```

3. We repeat the process for each cell type and store the results in a dictionary:

```
protein_tf=dict()
for ct in np.unique(adata.obs["cell_types"]):
    protein_tf[ct]=models[ct].get_tf_protein_cor()
```

4. A clustermap can be made for any cell type in the dictionary as follows:

```
ct="B"
plt.figure(dpi=400, figsize=(10,10))
tfs_to_plot=(adata[adata.obs["cell_types"]
==ct].obsm["tf_p_val"]<0.15).mean().sort_values(ascendin
g=False).head(15).index.tolist()

sns.clustermap(
     protein_tf[ct].loc[tfs_to_plot],
     cmap="bwr",vmin=-1, vmax=1
).fig.suptitle(ct+ ' Cells', y=1.05,fontsize=20)
```

## 4 Notes

1. The system of equations used to find the optimal TF-surface protein interaction matrix, **W**, can become prohibitively large in practice. The AR problem can be compressed by left multiplying the system of equations by $\mathbf{Y}^\mathbf{T}$, greatly reducing the

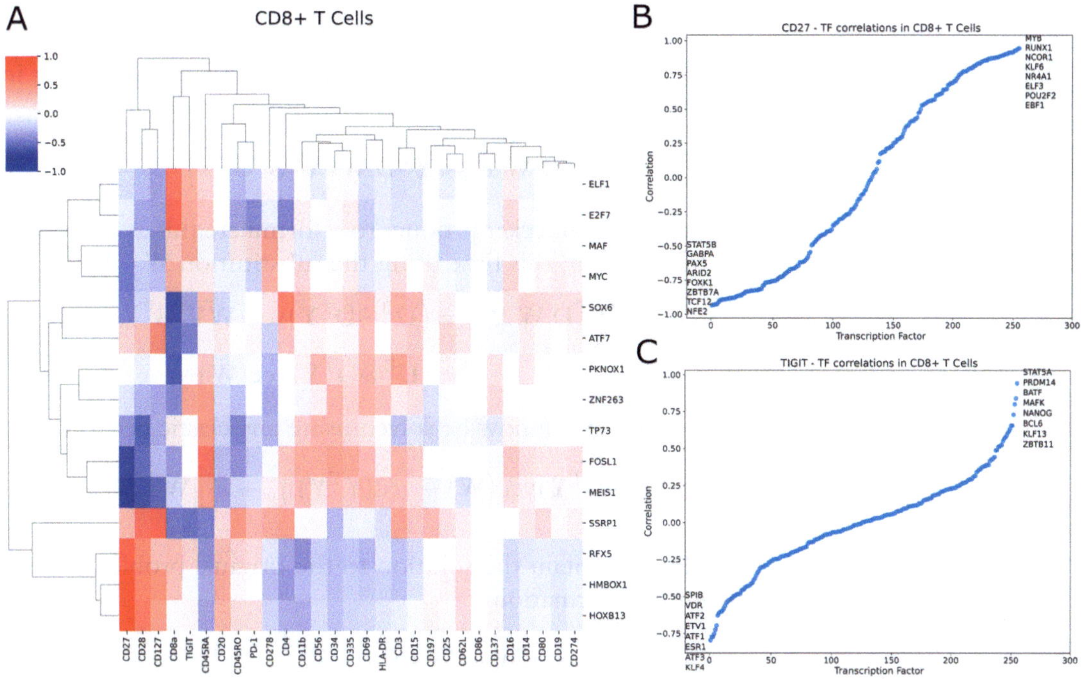


**Fig. 4** Relationships between inferred TF activities and surface receptor proteins. (**a**) Clustered heatmap showing the correlation between TF activities and surface protein expression in CD8+ T cells. (**b**) and (**c**) Sorted correlation plots for CD27 and TIGIT surface proteins in CD8+ T cells, showing the strongest positive and negative correlations. TFs are labeled

number of equations. The model now relates the measurements as:

$$\left(\mathbf{Y}^{\mathrm{T}}\mathbf{D}\right)\,\mathbf{W}\,\mathbf{P}^{\mathrm{T}} \sim \mathbf{Y}^{\mathrm{T}}\mathbf{Y}$$

With this transformation, we are now using the sum of the gene expression for each TF-target and the expression level of each protein to predict the similarity between cells.

2. AR can be solved by the following convex optimization problem, which uses the Kronecker product to reformulate the bilinear optimization problem into a form more suitable for standard optimization solvers:

$$\underset{\tilde{\mathbf{W}}}{\operatorname{argmin}} \left\| \left(\mathbf{P} \otimes \mathbf{Y}^{\mathrm{T}}\mathbf{D}\right)\operatorname{vec}\left(\tilde{\mathbf{W}}\right) - \operatorname{vec}\left(\mathbf{Y}^{\mathrm{T}}\mathbf{Y}\right) \right\|_F^2 + \lambda_2 \left\|\tilde{\mathbf{W}}\right\|_F^2 + \lambda_1 \left\|\tilde{\mathbf{W}}\right\|_1.$$

For large data sets, the Kronecker product $(\mathbf{P} \otimes \mathbf{Y}^{\mathrm{T}}\mathbf{D})$ becomes prohibitively large to compute and store in memory. The size problem can be drastically reduced by computing the truncated singular value decomposition (tSVD) of $\mathbf{P}$:

$$\left(\mathbf{U}^{\mathbf{P}}, \mathbf{S}^{\mathbf{P}}, \mathbf{V}^{\mathbf{P}}\right) = svd_t(\mathbf{P}.)$$

We use the tSVD to approximate the matrix multiplication in the objective function as follows:

$$\mathbf{Y}^{\mathbf{T}}\mathbf{D}\ \mathbf{W}\ \mathbf{P}^{\mathbf{T}} \approx \mathbf{Y}^{\mathbf{T}}\mathbf{D}\ \mathbf{W}\left(\mathbf{V}^{\mathbf{P}}\mathbf{S}^{\mathbf{P}}\mathbf{U}^{\mathbf{P}^{\mathbf{T}}}\right)$$

We use the Kronecker product to reformulate this as a standard quadratic program, this time using the substitution $\tilde{\mathbf{W}} = \mathbf{W}\mathbf{V}^{\mathbf{P}}$:

$$\begin{aligned}\text{vec}\left(\mathbf{Y}^{\mathbf{T}}\mathbf{D}\ \mathbf{W}\ \mathbf{P}^{\mathbf{T}}\right) &= \left(\mathbf{U}^{\mathbf{B}} \otimes \mathbf{D}^{\mathbf{T}}\mathbf{Y}\right)\text{vec}\left(\mathbf{W}\mathbf{V}^{\mathbf{P}}\mathbf{S}^{\mathbf{P}}\right) \\ &= \left(\mathbf{U}^{\mathbf{B}} \otimes \mathbf{D}^{\mathbf{T}}\mathbf{Y}\right)\text{vec}\left(\tilde{\mathbf{W}}\right)\end{aligned}$$

The problem can now be solved using any elastic net solver:

$$\underset{\tilde{\mathbf{W}}}{\text{argmin}} \left\| \left(\mathbf{U}^{\mathbf{B}} \otimes \mathbf{D}^{\mathbf{T}}\mathbf{Y}\right)\text{vec}\left(\tilde{\mathbf{W}}\right) - \text{vec}\left(\mathbf{Y}^{\mathbf{T}}\mathbf{Y}\right) \right\|_F^2 + \lambda_2 \left\|\tilde{\mathbf{W}}\right\|_F^2 + \lambda_1 \left\|\tilde{\mathbf{W}}\right\|_1$$

Finally, we obtain the solution to the original problem by the following transformation:

$$\mathbf{W} = \tilde{\mathbf{W}}\ \left(\mathbf{S}^{\mathbf{P}}\right)^{-1}\mathbf{V}^{\mathbf{P}^T}.$$

3. Once the model has been trained using a subset of the data, we can predict gene expression from any cell type, without further training, using only the surface protein expression of the cell. We calculate an initial estimate of the predicted gene expression as:

$$y_{\text{pred}} = \mathbf{D}\ \mathbf{W}\ p_{\text{test}}^T$$

The model was trained to predict similarities in gene expression, not gene expression itself. To account for the bias introduced by this difference, we project the initial prediction onto the span of the gene expression matrix used to train the model.

## Acknowledgments

This study was funded by support from the National Institutes of Health (R00 CA207871 and R35GM146989 to H.U.O., T15 LM007059-35 to A.S.) and the Innovation in Cancer Informatics Funds (to H.U.O.). Figures 1 and 2 were created with BioRender.com.

## References


1. Takahashi K, Yamanaka S (2006) Induction of pluripotent stem cells from mouse embryonic and adult fibroblast cultures by defined factors. Cell 126:663–676. https://doi.org/10.1016/j.cell.2006.07.024
2. Stoeckius M, Hafemeister C, Stephenson W et al (2017) Simultaneous epitope and transcriptome measurement in single cells. Nat Methods 14:865–868. https://doi.org/10.1038/nmeth.4380
3. Peterson V, Zhang K, Kumar N et al (2017) Multiplexed quantification of proteins and transcripts in single cells. Nat Biotechnol 35:936–939. https://doi.org/10.1038/nbt.3973
4. Buenrostro J, Wu B, Litzenburger U et al (2015) Single-cell chromatin accessibility reveals principles of regulatory variation. Nature 523:486–490. https://doi.org/10.1038/nature14590
5. Cusanovich DA, Daza R, Adey A et al (2015) Multiplex single cell profiling of chromatin accessibility by combinatorial cellular indexing. Science 348:910–914. https://doi.org/10.1126/science.aab1601
6. Baron CS, Barve A, Muraro MJ et al (2019) Cell type purification by single-cell transcriptome-trained sorting. Cell 179:527–542.e19. https://doi.org/10.1016/j.cell.2019.08.006
7. Ma X, Somasundaram A, Qi Z et al (2021) SPaRTAN, a computational framework for linking cell-surface receptors to transcriptional regulators. Nucleic Acids Res 49:9633–9647. https://doi.org/10.1093/nar/gkab745
8. Pelossof R, Singh I, Yang J et al (2015) Affinity regression predicts the recognition code of nucleic acid-binding proteins. Nat Biotechnol 33:1242–1249. https://doi.org/10.1038/nbt.3343
9. Osmanbeyoglu HU, Pelossof R, Bromberg JF, Leslie CS (2014) Linking signaling pathways to transcriptional programs in breast cancer. Genome Res 24:1869–1880. https://doi.org/10.1101/gr.173039.114
10. Osmanbeyoglu HU, Toska E, Chan C, Baselga J, Leslie CS (2017) Pancancer modelling predicts the context-specific impact of somatic mutations on transcriptional programs. Nat Commun 8:14249. https://doi.org/10.1038/ncomms14249
11. Garcia-Alonso L, Holland CH, Ibrahim MM, Turei D, Saez-Rodriguez J (2019) Benchmark and integration of resources for the estimation of human transcription factor activities. Genome Res 29:1363–1375. https://doi.org/10.1101/gr.240663.118
12. Rule A, Birmingham A, Zuniga C et al (2019) Ten simple rules for writing and sharing computational analyses in Jupyter notebooks. PLoS Comput Biol 15:e1007007. https://doi.org/10.1371/journal.pcbi.1007007
13. Virshup I (2022) Analyzing CITE-seq data. <https://scanpy-tutorials.readthedocs.io/en/latest/cite-seq/pbmc5k.html>

# Chapter 12

# Mass Spectrometry-Based Tissue Imaging of the Tumor Microenvironment


## Tyler Risom, Patrick Chang, Sandra Rost, and James Ziai



## Abstract

Multiplex ion beam imaging (MIBI) and imaging mass cytometry (IMC) enable highly multiplexed antibody (40+) staining of frozen or formalin fixed, paraffin-embedded (FFPE) human or murine tissues through detection of metal ions liberated from primary antibodies by time-of-flight mass spectrometry (TOF). These methods make detection of more than 50 targets theoretically possible while maintaining spatial orientation. As such, they are ideal tools to identify the multiple immune, epithelial, and stromal cell subsets in the tumor microenvironment and to characterize spatial relationships and tumor-immune status in either murine models or human samples. This chapter summarizes methods for antibody conjugation and validation, staining, and preliminary data collection using IMC or MIBI in both human and mouse pancreatic adenocarcinoma samples. These protocols are intended to facilitate use of these complex platforms in not only tissue-based tumor immunology studies but also tissue-based oncology or immunology studies more broadly.



**Key words** Multiplexing, Imaging mass cytometry (IMC), Multiplex ion beam imaging (MIBI), Metal-conjugated antibodies (MCA), Regions of interest (ROI), Frozen or formalin fixed, Paraffin-embedded (FFPE) tissues, Tumor microenvironment (TME), Pancreatic adenocarcinoma (PDAC), Genetically engineered mouse models (GEMM)


## 1 Introduction

Multiplexed immunohistochemistry makes possible the simultaneous visualization of multiple proteins with spatial context in a single tissue section. Primary antibodies conjugated either to enzymes or fluorophores can be applied to tissue slides with development of chromogenic products or fluorescence by introduction of organic compounds or light excitation, respectively. Alternatively, cyclic application of primary or secondary antibodies followed by imaging can be performed to enable higher levels of multiplexing as well as amplification [1–3]. However, in practice, these methods are limited by spectral overlap of chromogens or


Usha N. Kasid and Robert Clarke (eds.), *Cancer Systems and Integrative Biology*, Methods in Molecular Biology, vol. 2660, https://doi.org/10.1007/978-1-0716-3163-8_12,

**Table 1**
**Mass cytometry imaging platform specifications**

| | Imaging mass cytometry | Multiplex ion beam imaging |
|---|---|---|
| Tissue ablation method | Ultraviolet laser | Primary xenon ion beam |
| Sample types | Human (FFPE, frozen), mouse (FFPE, frozen) | Human (FFPE, frozen), mouse (FFPE, froze) |
| Slides required | Standard glass | Gold-coated |
| Maximum resolution | 1000 nm | 260 nm |
| Maximum no. targets per section (published) | 40+ | 40+ |
| Scan rate | 1 $mm^2$ / 1.5 h | 800 × 800 um / 36 min @750 nm res.<br>800 × 800 um / 2.4 h @550 nm res.<br>800 × 800 um / 4.6 h @350 nm res. |

fluorophores and can require multiple time-consuming rounds of staining, imaging, and deconvolution.

Recent chemistry has enabled design of metal-chelating polymers that can be conjugated to primary antibodies [4] which can then be applied to tissues. Subsequently, metal ions can be liberated by ablation and detected by time-of-flight (TOF) mass spectrometry enabling highly multiplexed staining of tissues in comparatively short time. Two platforms for this exist: multiplexed ion beam imaging (MIBI, Ionpath Inc.) and imaging mass cytometry (IMC, Standard BioTools). Both use TOF to detect secondary ions liberated from primary metal-conjugated antibodies (MCAs) but differ in certain technical components and capabilities (Table 1). Of note, MIBI employs a tunable ion beam that can be adjusted to vary tissue acquisition depth enabling multiple scans of the same area. Spot size can also be tuned to allow subcellular resolution or, with lower resolution, faster acquisition of regions of interest (ROI) or an entire tissue section. IMC and MIBI thus make possible acquisition of spatial data for a large number of proteins at single cell resolution and, in the case of MIBI, subcellular resolution. As such, the population and topography of tissue microenvironments can be more fully characterized to provide novel biological and, potentially, therapeutic insights in refractory disease indications.

Traditional chemotherapies [5], targeted therapies [6–8], and immunotherapies [9–11] have largely failed to significantly improve outcomes in pancreatic ductal adenocarcinoma (PDAC). Accordingly, increased focus has been placed on the stromal composition of PDAC as well as composition and location of immune cells within the tumor and stroma. Vimentin, a-SMA, and collagen I have been associated with epithelial–mesenchymal transition or progression of malignancy [12]. As well, prevalence of intratumoral CD4 and CD8 T cells has been associated with improved outcome [13, 14], but multiple immunosuppressive populations including cancer-associated fibroblasts (CAFs) and macrophages [15] and regulatory T cells [16, 17] must be accounted for to fully characterize a tumor's immune status and inform potential therapeutic combinations.

Here we describe development of a highly multiplexed immunohistochemistry panel for IMC and MIBI including antibody validation, staining, and data collection methods. The panel is comprised of 20 markers (Table 2) to identify major components of the tumor microenvironment in pancreatic cancer: T cells, B cells, tumor-associated macrophages, monocytes, dendritic cells, vasculature, stromal smooth muscle, fibroblasts, and tumor cells. While demonstrated here in the context of pancreatic cancer, the staining methods and markers are applicable to a wide array of tissue and tumor types as well as pure immunology indications.

**Table 2**
**Human and mouse tumor microenvironment profiling panel for PDAC**

| | Human panel | | | Mouse panel | | |
|---|---|---|---|---|---|---|
| Channel | Target | Clone | Conc. MIBI; IMC (ug/mL or ×) | Target | Clone | Conc. MIBI; IMC (ug/mL or ×) |
| 89Y | SMA | SP171 | 1; 7 | SMA | SP171 | 1; 5 |
| 113In | COL1A1 | E68AE | 3; 7 | COL1A1 | E68AE | 3; 5 |
| 115In | DsDNA | 35I9 | 2 (MIBI only) | DsDNA | 35I9 | 2 (MIBI only) |
| 115In | HH3 | D1H2 | 2× (MIBI only) | HH3 | D1H2 | 2× (MIBI only) |
| 142Nd | | | | | | |
| 143Nd | CD4 | EPR6855 | 1×; 2× | CD4 | RM4-5 | 2×; 2× |
| 144Nd | CD11c | EP1347Y | 1×; 1× | CD11c | D1V9Y | 1×; 1× |
| 145Nd | CD21 | SP186 | 1; 5 | CD21 | SP186 | 1; 5 |
| 151Eu | | | | CD49b | EPR17338 | 1×; 1× |
| 152Sm | CD31 | EP3095 | 1×; 1× | CD31 | D8V9E | 1×; 1× |
| 153Eu | Ki67 | D2H10 | 1×; 1× | Ki-67 | SP6 | 1×; 1× |
| 154Sm | CD14 | D7A2T | 1×; 2× | | | |
| 155Gd | | | | CD11b | EPR1344 | 1×; 2× |
| 156Gd | CD68 | D4B9C | 1×; 2× | F4/80 | D2S9R | 1×; 1× |

(continued)

**Table 2 (continued)**

| | Human panel | | | Mouse panel | | |
|---|---|---|---|---|---|---|
| Channel | Target | Clone | Conc. MIBI; IMC (ug/mL or ×) | Target | Clone | Conc. MIBI; IMC (ug/mL or ×) |
| 158Gd | CD8 | C8/144B | 1×; 2× | | | |
| 159 Tb | CD3 | MRQ-39 | 1×; 2× | CD3e | EPR22667-12 | 2×; 2× |
| 160Gd | PNAd-biotin | MECA79 | 0.5; 5 | PNAd-biotin | MECA79 | 0.5; 5 |
| 160Gd | Anti-biotin | 1D4-C5 | 1; 5 | Anti-biotin | 1D4-C5 | 1; 5 |
| 162Dy | CD44 | E7K2Y | 0.5; 7 | CD44 | E7K2Y | 0.5; 5 |
| 163Dy | Vimentin | D21H3 | 1×; 1× | Vimentin | D21H3 | 1×; 1× |
| 164Er | CD163 | D6U1J | 1; 7 | | | |
| 165Ho | PanCK | AE1/AE3 | 1×; 2× | PanCK | AE1/AE3 | 1×; 2× |
| 167Er | CD20 | L26 | 1×; 7× | B220 | RA3-6B2 | 1×; 2× |
| 170Er | CD90 | EPR3132 | 1; 13 | CD90 | EPR3132 | 1; 10 |
| 171Yb | | | | CD45 | D3F8Q | 0.5; 5 |
| 172Yb | HLA-DR | EPR3692 | 1×; 1× | | | |
| 174Yb | CK17 | D12E5 | 1; 10 | CK17 | D12E5 | 1; 10 |
| 175Lu | CD45 | 2B11 and PD7/26 | 1×; 1× | | | |
| 176Yb | NaK ATPase | EP1845Y | 1×; 1× | NaK ATPase | EP1845Y | 1×; 1× |
| 191Ir, 193Ir | Cell-ID intercalator | | 1:1000 (IMC only) | Cell-ID intercalator | | 1:1000 (IMC only) |

## 2 Materials

### 2.1 Tissue Slides

1. Tissue blocks—Archival FFPE tissue blocks of mouse and human pancreatic ductal adenocarcinoma (PDAC) tumors, as well as control tissues for marker validation:
   (a) LSL-KRAS$^{G12D}$-Pdx1-Cre-tg-p16INK4a-p19ARF murine PDAC model tumors ($n = 2$).
   (b) Human PDAC Tumor Resections ($n = 2$).
2. Tissue slides for sectioning

   MIBI—specially formulated high-purity gold-coated slides (Ionpath).

   IMC and chromogenic IHC—Standard positively charged glass slides (Epredia).

### 2.2 Tissue Processing

1. Histopathology oven.
2. Xylene.
3. Ethanol (200 proof).
4. Heat-induced epitope retrieval (HIER) solution.
   MIBI—Target Retrieval Solution pH 9 (Dako).
   IMC—EZ AR1 Elegance RTU (BioGenex) & EZ AR2 Elegance RTU (BioGenex).
5. HIER module.
   MIBI—LabVision PT Module (Thermo).
   IMC—EZ-Retriever System (BioGenex).
6. Block.
   MIBI—0.1% Fish Gelatin (Sigma-Aldrich) + 0.1% Triton X (Sigma-Aldrich) + 2% Normal Donkey Serum (Sigma-Aldrich) in TBST (Cell Marque).
   IMC—3% Bovine Serum Albumin (BSA, Cell Signaling Technology) in Maxpar PBS (Standard BioTools).

### 2.3 MCA Conjugation

1. Nanodrop 2000 microvolume spectrophotometer (Thermo Fisher).
2. MIBItag conjugation kit (for MIBI, Ionpath) or Maxpar Antibody Labeling Kit (for IMC, Standard BioTools).
3. Antibody Stabilizer (Candor).

### 2.4 Metal-Conjugated Antibody Cocktail

1. Antibody diluent.
   MIBI—3% Normal Donkey Serum (Sigma-Aldrich) in TBST (Cell Marque).
   IMC—0.5% Bovine Serum Albumin (BSA, Cell Signaling Technology) in Maxpar PBS (Standard BioTools).
2. Wash.
   MIBI: TBST (Cell Marque).
   IMC: Maxpar Water (Standard BioTools) and Maxpar PBS (Standard BioTools).
3. Counterstain.
   MIBI: None.
   IMC: Cell ID Intercalator-Ir (Standard BioTools).
4. Shandon Sequenza Staining Rack (Epredia).
5. Shandon Sequenza Coverplates (Epredia).

### 2.5 Tissue Preservation

1. 2% Glutaraldehyde solution (Electron Microscopy Sciences).
2. Barium-free PBS rinse.
3. TBS pH 8.5 rinse (Sigma-Aldrich).
4. Dehydration.
   MIBI: TBS ×3, H20 ×2, 70% EtOH, 80% EtOH, 95% EtOH ×2, 100% EtOH ×2.
   IMC: Maxpar Water (Standard BioTools), air-dry.

**2.6 Mass Cytometry Imaging Platforms**

1. Standard BioTools Hyperion Imaging System.
2. Ionpath MIBIscope.

## 3 General Methods

In this section we first present general methods regarding tissue selection, panel design, antibody conjugation, and MCA validation that are common to both the MIBI and IMC workflows (Figs. 1, 2, and 3). Adhering to these design principles will expedite panel design and workup and ensure the accuracy of mass spectrometry signal to antigen abundance. Step-by-step protocols for each platform will follow in the subsequent sections.

### 3.1 Tissue Selection

*3.1.1 Tissue Quality*

Tissue quality is one of the most influential factors on the robustness and uniformity of imaging results by MIBI and IMC. Like other immunohistochemical techniques, it is critical that cold ischemia times are minimalized in tissue harvesting and that fixation protocols are optimized for tissue density and size. While issues regarding overfixation such as autofluorescence are negligible in mass spectrometry imaging, issues regarding under-fixation are of

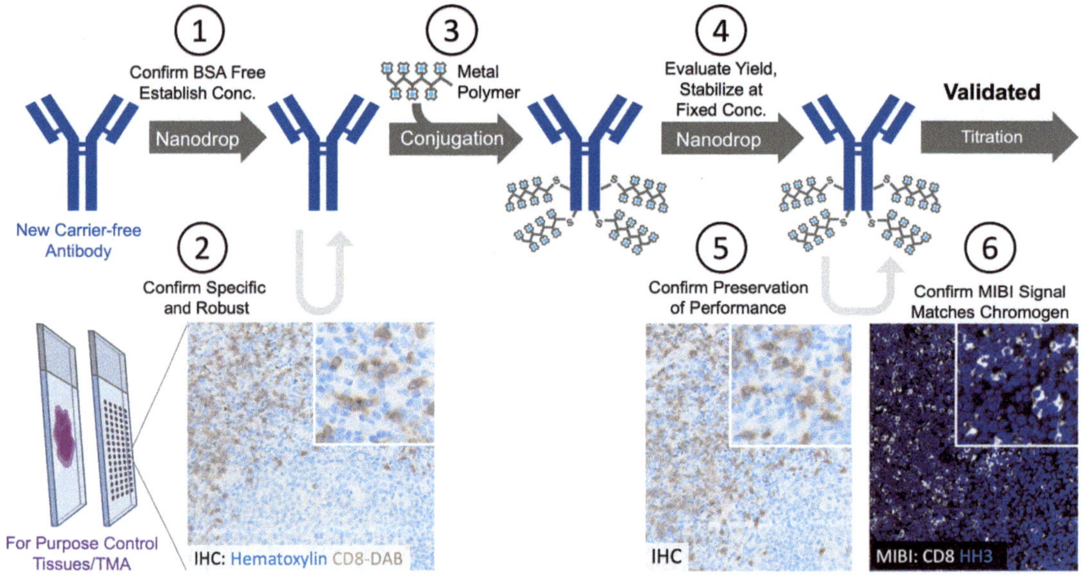


**Fig. 1** Workflow for metal-conjugated antibody validation. Schematic of the antibody conjugation and validation process. New carrier-free antibodies are first tested for IgG concentration using a spectrophotometer (1), where large derivations from the advertised concentration may indicate BSA contamination. Chromogenic IHC is then performed to assess antibody specificity and robustness (2). Antibodies are then conjugated to a metal-loaded polymer (3) and reassessed on a spectrophotometer for concentration post conjugation (4) to assess yield and to guide dilution in antibody stabilizer. MCA is then reassessed by chromogenic IHC (5) and by MIBI/IMC to ensure both chromogenic and MIBI/IMC signal replicates the unconjugated antibody

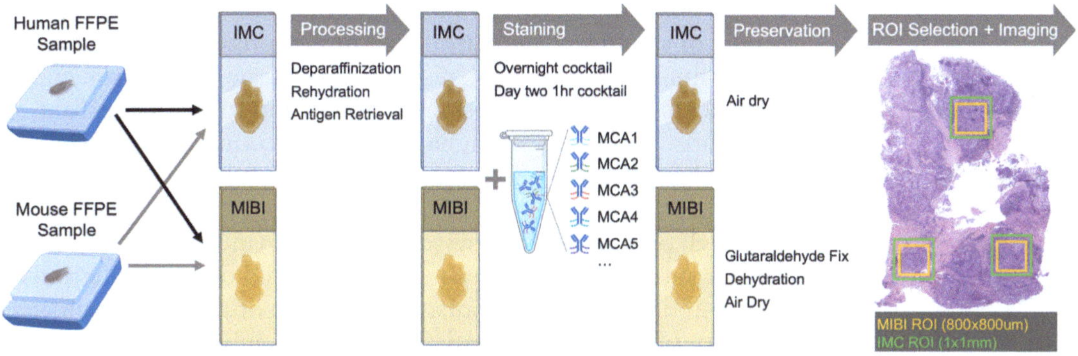


**Fig. 2** Tissue processing and imaging workflow. Schematic of the human and mouse PDAC tissue processing, staining, preservation, and imaging workflow for MIBI and IMC

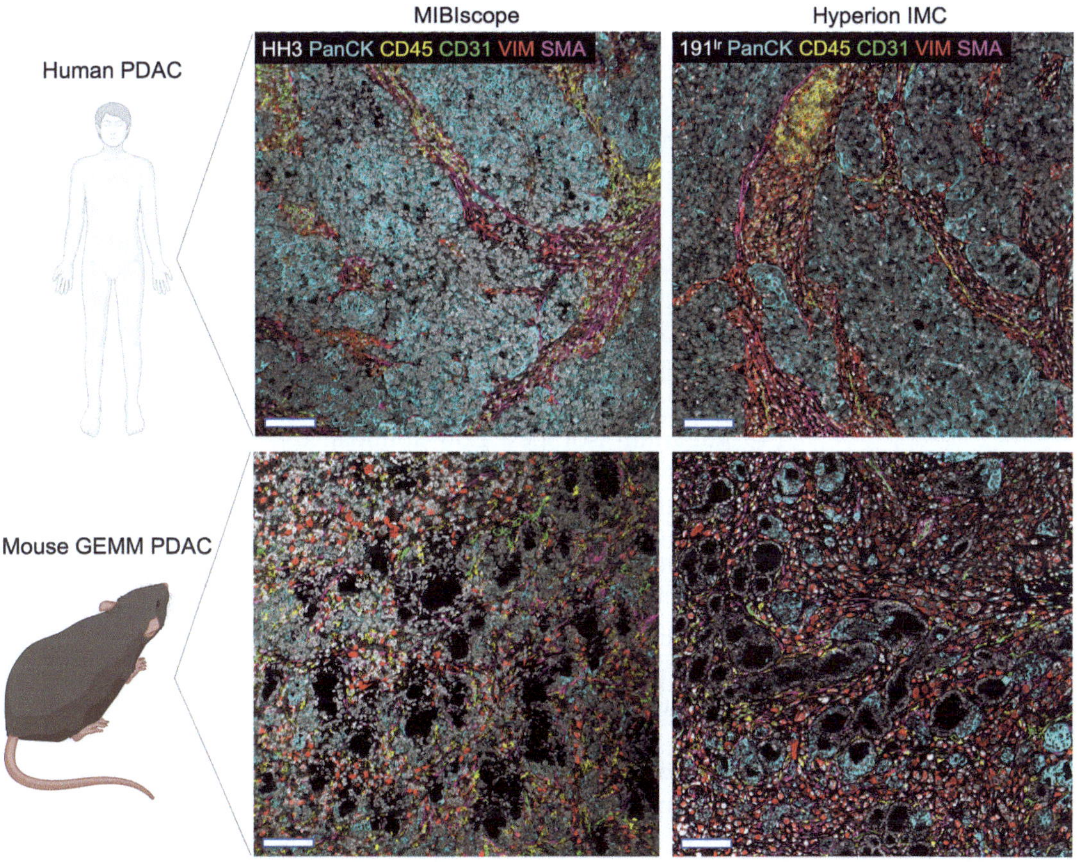


**Fig. 3** MIBI and IMC images of the PDAC tumor microenvironment. Representative images of the human and mouse tumor microenvironment of pancreatic adenocarcinoma by MIBI and IMC. The major cell lineage markers of PanCK (cyan, tumor), CD45 (yellow, immune), CD31 (green, endothelium), VIM (red, mesenchyme), SMA (pink, fibroblasts) are overlaid with nuclear markers (HH3 for MIBI, white, and 191-Iridium intercalator for IMC)

critical concern. Reduced fixation at the core of large specimens results in significant reductions in antigen preservation and resultant staining. Accordingly, dissection of tissue to uniform pieces prior to fixation can improve the uniformity of formalin fixation and subsequent MCA signal. ROI selection should consider regional tissue quality to produce the best results.

#### 3.1.2 Control Tissue Sourcing

The high plexing of these systems often demands a diverse array of tissue controls for marker validation by pathologists. While tonsil, lymph node, and other lymphatic organs can often help to validate a majority of markers, particularly immune cell lineage markers, tissue- or disease-specific markers often require the sourcing and inclusion of more exotic tissue specimens for marker validation. For this reason, the generation of project-specific marker validation tissue microarrays (TMAs) can combine a large array of diverse tissues that capture marker positivity and can expedite validation efforts.

### 3.2 Panel Design

Panel design in mass cytometry imaging refers to the placement of markers in the ~40 detected metal isotope positions in the mass spectrum. Considerations of marker abundance need to be weighed against machine sensitivity, organic interference, and produced organic metal adducts during ionization that result in predictable false-positive signal.

#### 3.2.1 Machine Sensitivity

MIBIscope and Hyperion IMC systems have highest TOF detection sensitivity in the 140–150 amu range of metals, including the 142–146, 148, and 150 neodymium isotopes, and 147 and 149 samarium isotopes. Low-abundance markers should be considered to be conjugated to these isotopes in order to take advantage of the highest TOF sensitivity. Similarly, when placing markers in positions outside of this maximal sensitivity window, one must consider the most robust antigens for positions like 89 yttrium, 113 and 155 indium, and 171–176 ytterbium, such as high-abundance nuclear antigens (HH3, DsDNA), highly expressed intermediate filaments (PanCytokeratin, vimentin), or highly abundant cell surface markers (HLA1, Na/K ATPase).

#### 3.2.2 Organic Interference

While we are most focused on the ionization and detection of the isotopic metals from our metal-conjugated antibodies, the ion beam or laser used to raster the sample in these technologies liberates a panoply of organic ions (e.g., H, O, OH) from the tissue as well as gold and tantalum from the coated slide in MIBI. These ions result in nonspecific noise that accumulates unequally across the mass spectrum. Channels that experience higher organic noise include 113/115In (MIBI), 141Pr, 157Gd, and 170Yb, and robust markers should be considered for placement in these locations as well.

#### *3.2.3 Organic Metal Adducts*

In addition to the nonspecific organic noise detailed above, the liberated metal ions from antibodies can interact with organic molecules including hydrogen, oxygen, and hydroxides to produce organic metal adducts, which will produce false-positive signal of the seed marker in the +1, +16, and +17, respectively (e.g., histone H3 signal from 149Sm will make a false signal in 150Nd, 175Lu, and 176Yb). In panel design, one must consider this and place bright, robust markers in positions where the organic-adduct signal falls outside of the detected spectrum or falls in +1/+16/+17 channels where spatially distinct marker expression is found (e.g., tumor marker vs stromal marker) so that this false signal can be easily addressed in analysis. The +1 false signal is the most profound and is akin to signal "bleed" in immunofluorescence staining; in line with the design principle previously mentioned, markers should be alternated in the panel by which tissue compartment they mark, as to avoid false signal from overlaying with real signal (e.g., tumor marker on 171, immune marker on 172, stromal marker on 173, repeat).

#### *3.2.4 Overnight Versus Day 2 Panel*

Most MCAs will provide optimal staining if incubated overnight at 4 °C; however, some very highly expressed antigens (e.g., DsDNA, HH3) have optimal detection when stained with an MCA for 1 h at 4 °C. In addition, secondary staining strategies such as PNAd-biotin + anti-biotin$^{160Gd}$ require the secondary antibody in the Day 2 stain.

### 3.3 Antibody Sourcing

#### *3.3.1 Commercial MCAs*

Metal-conjugated antibodies can either be sourced directly from the two companies that developed these platforms, Ionpath (MIBI) and Standard BioTools (IMC), or generated in house through the metal conjugation process (*see* Subheading 3.4). The available commercial conjugates include antibody clones that have been rigorously validated in past publications and by Ionpath/Standard BioTools pathologists and represent important resources to quickly build a high-plex panel.

#### *3.3.2 Unconjugated Antibodies*

For the selection of antibodies for in-house metal conjugation efforts, it is advantageous to similarly look for IHC-validated clones with a strong publication history of robust staining in FFPE tissue. Critical to the clone and vendor selection process is ensuring that the antibody is available carrier-free. Commercial antibodies almost ubiquitously contain antibody stabilizer to ensure shelf-life, and the most common stabilizer is Bovine Serum Albumin (BSA), a large protein often existing in high concentration. Due to this, executing the metal conjugation protocol on stabilized antibody solution results in metal conjugation of the BSA, not your target IgG molecule. Accordingly, purchasing carrier-free antibodies is critical to the generation of an in-house MCA library. Nowadays, many companies are aware of the growing demand for carrier-free antibodies and advertise specific "carrier-free" or "BSA- and azide-free" formulations of favorite antibody clones. In the case that one's favorite clone is not available as a carrier-free formulation, establishing a relationship with a company representative can usually help you source aliquots of the clone in a carrier-free format.

### 3.4 Antibody Conjugation

The conjugation process to generate MCAs relies on sourcing carrier-free antibodies, determining their concentration, and inputting the purified IgG into the conjugation kit protocols provided by Ionpath (MIBItag Conjugation Kit) or Standard BioTools (Maxpar X8 Antibody Labeling Kits).

*Please refer to the "MIBItag Conjugation Kit User Guide" (Ionpath) or "Maxpar Antibody Labeling User Guide" (Standard BioTools) for specific step-by-step instructions for antibody conjugation. Common steps are discussed below.*

#### *3.4.1 Establishing Input Material*

Carrier-free antibody concentration is first established using a Nanodrop 2000 microvolume spectrophotometer (Thermo Fisher). Concentration should match what is advertised by the distributor on the tube, usually 1 mg/mL, and large deviations from this value may indicate BSA contamination (*see* **Notes 6.1**).

#### *3.4.2 Antibody Purification*

The recommended 100 or 200 ug of input IgG is then purified in a 50 kDa centrifugal filter tube through numerous washes.

#### *3.4.3 Antibody Reduction*

The purified antibody is then reduced with TCEP (Sigma-Aldrich) in order to break cysteines in the FC domain and expose free thiol groups. The reduced antibody is then washed.

#### *3.4.4 Antibody: Metal Polymer Conjugation*

The reduced antibody is then incubated with a solution of metal-loaded polymer (MIBItag from Ionpath; Maxpar X8 polymer from Standard BioTools) with a maleimide group at the linkage site that reacts with the free thiol groups on the reduced antibody. For the Standard BioTools Maxpar kit, metal has to be previously loaded into the polymer, whereas MIBItags come ready for conjugation and only require solubilization.

#### *3.4.5 MCA Purification, Measurement, and Stabilization*

Following conjugation, the now metal-conjugated antibody is washed and purified in the same centrifugal filter. Following washing, MCA concentration is determined using a Nanodrop 2000 microvolume spectrophotometer (Thermo Fisher). The MCA is then diluted to 500 or 200 ug/mL using antibody stabilizer (Candor) and is ready for long-term storage.

### 3.5 Marker Validation

For the metal-conjugated antibodies sold and QC'ed by a third party, skip **step 1**.

1. Qualitative assessment of the chromogenic IHC staining of the carrier-free version of the antibody clone is done against an internally validated clone for the same marker. Move forward with the clones that match or exceed the validated clone in specificity and sensitivity.
2. After the metal conjugation (*see* Subheading 3.3) or when using purchased metal-conjugated antibodies, perform the same qualitative assessment for specificity and sensitivity using chromogenic IHC staining again with the internal validated clone.
3. Finally and most critically, test the MCA on the MIBI or IMC platform, and work with a pathologist to compare the imaging mass spectrometry detected signal with the chromogenic IHC staining from 1 to 2.

## 4 Methods for Multiplexed Ion Beam Imaging

### 4.1 Tissue Preparation (See Notes 6.2)

1. Bake slides at 70 °C for 10 min laying nearly flat, with the label end slightly elevated, then bake for 10 min standing up.
2. Take slides out and let cool to room temperature (RT), laying flat.
3. Submerge tissue on slides in xylene for 5 min.
4. Slides are then run through a deparaffinization and rehydration sequence of xylenes, alcohol, and water in a linear stainer (ST4200, Leica) for 30 s, 20 dips, each reagent as follows: xylene, xylene, xylene, 100% EtOH, 100% EtOH, 95% EtOH, 95% EtOH, 80% EtOH, 70% EtOH, $H_2O$, $H_2O$.
5. Antigen retrieval is performed using a pH 9 heat-induced epitope retrieval solution (DAKO), slides are submerged in 1× solution and incubated at 97 °C for 40 m, after which they are left to cool to 65 °C.
6. Slides are next mounted into the Sequenza Staining Rack (Epredia), alternatively, a PAP pen (Fisher) border can be made around the tissue to contain staining reagents.
7. Slides are then washed 2× with TBST Wash Buffer (Cell Marque).
8. Slides are then blocked with MIBI block containing diluted fish gelatin (Sigma-Aldrich) and normal donkey serum (NDS, Sigma-Aldrich) for 1 h RT.

#### 4.2 Antibody Staining

1. Prepare the overnight MCA cocktail by first adding 100 uL of 3% NDS to a low binding 1.5 mL tube.
2. For each tissue slide, 150 uL of MCA cocktail will be generated, add each MCA to the tube according to total volume and MCA titer.
3. After addition of each MCA, adjust total volume by adding more 3% NDS.
4. Add 130 uL of the overnight MCA cocktail to each slide in the Sequenza rack, ensure the lid is on and place rack at 4 °C overnight.
5. The next morning, generate the Day 2 staining cocktail. Add 100 uL to a low binding tube and add the appropriate amount of each Day 2 MCA.
6. Wash slides 1× with TBST wash buffer (Cell Marque).
7. Add 130 uL day 3 cocktail to each slide and incubate 4 °C for 1 h.
8. Wash slides 2× with TBST wash buffer (Cell Marque).

#### 4.3 Tissue Preservation

1. Remove slides from Sequenza rack and place slides in a ST4200 Linear Stainer rack.
2. Submerge slides in PBS pH 7.5 for 1 min RT.
3. Submerge slides in a 2% Glutaraldehyde (EMS) in PBS pH 7.5 solution for 5 min RT.
4. Slides are then run through a dehydration sequence on the ST4200 Linear Stainer for 30 s, 20 dips, each reagent as follows: TBS pH 8.5, TBS pH 8.5, TBS pH 8.5, $H_2O$, $H_2O$, 70% EtOH, 80% EtOH, 95% EtOH, 95% EtOH, 100% EtOH, 100% EtOH.
5. Remove the rack of slides from the linear stainer track, and place in an open area in the chemical hood to ensure even evaporation, allow to air dry for 1 h RT.
6. Slides can now be stored in a vacuum chamber, nitrogen cabinet, or alternative low-oxidation environment until ready to image.

## 5 Methods for Imaging Mass Cytometry

#### 5.1 Tissue Preparation (See Notes 6.3)

1. Bake the slides in a slide rack for 30 min or longer at 70 °C in the slide oven to ensure that all visible wax is removed.
2. Slides are then run through a deparaffinization and rehydration sequence of xylenes, alcohol, and water in a autostainer (Leica) for 2 min, eight dips, each reagent as follows: xylene, xylene,

xylene, 100% EtOH, 100% EtOH, 95% EtOH, 95% EtOH. And then leave in water until retrieved from the stainer.

3. Antigen retrieval is performed using the EZ-Retriever® System (BioGenex). The slides are first submerged in EZ-AR2 solution and incubated at 107 °C for 10 min. The slides are then moved to preheated EZ-AR1 solution and incubated at 107 °C for another 10 min. After that, slides are left in EZ-AR1 solution at room temperature for approximately 20 min to cool down to about 74 °C.
4. Wash the slides in Maxpar Water for 10 min.
5. Use a PAP pen to encircle the samples on the slide to conserve reagents.
6. Slides are blocked with freshly made 3% BSA in Maxpar PBS solution for 45 minutes at room temperature.

### 5.2 Antibody Staining

1. Prepare overnight MCA cocktail by adding calculated volume of antibodies at concentrations specific for the assay and bring volume up to 150 uL per slide in 0.5% BSA in Maxpar PBS.
2. Discard the blocking solution by tapping the slides on a kimwipe and absorb as much residual solution off the slides as possible using the tip of a kimwipe. Incubate overnight with MCA cocktail mix at 4 °C in a hydration chamber.
3. For secondary antibody incubation, take out slides and submerge into a slide mailer with Maxpar PBS for 8 min at room temperature after the overnight incubation. Prepare 5 ug/mL of secondary antibodies in 0.5% BSA in Maxpar PBS. Incubate the slides with the secondary antibody solution for 30 min at room temperature.

### 5.3 Counter Staining

1. After the incubation of primary antibodies (or secondary antibodies), prepare 0.2% Triton X-100 in Maxpar PBS, and use it to wash the slides for 8 min at room temperature.
2. Wash the slides in Maxpar PBS for 8 min with gentle agitation at room temperature twice.
3. Thaw Intercalator-Ir aliquot and dilute in Maxpar PBS (1: 1000). Spin the Intercalator-Ir aliquot down briefly before use.
4. Incubate slides in Intercalator-Ir for 30 min at room temperature.

### 5.4 Tissue Preservation

1. Wash the slides in Maxpar Water for 5 min at room temperature.
2. Air-dry the slides overnight at room temperature.
3. Keep slides in slide mailers with desiccants for long-term storage.

## 6 Notes

### 6.1 Antibody Conjugation Notes

1. If BSA contamination is suspected, you can run the antibody solution on an SDS PAGE gel and use anti-BSA antibodies to evaluate BSA presence in that lane.
2. BSA can be cleaned from contaminated antibodies using BSA-removal kits, however significant antibody input is required to get a robust yield of purified IgG.

### 6.2 MIBI Methods Notes

1. Try to use fresh-cut tissue sections, and maintain cut slides in oxygen-free, desiccated environment.
2. For heat-induced epitope retrieval, slides can be incubated in the HIER buffer in slide mailers, restricting total volume to 25 mL and conserving buffer.
3. When adding the Sequenza Coverplate to each slide, fill a reagent reservoir halfway with wash buffer, fully wet the slide and coverplate, then combine the two, pinching the two together according to instructions and keeping an eye on the formed liquid interface, ensuring it is uniform before loading into the rack.
4. Ensure no bubbles form in the coverplate/slide interface, remove slide and coverplate, rewet, and recombine as noted above if bubbles are observed.
5. After the dehydration series, set the slides in an area with high air circulation to ensure quick evaporation and to prevent reagent crystallization.
6. Once fixed and dried, slides have a >1 year shelf life if maintained in a low-oxygen, dry environment, enabling rescanning with equal yield.
7. Generate an H&E slide on a serial section to guide ROI selection, geographical landmarks can be identified in the MIBI-scope run setup to ensure accurate ROI capture.

### 6.3 IMC Methods Notes

1. Try to use fresh-cut tissue sections because tissue integrity and antigen stability for pre-cut sections may degrade over time.
2. For antigen retrieval in BioGenex microwave, when the microwave alarm goes off indicating the end of the first cycle, open the microwave door directly (alarm will stop), set the slide holder aside, pour the EZ-AR2 solution from the bucket with sensor space into a third bucket, pour the heated EZ-AR1 from the second bucket into the bucket with sensor space, emerge the slides into EZ-AR1 (bucket with sensor space), put both buckets with solutions into the microwave, and start the second cycle.

3. When preparing the antibody cocktail, add a small volume of individual antibodies into a larger volume of diluent. Spin down the antibodies with table top centrifuge briefly and pipette from the top of the tube to avoid antibody aggregates. Put the antibody cocktail mix on ice, and add it to samples within 2 h of preparation for best results.

## Acknowledgments

Tyler Risom and Patrick Chang contributed equally to this study.

### References


1. Eng J, Thibault G, Luoh SW et al (2055) Cyclic Multiplexed-Immunofluorescence (cmIF), a highly multiplexed method for single-cell analysis. Methods Mol Biol 2020:521–562
2. Golstev Y, Samusik N, Kennedy-Darling J et al (2018) Deep profiling of mouse splenic architecture with CODEX multiplexed imaging. Cell 74(4):968–981
3. Radtke A, Postolova E, Varlamova A et al (2022) A multi-scale, multiomic atlas of human normal and follicular lymphoma lymph nodes. Biorxiv. https://doi.org/10.1101/2022.06.03.494716
4. Majonis D, Herrera I, Ornatsky O et al (2010) Synthesis of a functional metal-chelating polymer and steps toward quantitative mass cytometry bioassays. Anal Chem 82:8961–8969
5. Conroy T, Hammel P, Hebbar M et al (2018) FOLFIRINOX or gemcitabine as adjuvant therapy for pancreatic cancer. N Engl J Med 379:2395–2406
6. Hecht JR, Lonardi S, Bendell J et al (2021) Randomized phase III study of FOLFOX alone and with pegilodecakin as second-line therapy in patients with metastatic pancreatic cancer (SEQUOIA). J Clin Oncol 39:1108–1118
7. Tempero M, Oh DY, Tabernero J et al (2021) Ibrutinib in combination with nab-paclitaxel and gemcitabine for first-line treatment of patients with metastatic pancreatic adenocarcinoma: phase III RESOLVE study. Ann Oncol 32:600–608
8. Van Cutsem E, Tempero MA, Sigal D et al (2020) Randomized phase III trial of pegvorhyaluronidase α with nab-paclitaxel plus gemcitabine for patients with hyaluronan-high metastatic pancreatic adenocarcinoma. J Clin Oncol 38:3185–3194
9. Balachandran VP, Beatty GL, Dougan SK (2019) Broadening the impact of immunotherapy to pancreatic cancer: challenges and opportunities. Gastroenterology 156:2056–2072
10. Le DT, Durham JN, Smith KN et al (2017) Mismatch repair deficiency predicts response of solid tumors to PD-1 blockade. Science 357: 409–413
11. Royal RE, Levy C, Turner K et al (2010) Phase 2 trial of single agent ipilimumab (anti-CTLA-4) for locally advanced or metastatic pancreatic adenocarcinoma. J Immunother 33:828–833
12. Zeisberg M, Neilson EG (2009) Biomarkers for epithelial-mesenchymal transitions. J Clin Invest 119:1429–1437
13. Carstens JL, Correa de Sampaio P, Yang D et al (2017) Spatial computation of intratumoral T cells correlates with survival of patients with pancreatic cancer. Nat Commun 8:15095
14. Knudsen ES, Vail P, Balaji U et al (2017) Stratification of pancreatic ductal adenocarcinoma: combinatorial genetic, stromal, and immunologic markers. Clin Cancer Res 23: 4429–4440
15. Väyrynen SA, Zhang J, Yuan C et al (2021) Composition, spatial characteristics, and prognostic significance of myeloid cell infiltration in pancreatic cancer. Clin Cancer Res 27:1069–1081
16. Clark CE, Hingorani SR, Mick R et al (2007) Dynamics of the immune reaction to pancreatic cancer from inception to invasion. Cancer Res 67(19):9518–9527
17. Wang X, Lang M, Zhao T et al (2017) Cancer-FOXP3 directly activated CCL5 to recruit FOXP3(+)Treg cells in pancreatic ductal adenocarcinoma. Oncogene 36(21):3048–3058

# Chapter 13

# Multiplexed Immunoassay Using Quantum Dots to Monitor Proteins Secreted from Single Cells at Near-Single Molecule Resolution

**Veena Y. Naveen, Tingwei Deng, Vanessa Herrera, and Jered B. Haun**

## Abstract

Single-cell secretion studies find important applications in molecular diagnostics, therapeutic target identification, and basic biology research. One increasingly important area of research is non-genetic cellular heterogeneity, a phenomenon that can be studied by assessing secretion of soluble effector proteins from single cells. This is particularly impactful for immune cells, as secreted proteins such as cytokines, chemokines, and growth factors are the gold standard for identifying phenotype. Current methods that rely upon immunofluorescence suffer from low detection sensitivity, requiring thousands of molecules to be secreted per cell. We have developed a quantum dot (QD)-based single-cell secretion analysis platform that can be used in different sandwich immunoassay formats to dramatically lower detection threshold, such that only one to a few molecules need be secreted per cell. We have also expanded this work to include multiplexing capabilities for different cytokines and employed this platform to study macrophage polarization under different stimuli at the single-cell level.

**Key words** Quantum dot, Multiplexing, Immunoassay, Single cell-secreted proteins, Cytokines, Macrophage polarization heterogeneity

## 1 Introduction

Single-cell analysis methods are important tools used in studies of phenotypic heterogeneity within cell populations [1, 2]. Such heterogeneity has implications on how cells respond to microenvironmental stimuli and is hence crucial in informing diagnostic decisions and therapeutic strategies. Immune cells, in particular, are known to exhibit a large degree of response heterogeneity [3–6]. Soluble effector proteins are markers of immune cell phenotypes, as well as driving factors of varied responses to stimuli [7–9]. Characterizing the secretion of cytokines from immune cells would give useful insight into the molecular regulators of non-genetic phenotypic variability and in turn provide knowledge

Usha N. Kasid and Robert Clarke (eds.), *Cancer Systems and Integrative Biology*, Methods in Molecular Biology, vol. 2660,
https://doi.org/10.1007/978-1-0716-3163-8_13,

on how these phenotypes can be induced and modulated in vivo [10, 11]. However, there are many challenges to monitoring secretion directly from single cells. Ideally, a platform for monitoring protein secretion from single cells would be one that can detect a large panel of proteins with high sensitivity, using standard methodologies that can be implemented easily by many research labs.

Enzyme-linked immunosorbent assays (ELISAs) are the gold standard for detecting secretion of soluble proteins, but these are bulk assays whose averaged results mask the heterogeneity within the populations being characterized. Microwell chamber platforms have been used for isolating single cells and interrogating protein secretion in sandwich immunofluorescence-based platforms [12–14]. However, detection threshold has historically been limited to ~10 pM (or ~3000 molecules per cell), which is two orders of magnitude higher than that of standard bulk ELISAs. Nanomaterial probes exhibit properties that confer numerous advantages to immunoassays over organic fluorophores. For example, quantum dots (QDs) emit a highly photostable luminescence signal, exhibit broad excitation and narrow emission spectra, and are negligibly susceptible to photobleaching [15, 16]. Moreover, QDs can be imaged using standard fluorescence microscopy with only slight modification with an appropriate filter cube that enables ultraviolet excitation. While QD-based assays have been reported that reached detection thresholds as low as 25 fM, these are bulk assays [17] and hence have the same limitations as standard ELISAs. Moreover, they failed to leverage the full suite of advantageous properties of QDs to further reduce detection thresholds or to explore the use of QDs in single-cell secretion assays.

Previously, we developed a QD-based sandwich immunoassay capable of characterizing secretion of the cytokine TNF-$\alpha$ from single macrophages at a resolution of ~180 aM, which corresponds to <1 molecule per cell in a standard microwell array [18]. The platform consisted of a polydimethylsiloxane (PDMS) microwell array in which single macrophage cells were isolated in picoliter microwells. Protein secretion was detected by overlaying the microwell arrays with a glass detection slide containing immobilized antibodies. The detection slides were then incubated with QD immunoconjugates (QD ICs) and secretion was quantified by fluorescence microscopy. QD ICs consist of tetrazine-modified quantum dots conjugated with TCO-modified detection antibodies. Detection sensitivity was enhanced by amplifying binding of protein molecules to the QD ICs. This was achieved by employing a novel bioconjugation method in which the detection antibody is modified with *trans*-cyclooctene (TCO), which is first used to label cells. Tetrazine-modified QDs can then be covalently reacted rapidly and specifically via the catalyst-free bioorthogonal cycloaddition reaction between TCO and tetrazine. This results in multiple QDs attaching each detection antibody, thereby enhancing

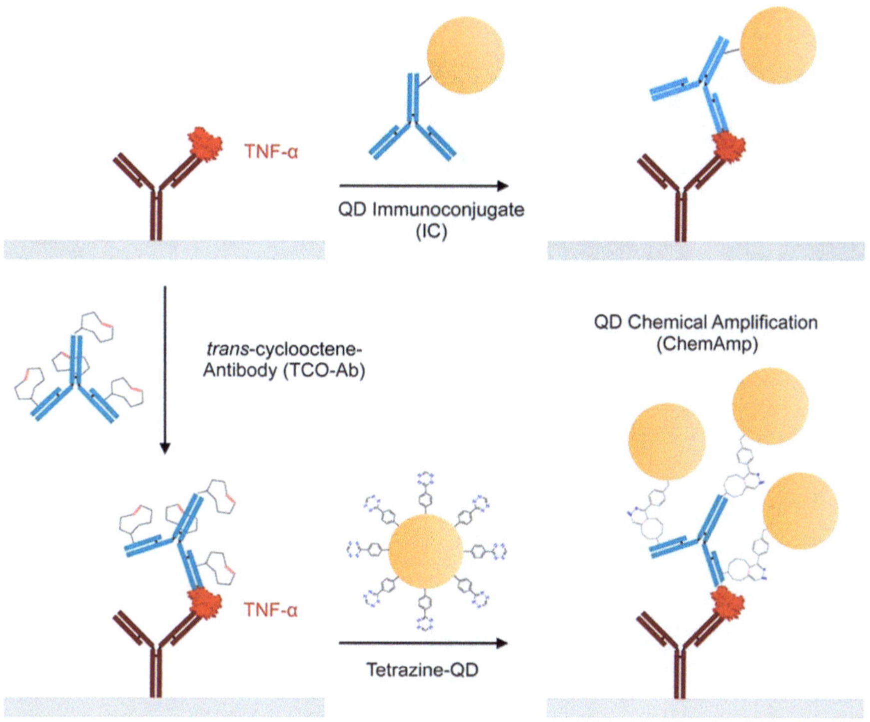


**Fig. 1** Schematic representation of the sandwich detection schemes used for detecting TNF-$\alpha$ (red). (Top) Standard immunoconjugate (IC) format in which the detection antibody (blue) is first attached to the QD. (Bottom) Chemical amplification (ChemAmp) technique in which the detection antibody is modified with trans-cyclooctene (TCO), bound to TNF-$_{\alpha}$, and covalently reacted with tetrazine-modified QDs via bioorthogonal cycloaddition reaction. (*Reprinted with permission from Reference* [18]. *Copyright 2019 Royal Society of Chemistry)*

detection sensitivity. A schematic representation of the sandwich construct is shown in Fig. 1. The performance of the assay was assessed and compared against that of a standard immunofluorescence assay that used an organic fluorophore. The results of this analysis are displayed in Fig. 2.

Subsequently, we developed a multiplexed version of this immunoassay using a multicolored QD panel [19]. This was achieved by adapting the previously described bioconjugation method to three additional targets, allowing the characterization of secretion of TNF-$\alpha$, MCP-1, TGF-$\beta$, and IL-10 (Fig. 3). The multiplexed assay was evaluated by conducting a similar study in which single macrophages were divided into three populations that were induced toward three different stimulation conditions: no stimulation, pro-inflammatory, and pro-healing. The performance indicators of the multiplexed assay are shown in Fig. 4. Finally, we used the Seurat pipeline to cluster single macrophages based on secretion profiles, resulting in seven primary groups (Fig. 5). We also performed principal component analysis and found that the

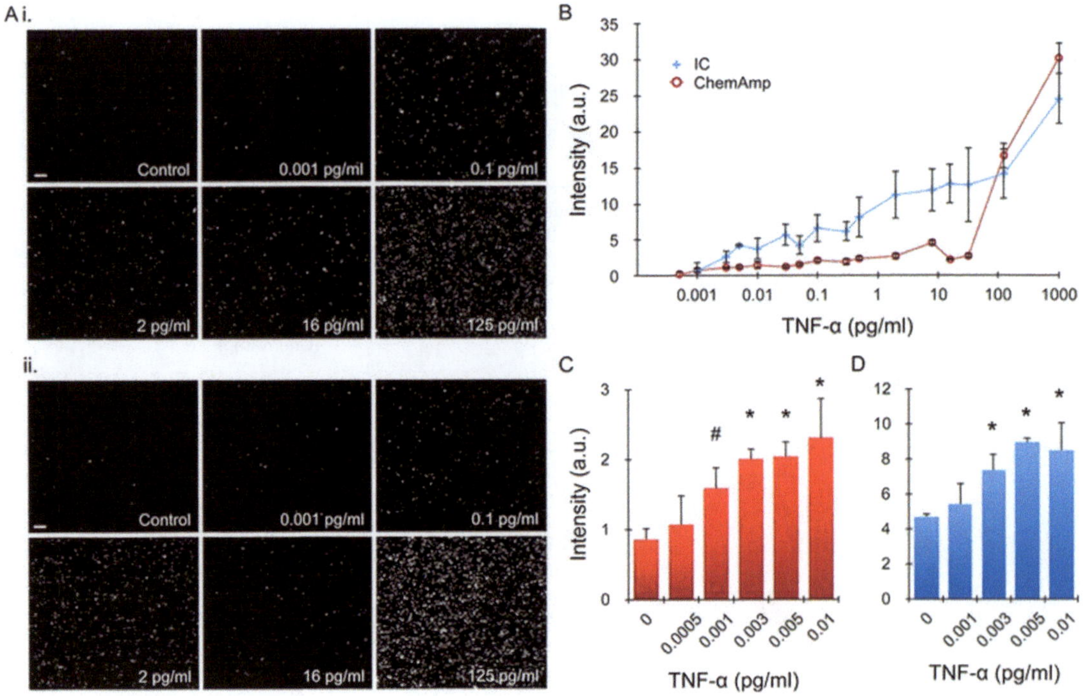


**Fig. 2** (**a**) ELISA results performed using biotinylated detection antibody, avidin-HRP, and TMB substrate. Detection threshold, defined as the lowest TNF-$_{\alpha}$ concentration at which signal was significantly above background, was ~5 pg/mL, or 300 fM. (**b**) Standard immunofluorescence assay performed using rhodamine-labeled detection antibody (TMR), with detection threshold at ~300 pg/mL, or 18 pM. (**c**) QD results for an IC, as well as the ChemAmp technique using TCO and PEG-TCO modified detection antibodies. (*Reprinted with permission from Reference* [18]. *Copyright 2019 Royal Society of Chemistry)*

clusters could be divided into primarily pro-inflammatory (clusters 1–3), pro-healing (clusters 4 and 5), or mixed (clusters 6 and 7) cytokine secretion profiles. However, the mixed profiles were predominantly associated with the pro-healing phenotype.

## 2 Materials

All chemicals and reagents were purchased from Sigma-Aldrich (St Louis, MO) unless otherwise noted.

### *2.1 Antibody Conjugation*

1. Monoclonal mouse capture/detection antibody pairs [BioLegend (San Diego, CA)]:
   (a) Human MCP-1 (capture: IgG1$\kappa$ clone 2H5, detection: IgG1$\kappa$ clone 5D3-F7).
   (b) Human TGF-$\beta$ (capture: IgG1$\kappa$ clone TW7-7H4, detection: IgG1$\kappa$ clone TW46H10).

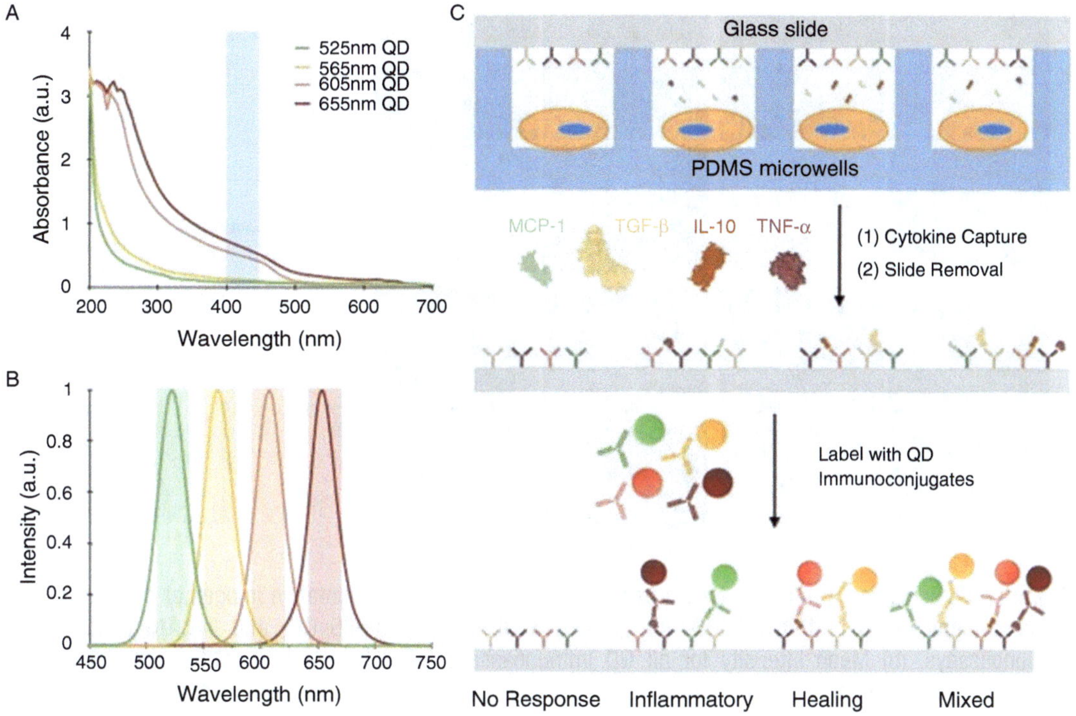


**Fig. 3** Multiplexed detection of soluble proteins using QDs. (**a**) Absorbance and (**b**) emission spectra for the four QDs in the detection panel. The same excitation window, from 400 to 450 nm (blue), was used for each QD. Emission filter windows are overlaid to indicate spectral overlap. (**c**) Schematic of single-cell secretion and detection assays. Inflammatory cytokines are MCP-1 (green, 525 nm QD) and TNF-$_{\alpha}$ (dark red, 655 nm QD). Healing/repair cytokines are TGF-$_{\beta}$ (orange, 565 nm QD) and IL-10 (red, 605 nm QD). (*Reprinted with permission from Reference* [19]. *Copyright 2022 American Chemical Society*)

(c) Human IL-10 (capture: IgG1κ clone JES3-9D7, detection: IgG1κ clone JES312G8).

(d) Human TNF-α (capture: IgG1κ clone MAb1, detection: IgG1κ clone MAb11).

2. NanoDrop 2000 spectrophotometer (Thermo Scientific).
3. Gyromini Nutating Mixer (LabNet International, Inc., NJ).

*2.1.1 Biotinylated Capture Antibodies*

1. Capture antibodies (see above).
2. Sulfo-NHS-LC-Biotin.
3. 1× Phosphate-buffered saline (PBS).
4. 0.1 M sodium bicarbonate (pH 8.4).
5. Zeba Spin desalting columns (Thermo Fisher, MA).

*2.1.2 TCO-Modified Detection Antibodies*

1. Detection antibodies (see above).
2. NHS-TCO (Click Chemistry Tools, AZ).
3. Dimethylformamide (DMF).
4. 1× PBS.

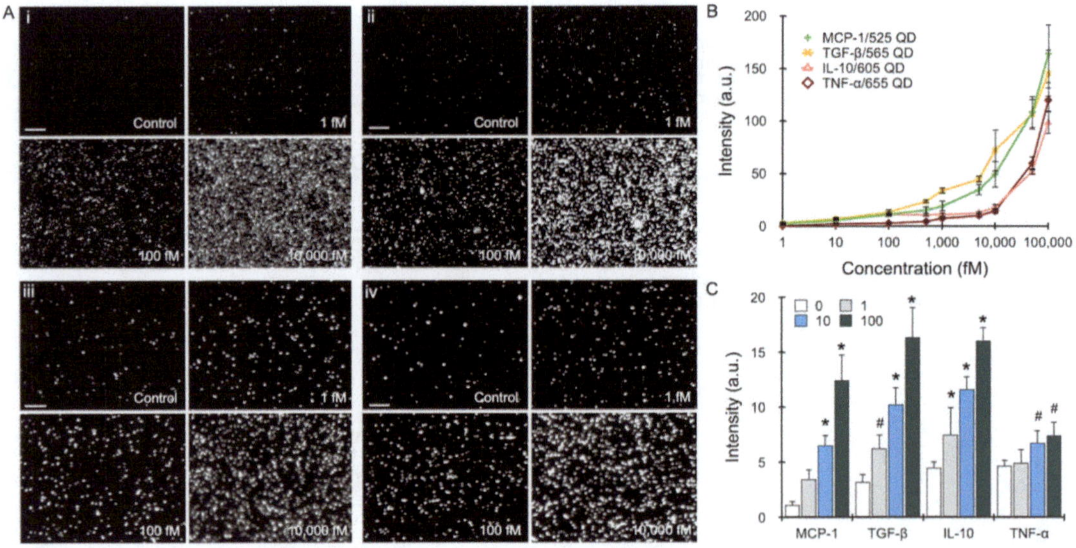


**Fig. 4** Characterization of immunoassays using different QDs. (**a**) Representative images at different protein concentrations for (i) MCP-1/525 QD, (ii) TGF-$_{\beta}$/565 QD, (iii) IL-10/605 QD, and (iv) TNFalpha/655 QD immunoassays. (**b**) Mean intensity for all QD immunoassays, after background subtraction. (**c**) Average intensity in the low-concentration range of 1100 fM, along with controls. Detection sensitivity was 1 fM for the MCP-1/525 QD, TGF-$_{\beta}$/565 QD, and IL-10/605 QD cases, and 10 fM for TNF-$_{\alpha}$/655 QD. Error bars represent the standard error from at least three independent experiments. # denotes $p < 0.05$ and * denotes $p < 0.01$ when compared to control samples. Scale bars are 30 μm. *(Reprinted with permission from Reference* [19]. *Copyright 2022 American Chemical Society)*

### *2.2 Tetrazine-Modified Quantum Dots*

1. Primary amine-terminated QDs (Qdot 525, 565, 605, and 655 ITK Amino PEG) (Thermo Fisher Scientific, Waltham, MA).
2. Tetrazine-NHS.
3. DMF.
4. 1× PBS.
5. 0.01 M sodium bicarbonate.
6. Amicon Ultra-4 centrifugal filtration systems with 100 kD MWCO (EMD Millipore).

### *2.3 QD Immunoconjugates*

1. Tz-modified QDs.
2. TCO-modified detection antibodies.
3. PBS containing 1% bovine serum albumin (PBS+).
4. Sephacryl S-400 gel filtration media (GE Healthcare).
5. AKTA Pure FPLC system (GE Healthcare).
6. Amicon Ultra-15 centrifugal filtration systems with 100 kD MWCO (EMD Millipore).

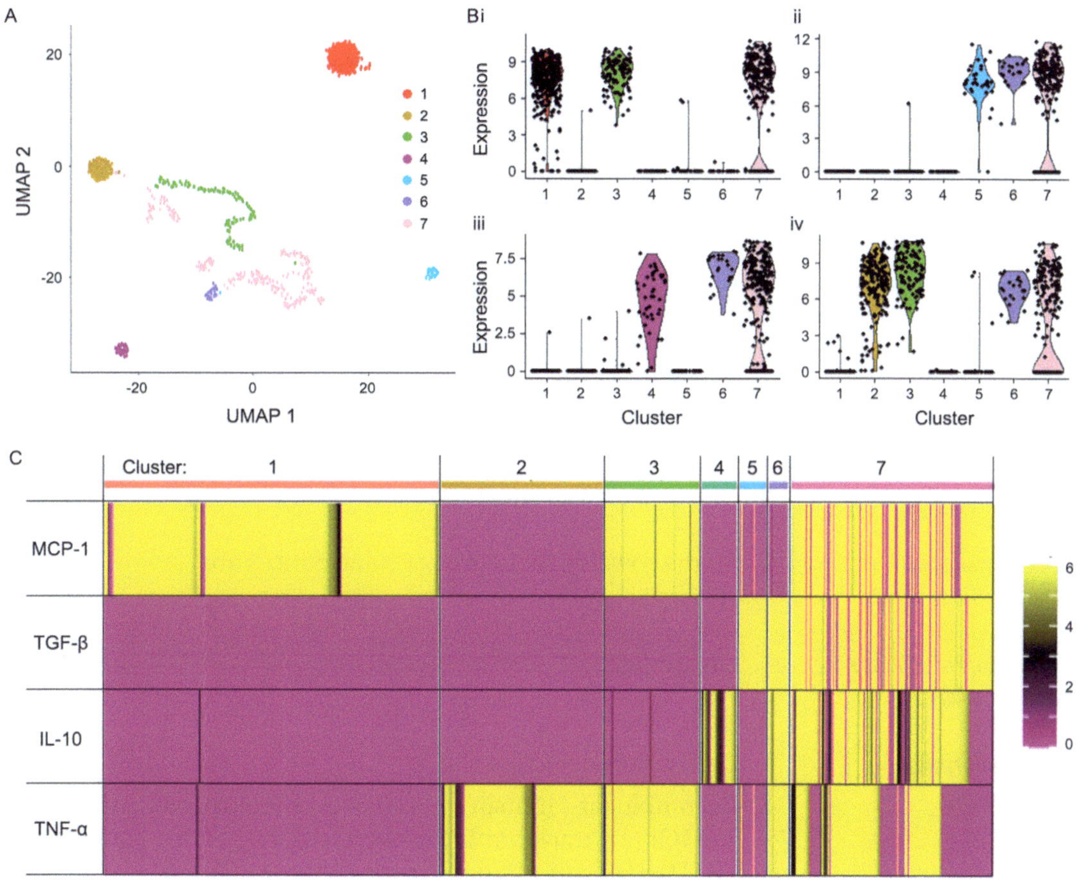


**Fig. 5** Bioinformatic analysis of secretion profiles. (**a**) UMAP projection showing the sizes and distribution of seven distinct clusters. (**b**) Violin plots showing secretion levels across clusters. The shape of the violin plot shows the probability density of expression levels for (i) MCP-1, (ii) TGF-$\beta$, (iii) IL-10, and (iv) TNF-$\alpha$. (**c**) Heatmap showing sizes and secretion profiles arranged by cluster. *(Reprinted with permission from Reference* [19]. *Copyright 2022 American Chemical Society)*

### *2.4 Fabrication of PDMS Microwell Arrays*

1. 3″ silicon wafer (University Wafer, MA).
2. SU-8 50 photoresist (MicroChem, MA).
3. Transparency mask containing clear rectangles (90 μm x 90 μm).
4. UV light (AB & M UV Flood Lamp Exposure System).
5. SU-8 developer.
6. Polydimethylsiloxane (PDMS) and curing agent (Dow Silicones, MI).
7. (Tridecafluoro-1,1,2,2-tetrahydrooctyl)trichlorosilane (Gelest, PA).
8. 70% ethanol.

### 2.5 Preparation of Detection Slides

1. Glass microscope slides (25 mm × 75 mm).
2. Staining dish.
3. Sulfo-SMCC [sulfosuccinimidyl 4-(N-maleimidomethyl)cyclohexane-1-carboxylate].
4. NeutrAvidin (Thermo Fisher Scientific, Waltham, MA).
5. Piranha solution (3% $H_2O_2$ and concentrated $H_2SO_4$ at a 1: 2 volume ratio).
6. Double-deionized water ($dH_2O$).
7. 4% (3-mercapto-propyl)-trimethoxysilane solution in 100% ethanol.
8. 100% ethanol.
9. PBS+.
10. Biotinylated capture antibodies.

#### 2.5.1 Calibration Detection Slides

1. 50 microwell silicone gasket (Grace Bio-Labs).

#### 2.5.2 Single-Cell Assay Detection Slides

1. Grease pen.

### 2.6 Calibration Experiments

1. Calibration detection slides.
2. Recombinant human monocyte chemoattractant protein 1 (MCP-1), transforming growth factor $\beta$ (TGF-$\beta$), interleukin 10 (IL-10), and tumor necrosis factor $\alpha$ (TNF-$\alpha$).
3. PBS+.

### 2.7 U-937 Cell Culture and Single-Cell Secretion Assays

1. Single-cell assay detection slides.
2. Pro-monocytic, human myeloid leukemia cell line U-937 (ATCC, Manassas, VA).
3. 4% Trypsin EDTA (Thermo Fisher Scientific, MA).
4. Phorbol 12-myristate 13-acetate (PMA).
5. RPMI 1640 medium supplemented with 2 mM L-glutamine, 10 nM HEPES, 1 mM sodium pyruvate, 4.5 g/L glucose, 1.5 g/L sodium bicarbonate, and 10% fetal bovine serum (Thermo Fisher, MA).
6. Lipopolysaccharide (LPS).
7. Interleukin 4 (IL-4).

### 2.8 Imaging and Analysis

1. Olympus X83 inverted microscope.
2. QD filter sets:
   (a) 400–450 nm single band exciter for all filter cubes.

  (b) 510–540 nm for QD525 and 590–620 nm for QD605 (Chroma, VT).
  (c) 550–580 nm for QD565 and 640–670 nm for QD655 (Semrock, NY).
3. 40× oil–immersion objective (NA 1.3, Olympus).
4. Orca–R2 CCD camera (Hamamatsu Photonics).
5. MManager control software.
6. ImageJ software with the Fiji package.

### 2.9 Phenotypic Analysis

1. R software environment with Seurat v3 installation.

## 3 Methods

### 3.1 Antibody Conjugations

#### 3.1.1 Biotinylated Capture Antibodies

1. React capture antibodies with 5 molar equivalents of sulfo-NHS-LC-Biotin in PBS and 0.1 M sodium bicarbonate:
   (a) Prepare the reaction solution (total volume: 1250 μL) by first combining 1000 μL capture antibody, 125 μL sodium bicarbonate, and 121.3 μL PBS in a 1.5 mL microcentrifuge tube. Briefly vortex the solution.
   (b) Add 3.7 μL sulfo-NHS-LC-Biotin to the solution. Briefly vortex the solution.
   (c) Incubate the reaction solution on a GyroMini for 3 h. Cover the tube with aluminum foil to minimize light exposure.
2. Purify the modified antibodies using a 5 mL Zeba Spin Desalting Column (7 K MWCO):
   (a) Buffer-exchange the column with 1× PBS and purify the sample according to the manufacturer's instructions.
   (b) Transfer the purified sample to a new 1.5 mL microcentrifuge tubes. Determine antibody concentration by absorption measurement using a NanoDrop 2000 spectrophotometer.
   (c) Store modified antibodies at 4 °C.

#### 3.1.2 TCO-Modified Detection Antibodies

1. React detection antibodies with 30 molar equivalents of NHS-TCO in PBS containing 10% DMF and 0.1 M sodium bicarbonate:
   (a) Prepare the reaction solution (total volume: 1000 μL) by first combining 500 μL capture antibody, 100 μL sodium bicarbonate, and 300 μL PBS in a 1.5 mL microcentrifuge tube. Briefly vortex the solution.
   (b) Add 98.66 μL DMF and 1.34 μL NHS-TCO. Briefly vortex the solution.

(c) Incubate the reaction solution on a GyroMini for 3 h. Cover the tube with aluminum foil to minimize light exposure.

2. Purify the modified antibodies using a 5 mL Zeba Spin Desalting Column (7 K MWCO):

(a) Buffer-exchange the column with 1× PBS and purify the sample according to the manufacturer's instructions.

(b) Transfer the purified sample to a new 1.5 mL microcentrifuge tubes. Determine antibody concentration by absorption measurement using a NanoDrop 2000 spectrophotometer.

(c) Store modified antibodies at 4 °C.

### 3.2 Tetrazine-Modified Quantum Dots

1. Vortex the QD stock solution. In subsequent steps, cover tubes containing QDs with aluminum foil to minimize light exposure.

2. Buffer-exchange and concentrate the QDs in PBS using an Amicon Ultra-4 centrifugal filtration system (100 kD MWCO):

(a) Add 50 μL of QD stock solution and 3 mL of PBS to the Amicon filter.

(b) Centrifuge at 4000 RCF for 7 min. Remove the liquid that passes through the filter. Mix the solute inside the filter using a pipette (to withdraw the particles settled at the bottom).

(c) Repeat **step 2** (b) 2–3 times (until the volume of the concentrated solute inside the filter is approximately 60 μL).

3. React 0.15 nmole of QDs with 500 molar equivalents of NHS-tetrazine in PBS containing 5% DMF and 0.01 M sodium bicarbonate:

(a) Add 790 μL of PBS to the concentrated QDs inside the Amicon filter. Mix the QD solution by pipetting inside the filter. Transfer the solution to a 1.5 mL microcentrifuge tube.

(b) Prepare the reaction solution (total volume: ~1 mL) by first adding 100 μL of 0.01 M sodium bicarbonate to the tube. Briefly vortex the solution.

(c) Add 7.52 μL Tz-NHS and 42.48 μL DMF. Briefly vortex the solution.

(d) Incubate the reaction at room temperature for 3 h on a GyroMini.

(e) Centrifuge at 4000 RCF for 7 min. Remove the liquid that passes through the filter. Mix the solute inside the filter using a pipette (to withdraw the particles settled at the bottom).

**Table 1**
**Sample calibration measurements for Qdot 605. Absorption was measured at 3 wavelengths near the peak of the excitation spectrum. The dilution of the sample was determined by taking the mean of the dilutions obtained from the three linear fits**

| Dilution (*x*) | Absorption (*y*) 580 nm | 590 nm | 600 nm |
|---|---|---|---|
| 0.5 | 0.12133 | 0.16767 | 0.154333 |
| 0.25 | 0.062 | 0.085 | 0.075 |
| 0.125 | 0.032 | 0.042 | 0.036 |
| 0.0625 | 0.017 | 0.02 | 0.016 |
| 0.03125 | 0.01 | 0.012 | 0.01 |
| Linear fit | $y = 0.238\times + 0.0023$ | $y = 0.3349\times + 0.0004$ | $y = 0.3116\times - 0.0021$ |

(f) Repeat **step 3** (e) 3 times.

(g) Transfer the purified Tz-modified QDs to a new 1.5 mL microcentrifuge tube.

(h) Determine the concentration of QDs by absorption measurements using a NanoDrop 2000 spectrophotometer and calibration using the QD stock solution. Sample calibration measurements for Qdot 605 are shown in Table 1.

### *3.3 QD Immunoconjugates*

1. React 0.15 nmole of Tz-QDs with 200 g of TCO-modified detection antibody in PBS+:

   (a) Calculate required volumes of Tz-QD and TCO-modified detection antibody using the respective sample concentrations and combine them in a 1.5 mL microcentrifuge tube.

   (b) Add PBS+ up to a final volume of 1 mL. Briefly vortex the solution.

   (c) Incubate the solution at room temperature for 3 h on a GyroMini.

2. Purify QD ICs using Sephacryl S-400 gel filtration media on an AKTA Pure FPLC system:

   (a) Immediately before purifying QD ICs, prepare the column by running a method with the following settings:

   - Column selection.
     - Show by technique: Gel Filtration.
     - Column type: Any.
     - Column volume: 80.000 mL.
     - Pressure limit pre-column: 1.00 MPa.

- Flow rate: 1.000 mL/min.
- Select "Control the flow to avoid overpressure".
- Inlet A: A1 (buffer: ultrapure water).
- Inlet B: B2 (buffer: 100% ethanol).
- Unit selection.
  - Method Base Unit: CV.
  - Flow Rate Unit: mL/min.
- Monitor settings.
  - UV variable wavelengths.
  - UV 1: 280 nm.

(b) To purify QD immunoconjugates, run a method with the same settings as in 3.3.2.(a), but with the following modifications:

- Column selection.
  - Column volume: 120.637 mL.
  - Pressure limit pre-column: 0.50 MPa.
- Inlet A: A2 (buffer: 1× PBS).
- Inlet B: B1 (buffer: 20% ethanol).
- Monitor settings.
  - UV variable wavelengths.
  - UV 1: 280 nm.
  - UV 2: 400 nm (for Qdots).

(c) Collect fractions eluted when peaks of both $A_{280nm}$ and $A_{592nm}/A_{545nm}/A_{510nm}/A_{640nm}$ appear on the chromatogram at the same time.

(d) Immediately after purifying QD ICs, clean the column by running a method with the same settings as in 3.3.2.

3. Concentrate the purified QD ICs using an Amicon Ultra-15 centrifugal filtration system:

(a) Transfer the fractions collected in **step 2** (c) to the Amicon filter.

(b) Centrifuge at 4000 RCF for 7 min. Remove the liquid that passes through the filter. Mix the solute inside the filter using a pipette (to withdraw the particles settled at the bottom).

(c) Repeat **step 3** (b) Until the solute inside the filter appears colored when visually inspected.

4. Transfer the concentrated QD ICs to a new 1.5 mL microcentrifuge tube. Determine antibody and QD concentrations by absorption measurement using a NanoDrop 2000 spectrophotometer and the same calibration curve used in **step 3.2.3**. (h).
5. Store QD ICs at 4 °C.

### 3.4 Fabrication of PDMS Microwell Arrays

1. Fabricate the silicon wafer:
   (a) Spin-coat a SU-8 50 photoresist onto a 3″ silicon wafer to get a uniform 80 mm thick layer. Bake at 95 °C for 2 h.
   (b) Position the wafer under the transparency mask and expose it to UV light, following MicroChem protocol.
   (c) After 10 min of post-exposure bake at 95 °C, immerse the wafer in SU-8 developer for 5 min to wash off unpolymerized photoresist.
   (d) Bake the clean and dried wafer at 200 °C for 30 min to allow SU-8 to cross-link completely.
2. Silanize the silicon wafer with 10 μL of (tridecafluoro-1,1,2,2-tetrahydrooctyl)trichlorosilane:
   (a) Place the wafer into a desiccator and add 10 μL of (tridecafluoro-1,1,2,2-tetrahydrooctyl)trichlorosilane into the desiccator equally on the top, bottom, left, and right sides.
   (b) Seal the desiccator, apply vacuum, and let the wafer sit for at least 6 h.
3. Prepare the polymer mixture:
   (a) Mix the PDMS and curing agent in a 10:1 ratio.
   (b) De-gas the mixture in a dessicator for at least 20 min.
4. Place the silicon wafer inside a dish and pour the polymer mixture onto it until it covers the wafer completely.
5. Cure the polymer overnight in an oven at 65 °C overnight.
6. Carefully remove the arrays from the silicon wafer by peeling them off using tweezers.
7. Sterilize the arrays by sonicating in 70% ethanol for 10 min.
8. Bake the arrays in an oven overnight to dry them.

### 3.5 Preparation of Detection Slides

1. Wash glass slides with Piranha solution (perform this step inside a fume hood):
   (a) Place the slides on a slide rack holder. Place the holder inside a staining dish.
   (b) Pour Piranha solution into the staining dish and slowly shake the dish. Incubate at room temperature for 15 min.
   (c) Briefly shake the rack slowly and incubate for another 15 min.
   (d) Transfer the slide rack to a new clean staining dish.

(e) Rinse the slides 10 times with $dH_2O$.

(f) Dry the slides in an oven for 1 h at 100 °C.

2. Silanize the glass slides:

(a) Transfer the staining dish containing the slides to a chemical hood.

(b) Pour the (3-mercapto-propyl)-trimethoxysilane solution into the staining dish until the slides are immersed. Incubate for 1 h at room temperature.

(c) Rinse the slides with 100% ethanol.

(d) Transfer the slides to a clean staining dish and incubate for 30 min in an oven at 100 °C.

(e) Transfer the slides to a polystyrene holder and store it in a dessicator.

3. Modify NeutrAvidin with maleimide using 10 molar equivalents of sulfo-SMCC:

(a) Prepare a 1 mg/mL NeutrAvidin solution in PBS according to the manufacturer's instructions.

(b) Prepare a 10 M solution of sulfo-SMCC in ultrapure water according to the manufacturer's instructions.

(c) Combine 1 mL of 1 mg/mL NeutrAvidin solution with 50 μL of 10 M sulfo-SMCC solution.

### 3.6 Calibration Experiments

1. Place a silicon gasket on the silanized slide.
2. Add 5 μL of 1 mg/mL maleimide-modified NeutrAvidin to each well. Incubate at room temperature for 2 h.
3. Wash the wells by adding 7 μL PBS to each well for 15 min. After this step, the slide can be stored overnight at 4 °C. If storing overnight and resuming later, then repeat this step before resuming.
4. Incubate antibodies at 2.5 g/mL: add 6 μL of biotinylated capture antibody to each well (one at a time for single-protein assay or all four together for multiplexed assay).
5. Incubate for 2 h at room temperature.
6. Wash the wells 3 times with 7 μL PBS.
7. Add 6 μL of purified protein in PBS+ (one at a time for single-protein assay or all four together for multiplexed assay) to each well at concentrations ranging from 1 fM to 100,000 fM. Incubate for 2 h at room temperature. Figure 6 shows a sample layout of protein concentrations in wells of a calibration experiment slide.
8. Incubate at room temperature for 30 min.
9. Wash 3 times with ice-cold PBS+.

A

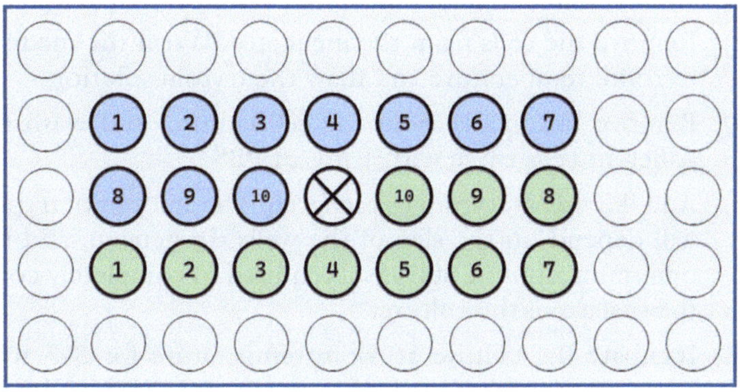


B

| Well label | Protein Concentration (fM) |
|---|---|
| 1 | 0 |
| 2 | 1 |
| 3 | 10 |
| 4 | 100 |
| 5 | 500 |
| 6 | 1000 |
| 7 | 5000 |
| 8 | 10000 |
| 9 | 50000 |
| 10 | 100000 |

**Fig. 6** (**a**) Sample layout of a calibration experiment slide. (**b**) Table showing protein concentrations corresponding to wells labeled 1–10 on the slide

10. Outline the region containing the wells at the bottom of the slide using a marker.
11. Carefully remove the gasket while ensuring that the region of the slide that contained the gasket does not get contaminated.
12. Incubate the slide with QD immunoconjugates at a final concentration of 20 nM (anti-IL-10 and anti-TNF-$\alpha$) or 100 nM (anti-MCP-1 and anti-TGF-$\beta$) for 30 min.
13. Wash 3 times with 7 μL PBS.
14. Cover the marked region of the slide with PBS+ and mount a cover slip. Blot away the excess PBS+.
15. Store the slides at 4 °C. Image them within 1 week.

### *3.7 U-937 Cell Culture and Single-Cell Secretion Assays*

1. Outline a 2 cm × 2 cm square region on the silanized glass slides using a grease pen.
2. Incubate the region in 125 μL of maleimide-modified NeutrAvidin for 2 h at room temperature.
3. Rinse the slide and block with 200 μL of PBS+ for 15 min at room temperature.
4. Incubate with 125 μL of biotinylated capture antibodies for 2 h at room temperature and wash with PBS+.
5. Differentiate U-937 Cells with 50 ng/mL PMA:
   - (a) Culture U-937 cells as recommended in the supplemented RPMI 1640 medium.
   - (b) Prepare a cell solution at a density of 500,000 cells/mL. Transfer 2 mL of the culture to each well of the well plate.
   - (c) Prepare a 0.01 mg/mL PMA solution. Add 10 $\mu$L of the solution to each well. Place the cells in an incubator for 48 h.
   - (d) Replace the PMA-containing media with fresh media. Place the cells in an incubator to rest for 24 h.

6. Seed differentiated U-937 cells onto PDMS microwells:
   (a) Remove the cells from the incubator. Warm the medium to room temperature and thaw the trypsin solution.
   (b) Remove the media from the cell culture, and wash the adherent cells twice with 1 mL of PBS.
   (c) Add 0.5 mL of trypsin to each well. The amount of trypsin will depend on the size of the wells. In general, add the minimum amount of trypsin required to completely cover the surface of the culture.
   (d) Incubate the culture at room temperature for 3–5 min. Check if the cells detach after incubation. If they do not detach, first tap the plate and gently pipette inside the well to help dislodge the cells. If cells are still attached, use a cell scraper to gently scrape the cells off.
   (e) Add 0.5 mL of media to neutralize the trypsin. For other types of well plates, add enough media for a 1:1 ratio of media to trypsin.
   (f) Collect the cell solution and centrifuge it at 120–130 RCF for 5 min.
   (g) Resuspend cells in media, and prepare the seeding solution at a density of 25,000 cells/mL.
   (h) Place the PDMS microwell arrays at the bottom of the wells of a 12-well plate, with the microwells facing up. Seed 1.5 mL of cells in each well.
   (i) Spin the plate at 700 rpm for 5 min to facilitate seeding. Incubate for 1 h to allow the cells to adhere to the bottom of the microwells.
   (j) After 1 h, use the cell scraper to remove cells outside of the microwells.
7. Stimulate the cells with 100 ng/mL LPS or 100 ng/mL interleukin 4 (IL-4):
   (a) Prepare a 400 ng/mL solution of LPS/IL-4. Add 0.5 mL of the solution to the plate wells containing microwell arrays with differentiated cells.
   (b) Briefly pipette directly over the microwells to ensure that the solution enters the wells.
   (c) Incubate the cells at room temperature for 10 min. Then aspirate the media.
8. Wet the detection slides slightly with LPS/IL-4-containing media from the well plate.
9. Gently remove the microwell arrays from the well plate using tweezers. Invert the arrays onto the marked region of the detection slides.

10. Place another glass slide on the exposed side of the microwell arrays. Seal the sandwich construct inside an acrylic housing. Perform this step gently and slowly to ensure proper cell viability.
11. Place the housing in a box and cover it with a damp paper towel to create a humidified chamber.
12. Incubate the setup for 24 h at 37 °C prior to imaging. After 24 h, carefully remove the sandwich construct from the housing.
13. Carefully separate the microwell arrays from the detection slides within an ice-cold bath of PBS+.
14. Incubate the slides with QD ICs at a final concentration of 20 nM (anti-IL-10 and anti-TNF-$\alpha$) or 100 nM (anti-MCP-1 and anti-TGF-$\beta$) for 30 min.
15. Wash 3 times with ice-cold PBS+.
16. Cover the marked region of the slide with PBS+ and mount a cover slip. Blot away the excess PBS+.
17. Store the slides at 4 °C. Image them within 1 week.

### 3.8 Imaging and Analysis

1. Image the glass slides using the Olympus X83 inverted microscope, 40× oil-immersion objective, and single–band QD filter sets.
2. Set the exposure time to 500 ms and capture images using the Orca–R2 CCD camera and MManager control software for at least five fields of view per sample.
3. Quantify mean fluorescence intensity using ImageJ software.
   (a) For single-cell studies, stitch the complete series of bright-field (cells), or fluorescence (protein) images were stitched together using the Fiji package and the grid/collection stitching plug-in.
   (b) Define microwell locations in the stitched fluorescence images by aligning with the stitched bright-field images using the built-in mask, selection, and ROI manager tools.
   (c) Manually select wells containing zero (empty wells), one (single-cell wells), or multiple cells, and store their ROIs.
   (d) Superimpose the defined ROIs for empty and single-cell wells on the fluorescence images from the detection glass slides.
   (e) Obtain mean intensity for each ROI using the built-in multimeasure tool in ImageJ.

4. Determine the detection limit by determining the intensity two standard deviations above the mean intensity for the empty wells.
5. Calibrate protein secretion for the positive population from intensity measurements obtained using purified proteins.

### 3.9 Phenotypic Analysis

The R commands that were used to perform each step listed under this method are shown below the respective subsection. The comments preceding each line indicate the step to which it corresponds.

1. Format the dataset into a Seurat counts matrix:
   (a) Import the Seurat package. Read in the file containing calibrated single-cell secretion data for M(−) cells as a dataframe.
   (b) Create an array of unique “IDs” for each cell, such that the ID keeps track of the cell’s stimulation condition.
   (c) Append the array of IDs to the dataframe.
   (d) Repeat **steps 1**. (a)-(c) for M(LPS) and M(LPS + IL-4) data.
   (e) Combine the three dataframes for the three stimulation conditions.
   (f) Remove the non-secretor cells from the combined dataframes.
   (g) Convert the concentrations from fM to pM.

```
# 1 (a) data_NS <- read.csv("IC_NS.csv", header = T) #
M(-)
# 1 (b)
IDs <- as.data.frame(1:length(data1[,1]))
IDs_NS <- apply(IDs, 1, function(x){paste("sc_NS_",
x, sep = ")}) #
M(-)
# 1 (c)
data_NS[,1] <- IDs
# 1 (d) Repeat 1 (a)-(c) for M(LPS) and M(LPS+IL4)
# 1 (e)
data_combined <- rbind(data_NS, data_LPS, data_LPS
+IL-4)
# 1 (f)
data_combined_secretors <- data_combined[rowSums(df)
==0,]
# 1 (g)
data_combined_secretors_pM <- data_combined_secre-
tors*0.001
```

2. Perform the analysis:
   (a) Create the Seurat object by using the dataframe obtained after **step 1**. (g) as the counts matrix.
   (b) Perform principal component and cluster analysis by following vignettes provided by Butler et al. in (10).

```
# 2 (a)
library(Seurat) library(ggplot2)
obj <- CreateSeuratObject(counts = t(data_combined_-
secretors_pM))
```

## 4 Notes

1. For sufficient protein recovery, ensure that the cap of the Zeba Spin Desalting Column is loosened before placing it in the centrifuge, as described in the manufacturer's instruction manual.
2. When preparing QD ICs, ensure that the molar ratio of detection antibodies to QDs in the reaction is at least 4:1. Excess antibodies are preferred to enable good binding, and a smaller ratio of antibodies to QDs favors multivalent binding. Furthermore, too few antibodies can cause QD-QD cross-linking.
3. QD ICs will be eluted from the gel filtration column after approximately 30–40% of the total column volume has been eluted.
4. Obtaining a very low concentration purified QD ICs may be a result of excessive aggregation due to cross-linking between TCO groups. One way to reduce aggregation is to reduce the molar ratio of NHS-TCO to antibody in the conjugation protocol. Aggregation can also be reduced by attaching TCO via PEG linkers. Finally, the molar ratio of antibody to QD can also be increased, as discussed in **Note 2**.
5. During the washing steps in the calibrations and single-cell secretion experiments, aspirate out the liquid by placing the glass tip gently on its side at the edge of the gasket wells or microwells.
6. To help seal the cover slip onto the glass slide, the cover slip can be outlined with transparent nail polish after it is placed on the slide, and the excess PBS+ is blotted away with a Kim-wipe or other absorbent material.

7. When seeding cells into the PDMS microwell arrays, the concentration of the seeding solution depends on the size of the wells. For larger wells, use a lower concentration to ensure good single-cell loading.
8. When selecting single-cell wells on the ROI manager in ImageJ, be sure to look for cells that are sitting on the sides of wells, as well as lysed cells. Lysed cells can be selected, if the well contained only a single cell.

## References


1. Proserpio V, Mahata B (2016) Single-cell technologies to study the immune system. Immunology 147:133–140
2. Gawad C, Koh W, Quake SR (2016) Single-cell genome sequencing: Current state of the science. Nat Rev Genet 17:175–188
3. Shalek AK, Satija R, Shuga J et al (2014) Single-cell RNA-seq reveals dynamic paracrine control of cellular variation. Nature 510:363–369
4. Patil S, Fribourg M, Ge Y et al (2015) Single-cell analysis shows that paracrine signaling by first responder cells shapes the interferon-beta response to viral infection. Sci Signal 8:ra16
5. Xue Q, Lu Y, Eisele MR et al (2015) Analysis of single-cell cytokine secretion reveals a role for paracrine signaling in coordinating macrophage responses to TLR4 stimulation. Sci Signal 8:ra59
6. Mcwhorter FY, Smith TD, Luu TU et al (2016) Macrophage secretion heterogeneity in engineered microenvironments revealed using a microwell platform. Integr Biol (Camb) 8:751–760
7. Chaplin DD (2010) Overview of the immune response. J Allergy Clin Immunol 125:S3–S23
8. Pettengill MA, Van Haren SD, Levy O (2014) Soluble mediators regulating immunity in early life. Front Immunol 5:457
9. Gessain G, Bleriot C, Ginhoux F (2020) Non-genetic heterogeneity of macrophages in diseases-A medical perspective. Front Cell Dev Biol 8:613116
10. Chen H, Ye F, Guo G (2019) Revolutionizing immunology with single-cell RNA sequencing. Cell Mol Immunol 16:242–249
11. Roy AL (2019) Transcriptional regulation in the immune system: One Cell at a time. Front Immunol 10:1355
12. Love JC, Ronan JL, Grotenbreg GM et al (2006) A microengraving method for rapid selection of single cells producing antigen-specific antibodies. Nat Biotechnol 24:703–707
13. Lu Y, Chen JJ, Mu L et al (2013) High-throughput secretomic analysis of single cells to assess functional cellular heterogeneity. Anal Chem 85:2548–2556
14. Lu Y, Xue Q, Eisele MR et al (2015) Highly multiplexed profiling of single-cell effector functions reveals deep functional heterogeneity in response to pathogenic ligands. Proc Natl Acad Sci U S A 112:E607–E615
15. Medintz IL, Uyeda HT, Goldman ER et al (2005) Quantum dot bioconjugates for imaging, labelling and sensing. Nat Mater 4:435–446
16. Resch-Genger U, Grabolle M, Cavaliere-Jaricot S et al (2008) Quantum dots versus organic dyes as fluorescent labels. Nat Methods 5:763–775
17. Hu M, Yan J, He Y et al (2010) Ultrasensitive, multiplexed detection of cancer biomarkers directly in serum by using a quantum dot-based microfluidic protein chip. ACS Nano 4:488–494
18. Herrera V, Hsu SJ, Rahim MK et al (2019) Pushing the limits of detection for proteins secreted from single cells using quantum dots. Analyst 144:980–989
19. Herrera V, Hsu SJ, Naveen VY et al (2022) Multiplexed detection of secreted cytokines at near-molecular resolution elucidates macrophage polarization heterogeneity. Anal Chem 94:658–668

# Chapter 14

# Isolation and Identification of Plasma Extracellular Vesicles Protein Biomarkers

**Michelle V. Lihon, Marco Hadisurya, Xiaofeng Wu, Anton Iliuk, and W. Andy Tao**

## Abstract

Extracellular vesicles (EVs) have emerged as a valuable source for disease biomarkers and an alternative drug delivery system due to their ability to carry cargo and target specific cells. Proper isolation, identification, and analytical strategy are required for evaluating their potential in diagnostics and therapeutics. Here, a method is detailed to isolate plasma EVs and analyze their proteomic profiling, combining EVtrap-based high-recovery EV isolation, phase-transfer surfactant method for protein extraction, and mass spectrometry qualitative and quantitative strategies for EV proteome characterization. The pipeline provides a highly effective EV-based proteome analysis technique that can be applied for EV characterization and evaluation of EV-based diagnosis and therapy.

**Key words** Extracellular vesicles (EVs), Exosomes, Plasma, Extracellular vesicles total recovery and purification (EVtrap), Mass spectrometry, Data-dependent acquisition (DDA), Data-independent acquisition (DIA), Automatic gain control (AGC) strategy, Gas-phase fractionated (GPF) spectral library, Spectronaut Pulsar search, Parallel reaction monitoring (PRM), Isobaric labeling, Tandem mass tag labeling (TMT labeling), Quantitative proteomics, Pharmacokinetics

## 1 Introduction

EVs are small lipid bilayer-enclosed particles, classified into two primary categories exosomes and microvesicles [1]. EVs are released by almost every cell type and are present in body fluids, including plasma, urine, and saliva [1]. Their primary role is intracellular communication by transporting various active biomolecules from producer cells to recipient cells [1, 2]. EVs are enriched in nucleic acids, lipids, and proteins that reflect the parent cell's current state, providing valuable real-time information on cellular proteome changes [3, 4]. Since there is evidence of EVs involved in several diseases like cardiovascular disease, cancer, autoimmune, neurodegenerative diseases, etc., their informational

Usha N. Kasid and Robert Clarke (eds.), *Cancer Systems and Integrative Biology*, Methods in Molecular Biology, vol. 2660, https://doi.org/10.1007/978-1-0716-3163-8_14,

cargo possesses diagnostic potential [5–8]. Due to EV's abundance in biofluids and its potential as a disease biomarker, there has been increased interest in exploring EVs as liquid biopsies to assess the therapeutic efficacy of a drug [9, 10]. Therefore, the proper isolation, identification, and quantitative strategy of EV proteome are required to characterize EVs further and explore their potential as disease biomarkers.

Proteomics profiling is a powerful tool to monitor and quantify changes in the proteome or specific disease biomarkers and discover novel disease biomarkers. There are two experimental proteomics methods for data acquisition and quantification, data-dependent acquisition (DDA) and data-independent acquisition (DIA). DDA selects N most abundant peptides for fragmentation, generating peptide-specific MS2 spectra. The DDA generated spectra are typically used for protein library generation and identification based on search engine algorithms applied to an existing protein database. In contrast, the DIA method is a powerful alternative to the DDA method for adequate qualitative and quantitative proteomics studies, perfect for monitoring and discovering disease biomarkers [11, 18]. DIA method consists of the fragmentation of all peptides resulting in complex MS2 fragmentation spectra that generate a spectral library to extract information from the highly rich data and allow for quantification at the MS2 level [11, 12]. Gas-phase fractionated (GPF) library is a DIA strategy for data acquisition, which has the advantage of generating an up-to-date protein library that is experiment-specific [13]. Like parallel reaction monitoring (PRM), GPF also provides targeted quantitative proteomics data for all peptides inside the scanning range, which helps produce a more comprehensive profile of the proteome.

Here is a method and strategy to isolate plasma EVs and analyze them by proteomic profiling. For the EV isolation, our group developed a novel chemical affinity magnetic bead-based (non-antibody) method called extracellular vesicles total recovery and purification (EVtrap). This method was designed to be fully compatible with LC-MS-based EV proteome analysis, providing high recovery and purity levels necessary for successful EV proteome analysis [14, 15]. For EV proteome quantification, we used mass spectrometry DDA and isobaric labeling (TMT labeling) to quantify changes in EV protein content (i.e., samples from individuals or cellular states). As an alternative to the DDA method, we also provide an optimized DIA label-free quantitative proteomic method that can increase coverage, sensitivity, and reproducibility for EV characterization [18]. In this chapter, we detail the procedure for effective plasma EV proteome analysis using EVtrap, phase-transfer surfactant-based EV lysis, protein extraction, and DDA and DIA mass spectrometry analysis (*see* Fig. 1 for the overall workflow).

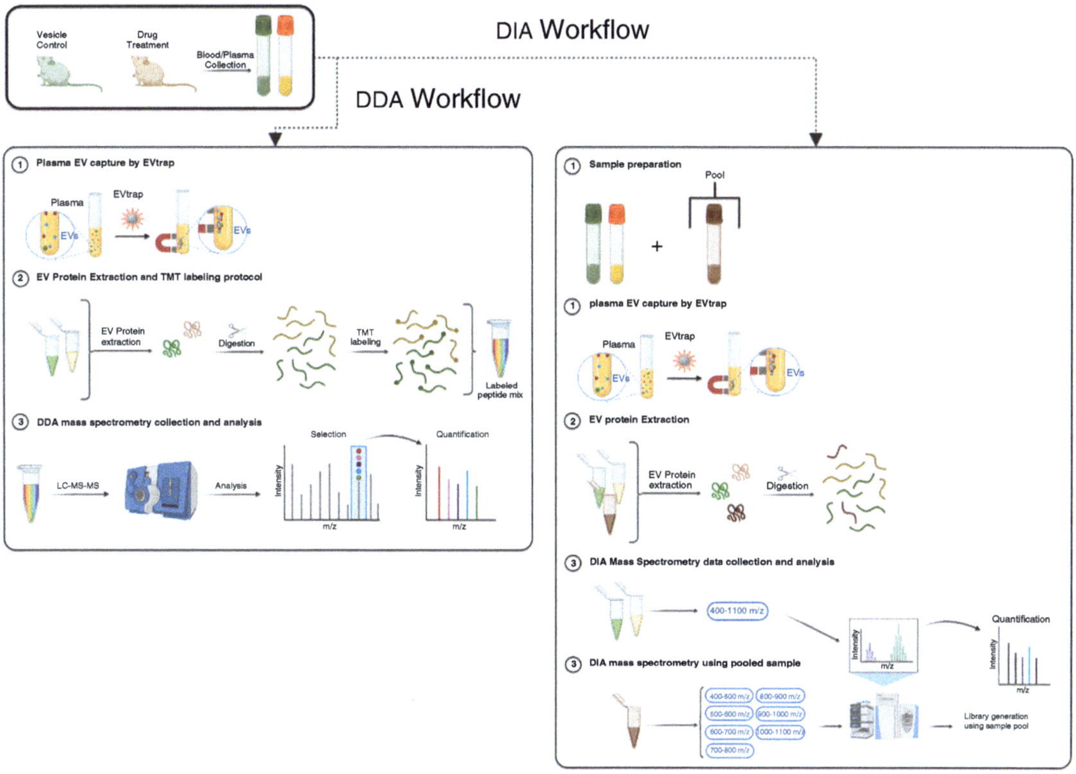


**Fig. 1** Workflow overview of DDA and DIA plasma EV proteomics analysis procedure

## 2 Materials

1. EVtrap EV enrichment beads (Tymora Analytical).
2. EVtrap elution solution (200 mM TEA in water) (**Note 1**. This solution is recommended to be prepared fresh).
3. 500 mM Tris-HCl stock buffer, pH 8.5. To prepare 100 mL, dissolve 6.05 g of Tris base in 80 mL of ultrapure water, then adjust pH to 8.5 with 5 M hydrochloric acid. Bring to a final volume of 100 mL by adding ultrapure water. Stock buffer can be stored at room temperature.
4. 120 mM SDC (sodium deoxycholate) stock solution. To prepare 10 mL, dissolve 500 mg of SDC in 10 mL ultrapure water. Stock solution can be stored at room temperature for up to 6 months.
5. 120 mM SLS (sodium lauroyl sarcosinate) stock solution. To prepare 10 mL, dissolve 352 mg of SLS in 10 mL of ultrapure water. Stock solution can be stored at room temperature for up to 6 months.

6. 200 mM TCEP (Tris(2-carboxyethyl)phosphine hydrochloride) stock solution. To prepare 100 μL, dissolve 6 mg of TECP in 100 μL of ultrapure water. Stock solution can be prepared ahead of time and stored for up to 6 months at − 20 °C. Freeze–thaw is not recommended; discard any remaining TCEP stock solution.
7. 300 mM CAA (chloroacetamide) stock solution. To prepare 100 μL, dissolve 3 mg of CAA in 100 μL of 50 mM TEAB. The stock solution must be prepared fresh and used immediately.
8. PTS lysis solution. To prepare 100 μL of PTS lysis solution, add the following stock solutions in that order below to 54 μL of ultrapure water immediately before use. Vortex after each added ingredient.

   10 μL 500 mM Tris–HCl stock solution, pH 8.5.

   10 μL 120 mM SDC (sodium deoxycholate) stock solution.

   10 μL 120 mM SLS (sodium lauroyl sarcosinate) stock solution.

   5 μL 200 mM TCEP (Tris(2-carboxyethyl)phosphine hydrochloride) stock solution.

   10 μL 300 mM CAA (chloroacetamide) stock solution
9. 50 mM TEAB (triethylammonium bicarbonate). To prepare 10 mL of 50 mM TEAB, dilute 500 μL 1 M TEAB stock and bring to a final volume of 10 mL by adding ultrapure water.
10. Bicinchoninic acid (BCA) protein assay kit (Life Technologies, Millipore-Sigma).
11. Proteomics grade trypsin (Millipore-Sigma, Promega).
12. Lys-C—lysyl endopeptidase (Wako Chemicals USA).
13. Acetonitrile.
14. Ethyl acetate.
15. Trifluoroacetic acid (TFA).
16. C18 loading solution (0.1% TFA, 3% acetonitrile in water).
17. Loading solution can be stored at room temperature for up to 6 months.
18. C18 elution solution (0.1% TFA, 50% acetonitrile in water).
19. Loading solution can be stored at room temperature for up to 6 months.
20. LC-MS loading solution (0.05% TFA, 3% acetonitrile or 0.1% formic acid, 3% acetonitrile in water).
21. 1.7 mL and 0.6 mL nonstick microfuge tubes.
22. Centrifuge capable of 16,000 × g speed.
23. Thermal shaker capable of reaching 95 °C.

24. Microplate reader for visible spectrum range (BioTek, Fisher Scientific).
25. Sep-Pak C18 column (Waters) or Top-Tip C18 tips (GlyGen).
26. Centrifugal vacuum concentrator (Labconco).
27. Magnetic separator rack (Promega, VWR).
28. Anhydrous acetonitrile (ACN, Thermo Fisher Scientific).
29. Perseus or MSStats software.
30. *Optional*: Tandem Mass Tag Reagent (Thermo Fisher Scientific).
31. *Optional*: Q-Exactive HF-X mass spectrometer (Thermo Fisher Scientific).
32. *Optional*: Ultracentrifuge capable of 100,000 × g speed.

## 3 Methods

### 3.1 Enrichment of Extracellular Vesicles (EVs)

1. Thaw the EV-containing plasma sample of interest (**Note 2**. We found that the minimum volume necessary for successful EV proteome analysis starts at 5–10 μL of plasma or serum. If possible, larger volumes of samples should be used. *Alternative approach*: If DIA analysis is going to be performed, an additional minimum of 3 μL per sample is needed for the GPF DIA library generation. The GPF DIA library sample needs to be combined as a sample pool to proceed to the next step).
2. Centrifuge plasma samples at 2500 × g for 10 min at room temperature and carefully collect the supernatant. This step is essential to remove platelets and apoptotic bodies from the plasma sample.
3. Dilute the supernatant (cleared plasma) 20-fold with the EVtrap plasma dilution buffer.
4. Add the EVtrap magnetic beads to the diluted supernatant (4 μL beads per 10 μL of original plasma) and incubate end-over-end for 30 min to 1 h.
5. Use a magnetic separator to separate the beads and remove the solution. The EVs will be bound to the EVtrap magnetic beads.
6. Wash the beads with the EVtrap plasma dilution buffer by vortexing the beads for 3–5 s. Then remove the wash solution using a magnetic separator.
7. Wash the beads with 1XPBS by vortexing the beads for 3–5 s. Then remove the wash solution using a magnetic separator.
8. Add 100–400 μL of EVtrap elution solution (200 mM TEA in water) to the beads (the elution volume should be enough to fully resuspend the beads and allow for efficient interaction, at least 3–5× original beads volume). Incubate for 10 min by vigorous shaking.

9. Collect the eluted sample (contains EVs) and dry the eluted sample completely in a vacuum centrifuge. It helps to freeze the eluted sample at −80 °C for 5–10 min before drying (**Note 3**. Ultracentrifugation is an alternative method for EVs isolation. Take the centrifuge pre-cleared plasma from **step 2** and centrifuge at 100,000 × g for 2 h at 4 °C. Remove and discard supernatant without disturbing the EV pellet. Add 1 mL of cold 1XPBS and to wash the EV pellet. Centrifuge the sample again at 100,000 × g for 2 h at 4 °C. Remove and discard the supernatant and use the pellet for EV lysis, as detailed in Subheading 3.2).

### 3.2 EV Lysis

1. Prepare fresh PTS lysis solution (**Note 4**. 100 μL of this solution per sample is enough for most EV samples).
2. Add the appropriate volume of lysis solution (e.g., 100 μL) to the EV pellet and resuspend the pellet by vortexing for several seconds.
3. Incubate at 95 °C for 5 min in the dark to lyse the EVs.
4. Cool the sample to room temperature and dilute it fivefold with fresh 50 mM TEAB or 50 mM ammonium bicarbonate (ensure pH is ~7.8–8.2 by using pH paper).
5. Prepare protein concentration standards following the BCA protein assay kit's instructions.
6. For each sample prepared in **steps 1–4**, prepare one nonstick microfuge tube to test the protein concentration, plus an extra tube for the blank sample.
7. Add 10 μL of each sample to the clean 0.6 mL tube. Also, add 10 μL of the PTS lysis solution diluted fivefold with 50 mM TEAB to the blank tube.
8. Prepare the protein concentration assay solution by mixing 1.25 mL of BCA assay reagent A with 25 μL of BCA assay reagent B (50:1 mix).
9. Add 200 μL of the BCA assay mix from step 8 into each concentration assay sample and blank tube. Vortex the samples.
10. Put all the assay tubes in a 37 °C water bath or a shaker for 30 min.
11. Transfer 200 μL of each protein concentration standards assay and samples into a microplate. Measure protein concentration of the EV lysate using the microplate reader at 562 nm. Subtract the reading from the blank tube and compare it to standards to calculate the lysate concentration.

### 3.3 Protein Digestion and Surfactant Removal

1. Add the appropriate amount of Lys-C (as suggested by the manufacturer) to the diluted sample and incubate at 37 °C for 3 h.
2. Add trypsin into the sample at a 1:50 ratio of trypsin to protein w/w (e.g., 1 μg trypsin per 50 μg protein to be digested). Digest overnight at 37 °C (best to digest 12–16 h).
3. Acidify the sample with 10% TFA to pH <3 (e.g., ~1% final TFA concentration) and vortex the sample.
4. Add equal parts of ethyl acetate according to the sample volume, and shake the sample vigorously for 2 min. Then spin down the sample at max speed (~16,000 × g) for 3 min.
5. Carefully pipette out the organic phase on top and discard it (**Note 5**. Ensure not to disturb the intermediate, which contains a portion of the peptides).
6. Dry the sample completely using a centrifugal vacuum concentrator to remove any remaining organic solvent.

### 3.4 Peptide Desalting

1. The sample can be desalted with a disposable Sep-Pak C18 column (Waters) or a disposable Top-Tip C18 tip (GlyGen).
2. Condition the column or the tip with one column volume of 0.1% TFA in 80% acetonitrile, followed by two volumes of 0.1% TFA in 3% acetonitrile. All conditioning steps can be done quickly by pushing the solution through the column using a pipette bulb or a tip.
3. Load the acidified sample onto the column/tip with a slow drip (see manufacturer's instructions) by letting gravity load the sample or by slow spin down in a centrifuge.
4. Wash column/tip thrice with 0.1% TFA in 3% acetonitrile each time. This step should be done quickly by pushing the solution through.
5. Elute the sample with one column volume of 0.1% TFA in 50% acetonitrile into a 1.7 mL nonstick microfuge tube. Do this step with a slow drip or low-speed centrifugation and collect the eluent.
6. Dry the sample completely using a centrifugal vacuum concentrator before proteome analysis (**Note 6**. To dry the sample faster, freeze the eluted samples at −80 °C for 5–10 min, and then place the sample in the centrifugal vacuum. Alternatively, the sample can be dried by using a lyophilizer).

### 3.5 Tandem Mass Tag (TMT) Labeling

1. Dissolve TMT reagents in anhydrous acetonitrile according to the manufacturer's instructions (e.g., dissolve 800 μg of labeling reagent in 41 μL of anhydrous acetonitrile).

2. To the dry desalted samples, resuspend in 25 μL in 50 mM TEAB and corresponding 50 μg of TMT reagent (**Note 7**. Apply the first channel of reporter ion at m/z 126 to the experimental control. The other channels can be occupied by experimental samples, e.g., samples corresponding to different individuals or different time points).
3. Incubate samples for 1 h at 25 °C. After incubation, quench the reaction by adding hydroxylamine and incubating for 15 min at 25 °C.
4. Combine all samples in a nonstick microfuge tube. Dry the sample completely using a centrifugal vacuum. Then desalt the sample following the steps described above in Subheading 3.4.
5. The desalted sample is ready for LC-MS analysis via data-dependent acquisition (DDA) (optimal resuspension solution is 0.05% TFA in 3% acetonitrile or 0.1% formic acid in 3% acetonitrile).

### 3.6 Alternative Approach: Mass Spec Analysis Via Data-Independent Acquisition (DIA)

1. The desalted sample is ready for LC-MS analysis (optimal resuspension solution is 0.05% TFA in 3% acetonitrile or 0.1% formic acid in 3% acetonitrile) running a linear 85-min gradient of 5–35% buffer B at a flow rate of 300 nL/min.
2. A Q-Exactive HF-X mass spectrometer (Thermo Fisher Scientific) was used for analysis. The mobile phase buffer consists of 0.1% formic acid in HPLC grade water (buffer A) and an eluting buffer containing 0.1% formic acid in 80% (vol/vol) acetonitrile (buffer B).
3. For each individual sample, utilize the mass spec in the data-independent mode, in which a full-scan MS (polarity, positive; scan range 390 to 1010 m/z with the resolution of 60,000; automatic gain control target (AGC), 1E6, maximum injection time, 60 ms; spectrum data type, centroid) followed by MS/MS with 8.0 m/z staggered-isolation windows schemes (polarity, positive; 15,000 resolution; normalized collision energy, 27%; AGC, 1E6, maximum injection time, 20 ms; loop count, 75; spectrum data type, centroid).
4. For the pooled sample, run the mass spec in the data-independent mode using the same setting as above with modification at the MS1 scan ranges ($n = 6$), 395–505, 495–605, 595–705, 695–805, 795–905, and 895–1005 m/z followed by MS/MS with 4 m/z staggered-isolation windows schemes [11].
5. Convert all RAW files into HTRMS files and enable MS2 demultiplexing using the HTRMS Converter for a faster Spectronaut search (Biognosys, v15, Switzerland) [21]. The effective isolation window is only 2 m/z after demultiplexing.

6. Generate the GPF spectra library from the pooled sample ($n = 6$) using Spectronaut Pulsar search with a default setting. Set the iRT reference strategy to the empirical iRT database with deep learning-assisted iRT regression as a backup.
7. Search the DIA data using the direct DIA or the GPF spectral library in Spectronaut (Biognosys) using a default setting.
8. Perform statistical analysis in Perseus [16] or MSStats software [17].

## 4 Additional Notes

1. There might not be a visible pellet if performing ultracentrifugation as an alternative to EVtrap for EV isolation. For this reason, leave at least 20–30 μL of solution from the bottom fraction when removing the supernatant.
2. PTS lysis solution might appear cloudy after the addition of TCEP. This seems to be expected; therefore, after adding TCEP vortexed solution immediately, the sample will become clear again.
3. To determine EV protein concentration accurately, it is essential to account for any interference from PTS components on the BSA assay. Use PTS lysis solution diluted fivefold with 50 mM TEAB as a blank and prepare protein concentration standards using diluted PTS (to mimic EV sample matrix). Also, to reduce interference of the sample, dilute the sample fivefold with 50 mM TEAB before the BCA assay.
4. After removing the ethyl acetate-based surfactant (*see* Subheading 3.3, **step 6**), it is not necessary to dry the sample; however, removing at least 80% of the volume guarantees a good sample recovery after the desalting step.
5. To improve digestion efficiency and higher protein/peptide identification number, use sequential digestion using Lys-C and trypsin.
6. Only proteomics-grade trypsin should be used for all proteomics experiments since it does not undergo autolytic activity (self-digestion) and has greater proteolytic specificity. Therefore, using lower-grade trypsin can be ineffective for proteomic studies.
7. If performing DIA workflow, direct DIA is an alternative strategy to GPF-DIA that does not require the preparation and processing of pooled sample for building an experimental-specific library, significantly reducing the instrument analysis time. For direct DIA analysis, mass spectrometer parameters can be set as described [19], full-scan MS (polarity, positive;

scan range 390 to 1010 m/z with the resolution of 60,000; automatic gain control target (AGC), 1E6, maximum injection time, 60 ms; spectrum data type, centroid), followed by MS/MS with 8.0 m/z staggered-isolation windows schemes (polarity, positive; 15,000 resolution; normalized collision energy, 27%; AGC, 1E6, maximum injection time, 20 ms; loop count, 75; spectrum data type, centroid).

8. For DIA data analysis, there is multiple software available, including Spectronaut [21], which you would have to purchase, or MaxDIA [20], which is free, among others.

## Acknowledgments

This project has been funded by NIH grants 3RF1AG064250 to W.A.T. and by Purdue Institute for Cancer Research (PICR).

### References

1. Bazzan E, Tinè M, Casara A et al (2021) Critical review of the evolution of extracellular vesicles knowledge: From 1946 to today. Int J Mol Sci 22:6417
2. Mallia A, Gianazza E, Zoanni B et al (2020) Proteomics of extracellular vesicles: update on their composition, biological roles and potential use as diagnostic tools in atherosclerotic cardiovascular diseases. Diagnostics 10:843
3. Mathilde M, Martin-Jaular L, Lavieu G, Théry C (2019) Specificities of secretion and uptake of exosomes and other extracellular vesicles for cell-to-cell communication. Nat Cell Biol 21: 9–17
4. Herrmann IK, Wood MJA, Fuhrmann G (2021) Extracellular vesicles as a next-generation drug delivery platform. Nat Nanotechnol 16:748–759
5. Lu M, DiBernardo E, Parks E et al (2021) The role of extracellular vesicles in the pathogenesis and treatment of autoimmune disorders. Front Immunol. https://doi.org/10.3389/fimmu.2021.566299
6. Hill AF (2019) Extracellular vesicles and neurodegenerative diseases. J Neurosci 39(47): 9269–9273. https://doi.org/10.1523/JNEUROSCI.0147-18.2019
7. Jansen F, Nickenig G, Werner N (2017) Extracellular vesicles in cardiovascular disease potential applications in diagnosis, prognosis, and epidemiology. Circulation Res 120:1649–1657. https://doi.org/10.1161/CIRCRESAHA.117.310752
8. Jabalee J, Towle R, Garnis C (2018) The role of extracellular vesicles in cancer: cargo, function, and therapeutic implications. Cell 7(8): 93. https://doi.org/10.3390/cells7080093
9. Rowland A, Ruanglertboon W, van Dyk M et al (2019) Plasma extracellular nanovesicle (exosome)-derived biomarkers for drug metabolism pathways: a novel approach to characterize variability in drug exposure. Br J Clin Pharmacol 85(1):216–226. https://doi.org/10.1111/bcp.13793
10. Useckaite Z, Rodrigues AD, Hopkins AM et al (2021) Role of extracellular vesicle-derived biomarkers in drug metabolism and disposition. Drug Metab Dispos 49:961–971
11. Pino L, Just SC, MacCoss MJ, Searle BC (2020) Acquiring and analyzing data independent acquisition proteomics experiments without spectrum libraries. Mol Cell Proteomics 19(7):1088–1103
12. Doerr A (2015) DIA mass spectrometry. Nat Methods 12:35. https://doi.org/10.1038/nmeth.3234
13. Searle BC, Swearingen KE, Barnes CA et al (2020) Generating high quality libraries for DIA MS with empirically corrected peptide predictions. Nat Commun 11:1548. https://doi.org/10.1038/s41467-020-15346-1
14. Iliuk A, Wu X, Li L et al (2020) Plasma-derived extracellular vesicle phosphoproteomics through chemical affinity purification. J Proteome Res 19:2563–2574

15. Wu X, Li L, Iliuk A, Tao WA (2018) Highly efficient phosphoproteome capture and analysis from urinary extracellular vesicles. J Proteome Res 17:3308–3316
16. Tyanova S, Temu T, Sinitcyn P et al (2016) The Perseus computational platform for comprehensive analysis of (prote)omics data. Nat Methods 13:731–740
17. Choi M, Chang CY, Clough T et al (2014) MSstats: an R package for statistical analysis of quantitative mass spectrometry-based proteomic experiments. Bioinformatics 30:2524–2526
18. Hadisurya M, Lee Z-C, Luo Z, et al (2023) Data-independent acquisition phosphoproteomics of urinary extracellular vesicles enables renal cell carcinoma grade differentiation. Mol Cell Proteomics 100536. https://doi.org/10.1016/j.mcpro.2023.100536
19. Bekker-Jensen DB, Bernhardt OM, Hogrebe A, et al (2020) Rapid and site-specific deep phosphoproteome profiling by data-independent acquisition without the need for spectral libraries. Nat Commun 11:1–12
20. Sinitcyn P, Hamzeiy H, Salinas Soto F, et al (2021) MaxDIA enables library-based and library-free data-independent acquisition proteomics. Nat Biotechnol 39:1563–1573
21. Bruderer R, Bernhardt OM, Gandhi T, et al (2015) Extending the limits of quantitative proteome profiling with data-independent acquisition and application to acetaminophen-treated three-dimensional liver microtissues. Mol Cell Proteomics 14:1400–1410

# Chapter 15

# Discovery of a Hidden Proinflammatory Signaling Proteome Using a Large-Scale, Targeted Antibody Microarray Platform

**Catherine Jozwik, Ofer Eidelman, and Meera Srivastava**

## Abstract

Dynamic post-translational processes regulate protein expression in eukaryotic cells. However, the processes are difficult to assess on a proteomic scale because protein levels actually reflect the sum of individual biosynthesis and degradation rates. These rates are presently hidden from the conventional proteomic technologies. We present here a novel and dynamic, antibody microarray-based time-resolved approach to simultaneously measure not only the total protein changes but also the rates of biosynthesis of low abundance proteins in the proteome of lung epithelial cells. In this chapter, we describe the feasibility of this technique by investigating the complete proteomic kinetics of 507 low abundance proteins in cultured cystic fibrosis (CF) lung epithelial cells using $^{35}$[S] methionine or $^{32}$[P] and the consequences of repair by gene therapy with [wildtype] CFTR. This novel antibody microarray-based technology identifies relevant, hidden proteins whose regulation by the CF genotype would never have been detected by simple measurements of total proteomic masses.

**Key words** Proteomics, Antibody microarray, Antibodies, Epithelial cells, Biomarkers, Bioinformatics

## 1 Introduction

In eukaryotic cells, there is a profound disconnect between genomic and proteomic information when samples are collected at one time point. In yeast, Gygi et al. [1] were among the first to document the lack of sufficient correlation between quantitative mRNA data and cognate protein expression. In a study on gal mutants in yeast, the overlap in mRNA and protein expression for hundreds of identified genes and proteins is only 10–15% [2]. Greenbaum et al. [3] found considerable overlap between categories of mRNA and proteins in yeast, but not for specific genes. This disparity is also found in mammalian systems. For example, in a myeloid cell system, Lian et al. [4] report that at single time points, mRNA and cognate proteins are not well correlated. Even in human platelets, which as anucleate cells have mRNA and protein synthetic machinery in the same compartment, there is almost no correlation

Usha N. Kasid and Robert Clarke (eds.), *Cancer Systems and Integrative Biology*, Methods in Molecular Biology, vol. 2660,
https://doi.org/10.1007/978-1-0716-3163-8_15,

between the order of abundance for messages and cognate proteins [5]. Our own experience with CF lung epithelial cells is consistent with these general conclusions. The simplest interpretation of these data is that the regulation of protein expression in eukaryotic cells is to a large extent post-translational. We therefore need to have an analytic system that is sufficiently sensitive and flexible to permit identification of changes in post-translational regulation of proteins on a massively parallel scale.

There are a number of useful methods available to detect the relative abundance of different proteins in two samples. For high abundance proteins, conventional 2-D gels can be stained with either silver or a fluorescent dye such as Sypro-Ruby and compared side-by-side. Silver has a low dynamic range and needs $10^8$ molecules in a spot for detection. A variant is the DIGE (dye-in-gel electrophoresis) method in which samples from the two conditions are labeled with different fluorescent dyes and combined for analysis in a single 2-D gel [6]. These dyes are only somewhat more sensitive than Sypro-Ruby or Silver but have a better dynamic range. Another method is the use of stable isotope dilution with either $^{15}$[N] or $^{13}$[C], in which cells are labeled to equilibrium and mixed together for 2-D gel separation and analysis by mass spectrometry [7]. Sensitivity is a problem here, too, since a "rule of thumb" is that to be detectable by mass spectrometry, the protein has to be seen by silver. Recently, the proteomes of both *Caenorhabditis elegans* and *Drosophila melanogaster* have been quantitatively compared in terms of relative abundance using metabolic incorporation of stable isotopes [8]. Finally, cells from two different conditions have been differentially labeled on SH groups with a reagent containing either deuterium or hydrogen atoms and then mixed together for isolation and measurement in the mass spectrometer [1]. This elegant method, called ICAT (isotope-coded affinity tag), permits one to measure relative total amounts of a given protein in complex mixtures, either directly as peptides in the mass spectrometer, or with intervening 2-D gel or free flow electrophoreses steps.

As alluded to above, a limitation to these methods, as conventionally practiced, is that one can learn only ratios rather than absolute amounts of a given protein in a sample. Furthermore, detection is limited by the sensitivity and resolution of the mass spectrometer. Many low abundance proteins such as transcription factors, ion channels, and signaling molecules were beyond the limit of detection. The incorporation of a radiolabel, with a known specific activity, as described by the Pulse-Chase Proteomics paradigm, can be a robust solution to these limitations if the system can be labeled to equilibrium. The dynamic range of $^{35}$[S], for example, is five logs and is substantially more sensitive than available mass labels. However, to detect whether a given protein in a proteome is synthesized or degraded at a different rate, simple mass spectrometry will not suffice. It is clear that some kind of "pulse-chase" method is needed.

The new antibody microarray technology is the key enabling platform for Pulse-Chase Proteomic analysis of low abundance signaling pathway proteins. This technology is intrinsically robust and dependable, since it is built on years of experience with antibody-based detection systems. Antibody microarrays, consisting of hundreds or thousands of antibodies on one slide, simply build up this mature technology into a massively parallel system. For detection, the technique employs the same optical instrumentation originally designed for detection of genomic microarrays [9–11]. The arrays have actually been used for a variety of purposes. One of the first uses was for high-throughput screening of phage display libraries [12]. Antibody microarrays have also been used to capture intact leukemia cells expressing specific antigens [13]. The whole proteome from yeast has been analyzed on an antibody microarray, with special attention paid to signal validation [14]. Nielson et al. [15] used antibody microarrays to profile the signaling pathway driven by the ErbB receptor tyrosine kinase. This particular application is of particular relevance to our intention to study the NFκB signaling pathway in CF cells.

To subject this new technology to a vigorous and biologically relevant test, we have examined the global kinetics of protein expression following gene therapy of cystic fibrosis (CF) lung epithelial cells with [wildtype] CFTR. This strategy allows for the rapid, sensitive, and specific identification and quantification of total amounts of all 509 proteins (antibody microarrays commercially available from Clontech (BD Biosystems)) recognized by the arrayed antibodies and for the simultaneous measurement of cognate rates of biosynthesis and degradation. We have also developed an internal standard-based approach to quantitation, which has reduced the average error for most proteins to less than 5%. The use of these microarrays for binding radiolabeled proteins has not been published previously and represents an important technical advance on our part. The key technical problem has been how to ratio the radiolabel measurement to the mass of the cognate protein. Our solution has been to label the cold protein with either Cy3 or Cy5 (two colors) and to label the radiolabeled protein with Cy2. This step yielded two more colors, one imaging the radiolabel on a phosphorimager and the other using Cy2 as a reference mass. By combining both experiments on two slides, and imaging in four colors, the data we get consists of the ratio of antigens in two conditions and the ratio of radiolabeled protein to cognate mass in the two conditions.

We have recently shown that the ΔF508-mutation in the CFTR gene, the principal cause of cystic fibrosis (CF), has profound effects on the rates of $^{35}$[S]methionine-based biosynthesis of ca. 15% of 194 proteins in the high abundance proteome of CF lung epithelial cells using 2-D gel electrophoresis and/or mass spectrometry [16]. Importantly, we found that the static expression levels were unaffected for most of these biosynthetically distinguished proteins [16]. We concluded from this initial study that a much larger-scale proteomic analysis, extending to the low abundance signaling proteome, might lead to a functionally relevant, comprehensive disease-specific CF proteome. Reaching the low abundance signaling proteome is of particular importance in CF because the disease phenotype includes an intrinsically hyperactivated TNFα/NFκB signaling pathway in the CF lung. However, extending the analysis to the low abundance signaling proteome, involving many hundreds of proteins associated with inflammation, clearly required that we develop and apply a much more sensitive, quantitative, and larger scale proteomic strategy. Our approach to this solution has been to perform pulse-chase experiments with $^{35}$[S] methionine or $^{32}$[P] and to use an antibody microarray platform to interrogate the system [8–13, 17, 18]. This strategy allows for the rapid, sensitive, and specific identification and quantification of total amounts of all proteins recognized by the arrayed antibodies and for the simultaneous measurement of cognate rates of biosynthesis and degradation.

## 2 Materials

### *2.1 Radioactively Labeled Cultured Cells*

1. 100 mm tissue culture dishes.
2. Appropriate tissue culture medium (complete, methionine-free, or phosphate-free).
3. In vitro translation grade $^{35}$[S]-methionine, 1000 Ci/mmol or [$^{32}$P]-orthophosphoric acid in water.
4. Phosphate-buffered saline.
5. Cell scrapers.
6. Lysis buffer (7 M urea, 4% CHAPS, 30 mM Tris-Cl, pH 8.5). A protease inhibitor like complete Protease Inhibitor (Roche Applied Science/MilliporeSigma, Burlington, MA) should be included (*see* **Notes 4.1.1–4.1.3**).

### *2.2 Protein Labeling*

1. Cy2, Cy3, and Cy5 Monoreactive Dye Packs (Cytiva, MilliporeSigma, Burlington, MA).
2. High-quality anhydrous dimethylformamide (MilliporeSigma, Burlington, MA).
3. 10 mM L-lysine (MilliporeSigma, Burlington, MA).
4. BCA protein assay (Thermo Fisher Scientific, Waltham, MA).

5. Pierce Dextran Desalting Spin Columns (Thermo Fisher Scientific, Waltham, MA).
6. Internal control extract (*see* **Notes 4.2.1–4.2.5**).

### 2.3 Antibody Microarray

1. Ab 500 Microarray (Takara Bio USA, Mountain View, CA) or other antibody microarray (*see* **Note 4.3.1**).
2. Ab Microarray Express Buffer Kit (Takara Bio USA, Mountain View, CA).
3. SecureSeal Incubation Chambers (Grace Bio-Labs, Bend, OR; the SA200 just fit the Clontech arrays, if you are unsure about your ability to center the incubation chamber over the array, use the larger SA500).
4. Fluorescent scanner with Cy2, Cy3, and Cy5 filter sets (Axon GenePix 4000B; Molecular Devices, Sunnyvale, CA).
5. Phosphorimager and phosphorimaging cassettes.

### 2.4 Bioinformatic Analysis of Antibody Microarrays

1. Fluorescent scanner analysis program (GenePix 6.0; Molecular Devices, Sunnyvale, CA).
2. Microsoft Excel or other spreadsheet program.

## 3 Methods

### 3.1 Radioactive Labeling

#### 3.1.1 $^{35}$[S]-Methionine Radioactive Labeling

1. Plate cells in 100 mm tissue culture dishes and grow to 80% confluency in complete culture medium.
2. Aspirate medium and wash with 10 mL methionine-free culture medium.
3. Aspirate wash medium and add 3 mL methionine-free medium to dishes.
4. Incubate in $CO_2$ incubator for 30 min.
5. Add 200 μCi in vitro translation grade $^{35}$[S]-methionine, mix.
6. Incubate in $CO_2$ incubator for 1 h, swirling dish to mix at after 30 min.
7. Aspirate radioactive medium. Wash with 10 mL complete culture medium.
8. Aspirate wash medium and add 10 mL complete medium. Incubate for desired "chase" times, processing samples as below at desired intervals (*see* **Note 4.1.1**).
9. Aspirate medium and wash with 10 mL PBS. Aspirate and repeat PBS wash. Aspirate final wash.
10. Add 400 μL lysis buffer. Scrape cells with cell scraper and transfer lysate to a microcentrifuge tube (*see* **Note 4.1.2**).

11. Incubate at 4 °C with end-over-end mixing for 10 min.
12. If lysate is viscous, sonicate in 5 s bursts with small tip until viscosity is eliminated.
13. Centrifuge lysates at 12,000 × g for 10 min. Transfer supernatant to new microcentrifuge tube.
14. Measure protein concentration using the protein assay of choice (*see* **Note 4.1.3**).
15. Store lysates at −80 °C.

*3.1.2 $^{32}$[P]-Orthophosphate Radioactive Labeling*

1. Plate cells in 100 mm tissue culture dishes and grow to 80% confluency in complete culture medium.
2. Aspirate medium and wash with 10 mL phosphate-free culture medium.
3. Aspirate wash medium and add 3 mL phosphate-free medium to dishes.
4. Incubate in $CO_2$ incubator for 30 min.
5. Add 750 μCi $^{32}$[P]-orthophosphate, mix.
6. Incubate in $CO_2$ incubator for 15–90 min, swirling dish to mix at 30 min intervals. Process plates as below at desired times (*see* **Note 4.1.1**).
7. Aspirate medium. Wash with 10 mL PBS. Aspirate and repeat PBS wash. Aspirate final wash solution.
8. Add 400 μL lysis buffer (*see* **Note 4.1.2**). Scrape cells with cell scraper and transfer lysate to a microcentrifuge tube.
9. Incubate at 4 °C with end-over-end mixing for 10 min.
10. If lysate is viscous, sonicate in 5 s bursts with small tip until viscosity is eliminated.
11. Centrifuge lysates at 12,000 × g for 10 min. Transfer supernatant to new microcentrifuge tube.
12. Measure protein concentration using the protein assay of choice (*see* **Note 4.1.3**).
13. Store lysates at −80 °C.

### 3.2 Protein Labeling

1. Reconstitute the Cy3 and Cy5 dyes by adding anhydrous DMF according to manufacturer's instructions.
2. Confirm that cell lysate pH is ~8.5 by spotting a small amount on pH paper. If the pH is not 8.5, adjust the pH of the sample (*see* **Note 4.2.1**).
3. Add 100 μg each unlabeled experimental sample (or pooled sample; *see* **Note 4.2.2**) to microcentrifuge tubes.
4. Add 100 μg each radioactively labeled sample (or pooled sample; *see* **Note 4.2.2**) to microcentrifuge tubes.

5. Add enough of the internal standard extract for all arrays (100 μg per array) to a microcentrifuge tube (*see* **Notes 4.2.3** and **4.2.4**).
6. Add 800 pmol Cy3/100 μg of protein to the internal standard tube, 800 pmol Cy5/100 μg to each experimental sample tube, and 800 pmol Cy2/100 μg to each radiolabeled sample tube.
7. Mix vigorously by pipetting.
8. Briefly spin the sample in a microcentrifuge and incubate on ice for 30 min, in the dark.
9. Add 2 μL 10 mM lysine/100 μg protein to each tube to stop reaction.
10. Mix vigorously by pipetting. Briefly spin the sample in a microcentrifuge.
11. Incubate on ice for 10 min, in the dark.
12. Remove unbound dye using a Protein Desalting Spin Column.
    (a) Prepare needed amount of 1× desalting buffer.
    (b) Centrifuge columns (attached to 1.5 mL microcentrifuge tubes) at 1500 × g for 2 min to remove storage buffer.
    (c) Add 400 μL 1× desalting buffer to each column and centrifuge at 1500 × g for 2 min. Discard flow-through.
    (d) Repeat **step C**.
    (e) Attach a clean microcentrifuge tube to each column.
    (f) Apply the labeled protein extract to the column. Allow sample to penetrate the column.
    (g) Centrifuge at 1500 × g for 2 min.
    (h) Detach and store the collection tubes on ice.
13. Measure the protein concentration using the BCA assay. Subtract the absorption of Cy2, Cy3, and Cy5 dyes at 562 nm by preparing a tube that contains the labeled sample in place of the BCA reagent. Calculate $\Delta OD_{562} = [OD_{562\{\text{protein sample}\}} - OD_{562\{\text{protein blank}\}}]$. Use this value as the blank when calculating the protein concentration.
14. Estimate the average number of dye molecules covalently attached to each protein.
    (a) Measure Cy2, Cy3, and Cy5 absorbance at 510, 552, and 650 nm, respectively. Use the Cy2, Cy3, and Cy5 molar extinction coefficients (Cy2 $\lambda_{510} = 150{,}000\ M^{-1}\ cm^{-1}$; Cy3 $\lambda_{552} = 150{,}000\ M^{-1}\ cm^{-1}$; Cy5 $\lambda_{650} = 250{,}000\ M^{-1}\ cm^{-1}$) to calculate the micromolar concentration of Cy2, Cy3, and Cy5.
    (b) Determine micromolar concentration of protein. Assume an average molecular weight in extracts of 60 kDa.

(c) Calculate the average number of dye molecules per protein. The dye/protein ratio should be less than 6, ideally between 2 and 4 (*see* **Note 4.2.5**).

### 3.3 Antibody Microarray

1. Set up incubation tray (one four-chambered tray per two arrays with an incubation and wash chamber for each array). Add 5 mL incubation buffer to each incubation chamber.
2. Combine each labeled sample with the appropriate amount of labeled internal standard in a microcentrifuge tube.
3. Transfer labeled protein from the tubes to the appropriate incubation chamber.
4. Incubate the tray at room temperature for 10 min with gentle rocking.
5. Decant storage buffer from antibody microarrays contained in green-capped shipping/storage tubes.
6. Add 30 mL stock incubation buffer. Invert tube 10 times.
7. Decant the stock incubation buffer. Add 20 mL incubation buffer. Invert tube 10 times and decant buffer.
8. Record slide lot numbers and assign arrays to samples.
9. Remove arrays from storage tube and place, face-up, in the appropriate incubation chamber that contains the labeled sample and internal standard.
10. Incubate the arrays at room temperature for 40 min with gentle rocking. Every 10 min, pry up one end of the slide with a pipette tip and rock the tray gently.
11. Add 5 mL Wash Buffer A to each wash chamber. Transfer slides to wash chambers.
12. Incubate the tray at room temperature for 5 min with gentle rocking.
13. Decant wash buffer and add 5 mL Wash Buffer B.
14. Incubate the tray at room temperature for 5 min with gentle rocking.
15. Decant wash buffer and add 5 mL Wash Buffer C.
16. Incubate the tray at room temperature for 5 min with gentle rocking.
17. Transfer each slide (slide label facing down) to a 50 mL conical tube filled with $_{d}H_2O$.
18. Decant $_{d}H_2O$.
19. Transfer slides to clean green-capped vial supplied with arrays.
20. Centrifuge slides at 1000 × g for 5 min at room temperature.
21. Remove slides from vial and protect from light until scanning.

22. Scan slides within 24 h of drying. GenePix 4000B recommended settings: Cy3 532 nm, PMT = 550 V, laser power = 33%; Cy5 635 nm, PMT = 670 V, laser power = 33%. Use BSA control spots to adjust scanner settings. BSA should fluoresce at 2500–30,000 fluorescence units (FU). If the control spots are greater than 50,000 FU, adjust PMT and laser power to obtain a signal within 2500–30,000 FU.
23. Use phosphorimager and cassettes to detect radioactive signals after scanning for CyDye signal (*see* **Note 4.3.1**).

### 3.4 Bioinformatic Analysis of Antibody Microarray

1. Use array analysis software program to determine the intensity of each spot.
2. Download the Ab500 Microarray workbook (an Excel file) that corresponds to the microarray lot number from the "Online Tools" at www.clontech.com. Please note that this number differs from the lot number of the kit; the slide lot number is found on the label affixed to the slide. This workbook contains the names and coordinates for the antibodies on the microarray (*see* **Note 4.4.1**).
3. Cut and paste the fluorescence data intensity for each antibody into the workbook.
4. Flag all spots with a signal-to-noise ratio (SNR) $<2$ or those that are below background (*see* **Notes 4.4.2** and **4.4.3**).
5. Determine a cohort of good spots (i.e., those spots not flagged in all of the arrays to be compared; *see* **Note 4.4.4**).
6. Calculate the median for the cohort on each array.
7. Normalize all spots on the array regardless of cohort inclusion by dividing the spot intensity by the cohort median for the respective array.
8. Correct for inter-array variability by dividing the Cy5 (experimental samples) values obtained for each spot in **step 6** by the Cy3 (internal standard) values obtained in **step 6**.
9. Calculate the average ratio for each antibody and experimental group using the ratios obtained in **step 6**.
10. Use these average ratios for each antibody to identify differentially expressed proteins in the experimental groups.
11. Calculate specific activity of radioactively labeled spots by dividing $^{35}$[S]methionine or $^{32}$P counts by Cy2 levels. Specific activity corresponds to cpm/mg protein (*see* **Note 4.4**).
12. Calculate synthesis rates for each spot as $^{35}$[S] methionine counts/unit time at the end of the pulse. Calculate degradation rates in units of $h^{-1}$ as the logarithmic slope of the time dependence of the specific activity at each given time divided

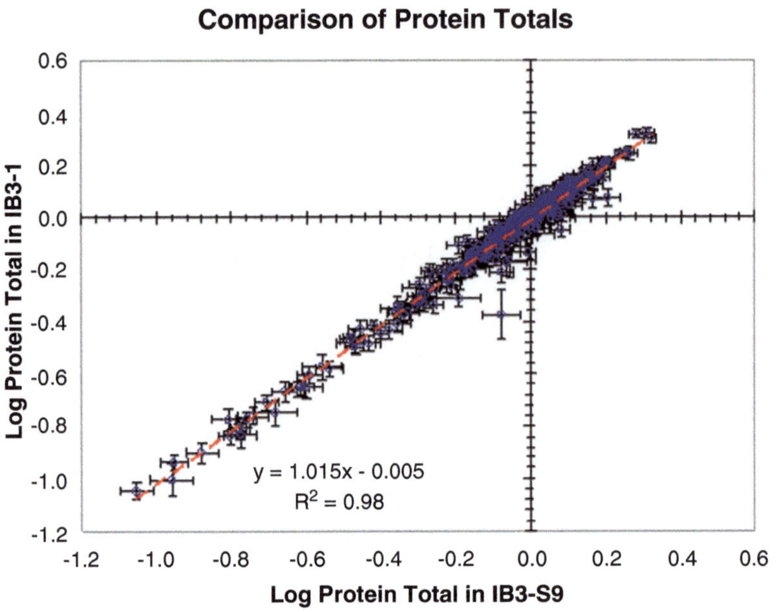


**Fig. 1** Comparison of protein totals in IB3-1 CF lung epithelial cells and in the wt-CFTR corrected IB3-S9 cells. Values on both axes are relative to the respective median. Each spot depicts the average ± SEM of 6 fluorescent determinations for each one of the cells. The red line represents the linear regression of the data

by the specific activity at the end of the pulse (*see* **Note 4.4.3** and Figs. 1 and 2).

13. Test for statistical significance (e.g., Student's t-test, SAM, etc.; *see* **Note 4.4.4**).
14. Identify differentially expressed/modified proteins.
15. Correlate differentially expressed/modified proteins with affected pathways and cellular processes using gene ontology and pathway analysis programs (*see* **Note 4.4.5**).
16. Validate differentially expressed proteins.

## 4 Notes

### 4.1 Radioactive Labeling

4.1.1. "Chase" times for metabolic labeling using $^{35}$[S] can and should be tailored to the expected half-life of the proteins recognized by the microarray. We typically "pulse" with $^{35}$[S] for 1–12 h, taking samples every 3 h. Likewise, "pulse" times should be tailored to fit expected modification time of the pathways of interest. We typically "pulse" with [$^{32}$P] for 15–90 min, taking samples at 15, 30, 60, and 90 min.

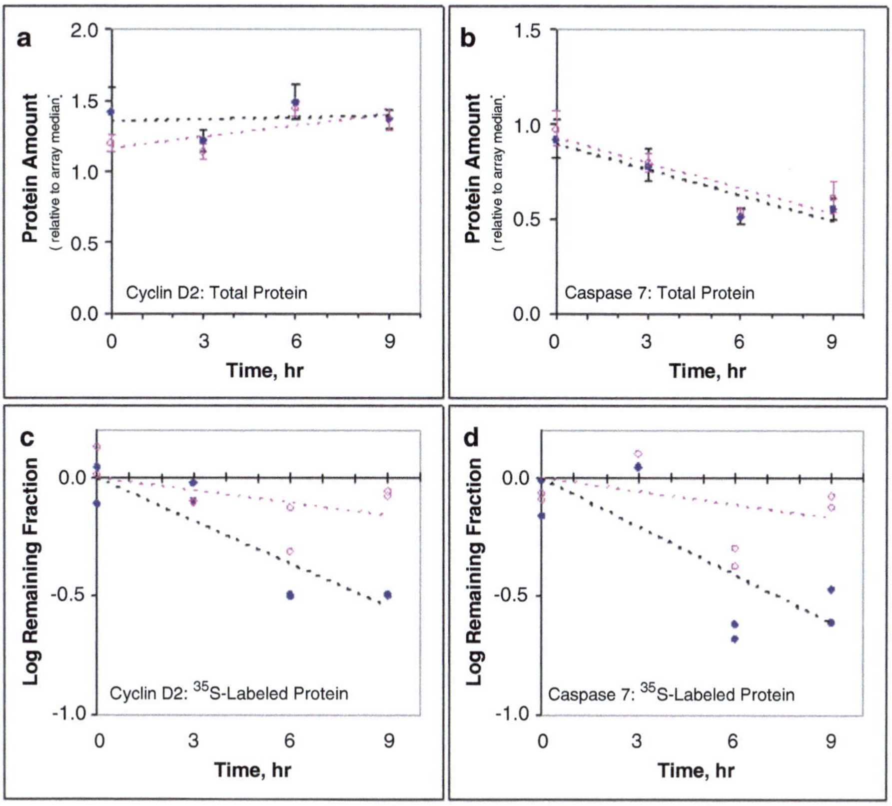


**Fig. 2** Kinetics of expressed protein and of labeled protein for two example proteins on the microarray. *Top*: Total protein levels for the two example proteins, cyclin D2 (**a**), which shows little variability during the 9 h chase; and caspase 7 (**b**), which shows a pronounced decrease. Error bars are standard errors of the mean for 6 determinations. Total and labeled protein levels were measured in CF lung epithelial line IB3-1 (blue filled symbols) and in the wt-CFTR corrected IB3-S9 cells (red open symbols). *Bottom*: Protein degradation rates were determined for cyclin D2 (**c**) and caspase 7 (**d**) by following the time course of the remaining fractions of [$^{35}$S] methionine-labeled of the respective protein. Data from duplicate spots are shown on logarithmic scale. Dotted lines depict linear regressions for each case

4.1.2. Alternate lysis buffers may be used. However, it is important to avoid primary amines and reducing agents in preparation of the samples as these can interfere with dye attachment.

4.1.3. Ideally, lysates should be between 1 and 20 mg/mL but 5–10 mg/mL is optimal.

### 4.2 CyDye Labeling

4.2.1. Sample pH can be increased through the careful addition of lysis buffer (pH 9.5 instead of 8.5) or 50 mM NaOH.

4.2.2. Experimental samples of the same type (controls, treated, etc.) may also be pooled to reduce the number of arrays needed.

4.2.3. Antibody microarray results are usually given in terms of ratios between two samples (e.g., "experiment" and "control"). However, given the printing and calibration issues inherent with antibody microarrays, this ratio'ing approach precludes statistically valid inter-array comparisons of the individual protein concentrations. We have developed a different strategy to control for inter-array variability. We use a benchmark mixture that is applied to every array as an internal standard; this corrects for any differences in antibody activities as well as printing imperfections. The amount of protein bound to each spot is determined relative to the respective protein in the internal standard. Since all samples in the study are compared to the same internal standard, this semi-quantitative approach permits the comparison of multiple samples. All data for the same antibody on multiple arrays can then be used to calculate the average and standard deviation for a given population (such as patients or control groups) in a parallel fashion. The composition of the internal standard is not important. Ideally, it would be a mixture of all proteins on the array at a known concentration. However, this is not practical, so the internal standard should consist of a protein sample that contains all of the proteins represented on the microarray. We use the same combination of cell culture and tissue extracts for every array run in our laboratory.

4.2.4. This tube should contain enough internal standard extract for all of the arrays.

4.2.5. A ratio higher than 6 can interfere with antibody binding.

### 4.3 Antibody Microarray

4.3.1. Available antibody microarrays are constantly changing. Check out commercially available arrays on the web or recent review articles such as the recent review by Chen et al. [19] or consider printing your own. Use the microarray buffers recommended by the manufacturer.

### 4.4 Bioinformatic Analysis

4.4.1. This workbook also contains an analysis worksheet that identifies differentially expressed proteins using the conventional approach using reciprocal CyDye labeling. In this approach, two samples are labeled with either Cy3 or Cy5 and mixed together for multiplex binding to a given antibody feature on the microarray. If there is more Cy3- or Cy5-labeled protein in the mixed sample, the spot will be either green or red, respectively. If the samples are present in equal amounts, the mixture will show as yellow. However, because of protein-specific differences in quantum yields and labeling efficiencies for Cy 3 and Cy5, a parallel set of two samples must be labeled reciprocally with the opposite dye. The geometric average of the two relative levels of Cy3 and Cy5 occurring at a given feature under the two conditions are then averaged, and the ratio is then used to express the *relative* abundance of the given protein in the two conditions. This strategy, while logical on its face, yields data that is intrinsically noisy. This approach also carries with it the limitation that the data are always ratios. In addition to statistical issues with ratios, this also means that one cannot easily compare one assay with another assay performed the next day or the next year. This is especially critical when trying to establish statistical differences between patient and control groups in clinical samples where each patient and each control may be different from the group as a whole. We believe that in using one dye for an internal standard, we circumvent the issues with ratios while also correcting for printing irregularities. Therefore, we only use the workbook to obtain the coordinates/names for the antibodies on the array. A complimentary approach, which we believe has significant advantages, is to treat each dye signal as an independent variable (rank order method). Rank order methods, either rank-sum or rank-product tests, are frequently used in clinical studies.

4.4.2. GenePix calculates SNR, otherwise SNR ~ net counts/SD of the background.

4.4.3. The simplest representation for the relationship between synthesis and degradation rates for a given protein "*n*" at any time *T* is as follows:

$$[P_n(T)] = \int_0^T \{S_n - D_n \cdot P_n(t)\} \cdot dt \qquad (1)$$

where $S_n$ is the rate of biosynthesis; $D_n$ is the rate of degradation; and $P_n(t)$ is the concentration of protein "*n*" at time *t*. If the total protein is changing over time, this equation becomes:

$$\frac{d}{dt}[P_n(T)] = S_n - D_n \cdot P_n(T) \qquad (2)$$

However, if that protein's concentration does not change over time, then:

$$S_n = D_n \cdot P_n \qquad (3)$$

which implies that there should be a (protein-dependent) correlation between the measured synthesis rate and the degradation rate. Thus, the difference between the synthesis rates in IB3-1 and IB3-S9 should be correlated with the respective difference in degradation rates. In particular, at least for those proteins for which the concentrations in both cell types are the same (*see* Fig. 1), the sign of the difference in synthesis rate should be the same as the sign of the difference in degradation rate, so that such points should all be in either the first or third quadrants. Conversely, points in the second and fourth quadrants might arise from protein whose total mass change with time or which have different total mass in the two cell types.

4.4.4. Recently, it has become apparent that reliance on the classical t-test for determining significance of a difference is misleading when dealing with microarray data. The statistical problem is associated with the need for the number of experimental replicates on the same order as the number of spots on the microarray. However, the individual microarrays are expensive, and it is experimentally impossible to do so many experiments in the first place. Tusher et al. [20], building on the concept of a false discovery rate, have suggested an alternative strategy in the statistical analysis of microarrays (SAM) algorithm. Their strategy has been to take data associated with each condition and to calculate statistical parameters based on many permutations of the data. The result is the ability to calculate the statistical validity of the difference based on a local false discovery rate (FDR) and an associated $q$-statistic. The $q$-value gives an estimate of the probability of falsely identifying a protein as "significant" within all the group of proteins that have $q$-values lower than that for the protein in question. The local FDR gives a measure of the probability that a given feature is identified as significant by random chance. The apparent advantage of combining non-parametric inference from rank orders and the random permutation steps in the SAM algorithm have been interpreted as an example of "borrowed power," because the resultant study seems to be more locally powered than might have been calculated from a the standard approach.

4.4.5. There are many excellent online resources for gene ontology analysis and the identification of functional protein classes. These include NAIAD DAVID (http://niaid.abcc.ncifcrf.gov/home.jsp) and NCI GOMiner (http://discover.nci.nih.gov/gominer/). Unfortunately, the best pathway analysis software must be purchased (Ingenuity, GeneGO).

## References


1. Gygi SP, Rist B, Gerber SA et al (1999) Quantitative analysis of complex mixtures using isotope-coded affinity tags. Nat Biotechnol 17:994–999
2. Ideker T, Thorsson V, Ranish JA et al (2001) Integrated genomic and proteomic analyses of a systematically perturbed metabolic network. Science 292:929–934
3. Greenbaum D, Jansen R, Gerstein M (2002) Analysis of mRNA expression and protein abundance data: an approach for the comparison of the enrichment features in the cellular population of proteins and transcripts. Bioinformatics 18:585–596
4. Lian Z, Kluger Y, Greenbaum DS et al (2002) Genomic and proteomic analysis of the myeloid differentiation program: global analysis of gene expression during induced differentiation in the MPRO cell line. Blood 100:3209–3220
5. McRedmond JP, Park SD, Reilly DF et al (2004) Integration of proteomics and genomics in platelets: a profile of platelet proteins and platelet-specific genes. Mol Cell Proteomics 3(2):133–144
6. Patton WF (2000) A thousand points of light: the application of fluorescence detection technologies to two-dimensional gel electrophoresis and proteomics. Electrophoresis 21:1123–1144
7. Sanford EJ, Smolka MB (2002) A field guide to the proteomics of post-translational modifications in DNA repair. Proteomics 22(15–16): e2200064. https://doi.org/10.1002/pmic.202200064
8. Krijgsveld J, Ketting RF, Mahmoudi T et al (2003) Metabolic labeling of C. elegans and D. melanogaster by quantitative proteomics. Nat Biotechnol 21:927–931
9. Huang RP (2001) Detection of multiple proteins in an antibody-based protein microarray system. J Immun Methods 255:1–13
10. Haab BB (2003) Methods and applications of antibody microarrays in cancer research. Proteomics 3:2116–2122
11. Glokler J, Angenendt P (2003) Protein and antibody microarray technology. J Chromatogr B Analyt Technol Biomed Life Sci 797:229–240
12. de Wildt RMT, Mundy CR, Gorick BD, Tomlinson IM (2000) Antibody arrays for high throughput screening of antigen-antibody interactions. Nat Biotechnol 18:989–994
13. Belov L, Huang P, Barber N et al (2003) Identification of repertoires of surface antigens on leukemias using an antibody microarray. Proteomics 3:2147–2154
14. Michaud GA, Salcius M, Zhou F et al (2003) Analyzing antibody specificity with whole proteome microarrays. Nat Biotechnol 21:1509–1512
15. Nielsen UB, Cardone MH, Sinskey AJ et al (2003) Profiling receptor tyrosine kinase activation by using Ab microarrays. Proc Natl Acad Sci U S A 100(16):9330–9335
16. Pollard HB, Eidelman O, Jozwik C, Huang W, Srivastava M et al (2006) de Novo biosynthetic profiling of high abundance proteins in cystic fibrosis lung epithelial cells. Mol Cell Proteomics 5:1628–1637
17. Pollard HB, Eidelman O, Srivastava M, Jozwik C et al (2007) Protein microarray platforms for clinical proteomics. Proteomics Clin Appl 1(9): 934–952
18. Srivastava M, Eidelman O, Jozwik C et al (2006) Antibody microarray platform for serum proteomics of cystic fibrosis. Mol Genet Metab 87:303–310
19. Chen Z, Dodig-Crnković T, Schwenk JM, Tao SC (2018) Current applications of antibody microarrays. Clin Proteomics 15:7–22
20. Tusher VG, Tibshirani R, Chu G (2001) Significance analysis of microarrays applied to the ionizing radiation response. Proc Natl Acad Sci U S A 98:5116–5121

# Chapter 16

# Multiplex Fluorescent Immunohistochemistry for Preservation of Tumor Microenvironment Architecture and Spatial Relationship of Cells in Tumor Tissues

**Jake J. McGue, Jacob J. Edwards, Brian D. Griffith, and Timothy L. Frankel**

## Abstract

The tumor microenvironment (TME), composed of immune cells, antigens, and local soluble factors, is integral to cancer development and progression. Traditional techniques such as immunohistochemistry, immunofluorescence, or flow cytometry limit the analysis of spatial data and cellular interactions within the TME, as they are restricted to colocalization of a small number of antigens or the loss of tissue architecture. Multiplex fluorescent immunohistochemistry (mfIHC) allows for detection of multiple antigens within a single tissue sample, providing a more comprehensive description of tissue composition and spatial interactions within the TME. This technique utilizes antigen retrieval, application of primary and secondary antibodies, followed by a tyramide-based chemical reaction to covalently bind a fluorophore to an epitope of interest and, eventually, stripping of the antibodies. This allows for multiple rounds of antibody application without concern for species cross-reactivity, as well as signal amplification which abrogates the autofluorescence that frequently plagues analysis of fixed tissues. As such, mfIHC can be used to quantify multiple cellular populations and their interactions, in situ, unlocking key biologic data that was previously unavailable. This chapter provides an overview of the experimental design, staining, and imaging strategies using a manual technique in formalin-fixed paraffin-embedded tissue sections.

**Key words** Immunofluorescence (IF), Multiplex fluorescent immunohistochemistry (mfIHC), Multispectral imaging, Spatial relationships in situ, Tumor microenvironment architecture, Cellular interactions, Cellular engagement, Tumor microenvironment (TME), Tumor immunology, The AKOYA OPAL system

## 1 Introduction

The tumor microenvironment (TME) is composed of tumor cells and stroma, consisting of non-tumor cells and soluble factors such as cytokines and chemokines, all of which work to shape the biology of cancer. Each tumor has a unique immune signature, where the balance of proinflammatory and immunosuppressive signaling

Usha N. Kasid and Robert Clarke (eds.), *Cancer Systems and Integrative Biology*, Methods in Molecular Biology, vol. 2660, https://doi.org/10.1007/978-1-0716-3163-8_16,

influences the overall immune state of the tumor [1]. This tumor immune microenvironment is integral in shaping tumor biology including progression, metastasis, and response to treatment [2–4].

Prior techniques used to describe cells within the TME, including immunohistochemistry (IHC), immunofluorescence (IF), and flow cytometry, have significant limitations with regard to assessing spatial cellular relationships. IHC, long considered the standard for diagnosis of tumor pathology, uses formalin-fixed paraffin-embedded (FFPE) tissue sections stained typically with a single antibody, followed by horseradish peroxidase (HRP) bound secondary antibody and chemical reaction to identify an antigen within the visible spectrum [5]. Limitations of IHC include colocalization of only one or two antigens given significant overlap within the visible spectrum, low sensitivity for poorly antigenic targets and high inter-observer variability [5–8]. IF allows for better detection and colocalization of a wide variety of antigens due to its use of the non-visible spectrum and can identify more difficult to detect antigens. However, as each antibody typically is attached to a single fluorophore, IF sensitivity is limited by tissue autofluorescence, as well as significant spectral overlap, cross species reactivity, and photobleaching [9]. Flow cytometry, often considered the gold standard for describing the immune microenvironment, involves processing fresh tissue into a single cell suspension with labeled fluorescent antibodies. Limitations of this technique include the need for significant compensation of signals and most importantly loss of tissue architecture and spatial information during processing [10, 11].

Multiplex fluorescent immunohistochemistry (mfIHC) is a technique that utilizes multispectral imaging analysis and signal amplification to allow for detection of multiple antigens in a single tissue sample while subtracting noisy autofluorescence. This approach allows for a comprehensive overview of tissue composition, subpopulation densities, and cellular interactions, as the multispectral image produced can identify multiple cellular populations and their spatial relationships, in situ [12]. During mfIHC, FFPE tissue is adhered to a charged slide and after antigen retrieval, undergoes staining with a primary antibody to the antigen of interest, followed by a secondary antibody with an HRP chemical tag, similar to traditional IHC. After each single round of staining, a tyramide-based chemical reaction covalently binds multiple fluorophores to the antigen of interest amplifying the signal [7]. Subsequently, heating of the slide removes the previously applied primary and secondary antibody, leaving the fluorescent tag bound to the epitope, allowing for additional rounds of antibody application without concern for cross-reactivity [7, 13]. Panel design using multiple fluorophores with minimal spectral overlap limits the need for manual compensation. Additionally, the strength of antigen-specific fluorescence after fluorophore signal amplification allows for subtraction of autofluorescence. Current mfIHC systems allow for analysis of up to eight antigens and a nuclear counterstain, and programs designed to analyze large data sets allow for assessment of cell locations and interactions among multiple cellular phenotypes, providing crucial spatial context to cells in the TME [12, 14, 15]. However, mfIHC can be limited by considerable time needed for optimization and decreased specificity as overactive tyramide deposits may lead to off target staining [12]. Similarly, as multiple systems and protocols exist for mfIHC, variability in procedures can limit reproducibility [12, 16]. Herein is a detailed methodology of experimental design, staining, and imaging strategies for mfIHC to aid analysis of spatial context of cellular interactions using a manual technique for FFPE tissue using the AKOYA OPAL system.

## 2 Materials

### 2.1 Wash Buffers

1. Tris-buffered saline (TBS;10×): 1.5 M NaCl, 0.1 M Tris-HCL, pH 7.4.
2. 1× TBS containing 0.1% Tween-20 (TBST).

### 2.2 Antigen Retrieval

1. AR6 buffer: prepare working 1× AR6 buffer from stock solution using deionized water. Store at room temperature (*see* **Note 1**).
2. AR9 buffer: prepare working 1× AR9 buffer from stock solution using deionized water. Store at room temperature (*see* **Note 1**).

### 2.3 Staining Reagents

1. Primary antibody: Dilute primary antibodies in antibody dilutant according to company IHC-P recommendations. Store at 4 °C.
2. 1× Opal polymer anti-Ms + Rb HRP secondary antibody.
3. Opal fluorophore stocks (480, 520, 570, 620, 690, 780, TSA DIG): Reconstitute all Opal stocks except 780 by adding 75 μL DMSO to the lyophilized Opal fluorophores. Vortex and allow the lyophilizate to dissolve for 10 to 15 min at room temperature. Vortex again and centrifuge the Opal fluorophore stock at 10 × g for 10 s at room temperature. Store in the dark at −20 °C. Reconstitute 780 in the same way as described above instead using 300 μL of deionized water in place of DMSO.
4. 4′,6-Diamidino-2-phenylindole (DAPI): Add 3 drops of DAPI to 1 mL of TBS. Vortex and store in the dark at 4 °C.
5. Fluorophore dilutant.
6. Antibody dilutant.

### 2.4 Special Equipment

1. Heat-resistant slide containers (10 total recommended).
2. Plastic slide holder.
3. Immunohistochemistry staining tray (humidified chamber).
4. Micro hybridization oven.
5. Microwave equipped with inverter technology.
6. Vacuum with glass pipette attachment.

## 3 Methods

### 3.1 Slide Preparation and Antibody Optimization

1. Prepare formalin-fixed, paraffin-embedded (FFPE) tissue onto charged slides
   (a) Cut blocks of FFPE tissue to 5 μm and adhere to charged slides (*see* **Note 2**).
   (b) Allow the slides to dry overnight at 37 °C. Keep the slides flat with the tissue side up.
   (c) Store the slides in a slide box at 4 °C until ready for use.
2. Determine the optimal antibody concentrations for multiplex by conventional immunohistochemistry (IHC):
   (a) Perform the IHC on tissue with an abundance of cells specific for each antibody evaluated.
   (b) Complete the experiment with the company recommended antibody concentration, as well as dilutions above and below the recommended concentration.
   (c) Visualize slides and note the most specific and sensitive concentration. This is the optimal antibody concentration to be used for the multiplex.

### 3.2 Staining Preparation

1. Preparation of wash buffer and antigen retrieval solutions
   (a) Prepare 0.1% TBST wash buffer by diluting 1 L of 10× Tris-buffered saline (TBS) in 9 L of deionized water. Add 10 mL Tween 20 and mix well.
   (b) Prepare antigen retrieval solutions (pH 6 and pH 9) by diluting to 1× with deionized water.
2. Preparation of FFPE slides
   (a) Deparaffinize slides in a hybridization oven. Bake slides at 60 °C for 1 h lying flat or vertically in a heat-resistant slide rack. Remove slides from the oven and allow to cool for at least 10 min.
   (b) Rehydrate slides through the following series: xylene in triplicate, followed by a single submersion in 100% ethanol, 95% ethanol, and 70% ethanol. Submerge the slides for 10 min during each step of the series.

(c) Wash the slides for 2 min in deionized water, followed by 30 min in neutral buffered formalin. Complete the rehydration by washing slides in deionized water for another 2 min (*see* **Note 3**).

3. Antigen retrieval

(a) Place the slide rack into a heat-resistant box filled with either pH 6 or pH 9 antigen retrieval buffer (*see* **Note 4**).

(b) Triple fold plastic wrap and cover the heat resistant box, secure tightly with a rubber band.

(c) Insert the heat-resistant box at the edge of the microwave plate. Microwave the slides for 45 s at 100% power, followed by 15 min at 20% power (*see* **Note 5**).

(d) Remove the heat-resistant box from the microwave and use the box cover to vent about 1/3 of the opening. Allow slides to cool for at least 20 min (*see* **Note 6**).

4. Preparation of primary antibody and Opal fluorophore working solutions

(a) Use the optimal antibody concentration determined by IHC. Dilute the primary antibody in antibody dilutant, approximately 150 μL per slide (*see* **Note 7**).

(b) Prepare the Opal fluorophores in the following manner: (*see* **Note 8**).

- Opal 480, 520, 570, 620, and 690: Dilute stock solutions 1:50 to working concentration with AMP dilutant.
- Opal 780: Dilute stock solution 1:25 to working concentration with antibody dilutant.
- TSA Dig: Dilute stock solution 1:100 to working concentration with AMP dilutant.

### 3.3 Slide Staining

1. Washing and peroxidase blocking

(a) Wash the slides in a slide box for 2 min with deionized water followed by TBST for 2 min in a separate slide box.

(b) Remove one slide at a time and dry around the tissue using a delicate task wipe. Trace around the outside of the tissue with a hydrophobic barrier pen. Place the slide in a humidified chamber and use a disposable pipette to transfer TBST onto the tissue to avoid tissue drying. Be careful not to touch the tissue with the hydrophobic pen or delicate task wipe. Repeat for the remaining slides.

(c) Gently vacuum the TBST off the tissue on each slide by placing the vacuum tip at the edge of the hydrophobic barrier. Quickly apply 3 to 4 drops of peroxidase blocking solution to the tissue and incubate for 10 min at room temperature.

2. Primary antibody application
   (a) Vacuum the blocking solution off the tissue and quickly add approximately 150 μL of working primary antibody solution per slide. Incubate in the humidified chamber for 1 h at room temperature.
   (b) Take each slide and remove the primary antibody solution by tapping the side of the slide against a stack of paper towels. Place the slides into a slide box with fresh TBST for 2 min for three washes (*see* **Note 9**).
3. Secondary antibody application
   (a) Omitting the hydrophobic pen step, load the humidified chamber as detailed in Subheading 3.3, **step 1(b)**.
   (b) Gently vacuum the TBST off the tissue on each slide and quickly add 3 to 4 drops of secondary antibody (mixture of rabbit and mouse secondary HRP-conjugated antibodies). Incubate in the humidified chamber for 10 min at room temperature (*see* **Note 10**).
   (c) Remove excess secondary antibody by tapping slides onto a stack of fresh paper towel and place slides into a slide box with fresh TBST. Wash three times for 2 min each.
4. Fluorophore application
   (a) Load slides into the humidified chamber.
   (b) Apply approximately 100 μL of fluorophore working solution and incubate in the humidified chamber for 10 min at room temperature.
   (c) Remove excess fluorophore solution and wash slides with fresh TBST for 2 min in triplicate.
5. Antibody removal
   (a) Microwave the slides as detailed in Subheading 3.2, **step 3** for removal of antibodies and in preparation of the next primary antibody staining.
   (b) At this point, the protocol may be paused by leaving the slides submerged in antigen retrieval buffer overnight at room temperature. Cover the slide box to prevent evaporation.
   (c) After determining multiplex positions as detailed in Subheading 3.4, repeat Subheading 3.3, **steps 1–5** for the remaining antibody–fluorophore pairs in the appropriate positions.
6. DAPI Application and Mounting
   (a) Prepare the 4′,6-diamidino-2-phenylindole (DAPI) working solution by adding three drops of DAPI into

1 mL of TBS. Vortex to ensure homogenization and store in the dark at 4 °C.

(b) Remove the final antibody application with pH 6 antigen retrieval buffer. Wash slides in deionized water for 2 min, followed by TBS for 2 min.

(c) Load slides into the humidified chamber and apply approximately 150 μL of working DAPI solution to each slide. Incubate for 10 min at room temperature.

(d) Remove excess DAPI by tapping slides onto a stack of fresh paper towel. Place slides into a slide box with fresh TBS for 2 min, followed by deionized water for 2 min.

(e) Place slides in fresh xylene for no longer than 2 min to remove the hydrophobic barrier.

(f) Mount coverslips with an antifade mountant. Be careful not to trap air bubbles over the tissue and allow slides to dry overnight.

### 3.4 Multiplex Workflow

1. Complete monoplexes with each primary antibody to determine the best position (order) for the multiplex

   (a) Prepare an equal number of tissues slides as the number of antibodies in the anticipated multiplex. For example, if staining for three antibodies in the multiplex, prepare three slides for each of the individual antibodies. Each slide should contain tissue with an ample amount of epitope for each optimized primary antibody.

   (b) For each individual antibody, perform a monoplex stain with the antibody in a unique position, following the staining procedures outline in Subheading 3.3, **steps 1–6**. For instance, if using three antibodies, each slide would have one primary antibody applied in one of the three unique positions. Use antibody and fluorophore dilutant on slides that do not require working solutions for that round (Fig. 1).

2. Image and analyze the monoplex

   (a) Images may be captured using any fluorescent microscope capable of recording emission at the appropriate wavelength for the fluorophores used. Often analysis software is dependent on the machine used for capture, so this should be considered.

   (b) Evaluate each monoplex slide by looking at the fluorescent intensity of the stained marker and the anticipated cells of interest.

   (c) Compare the intensity of the antigen of interest to the background. Positions in which the intensity is at least

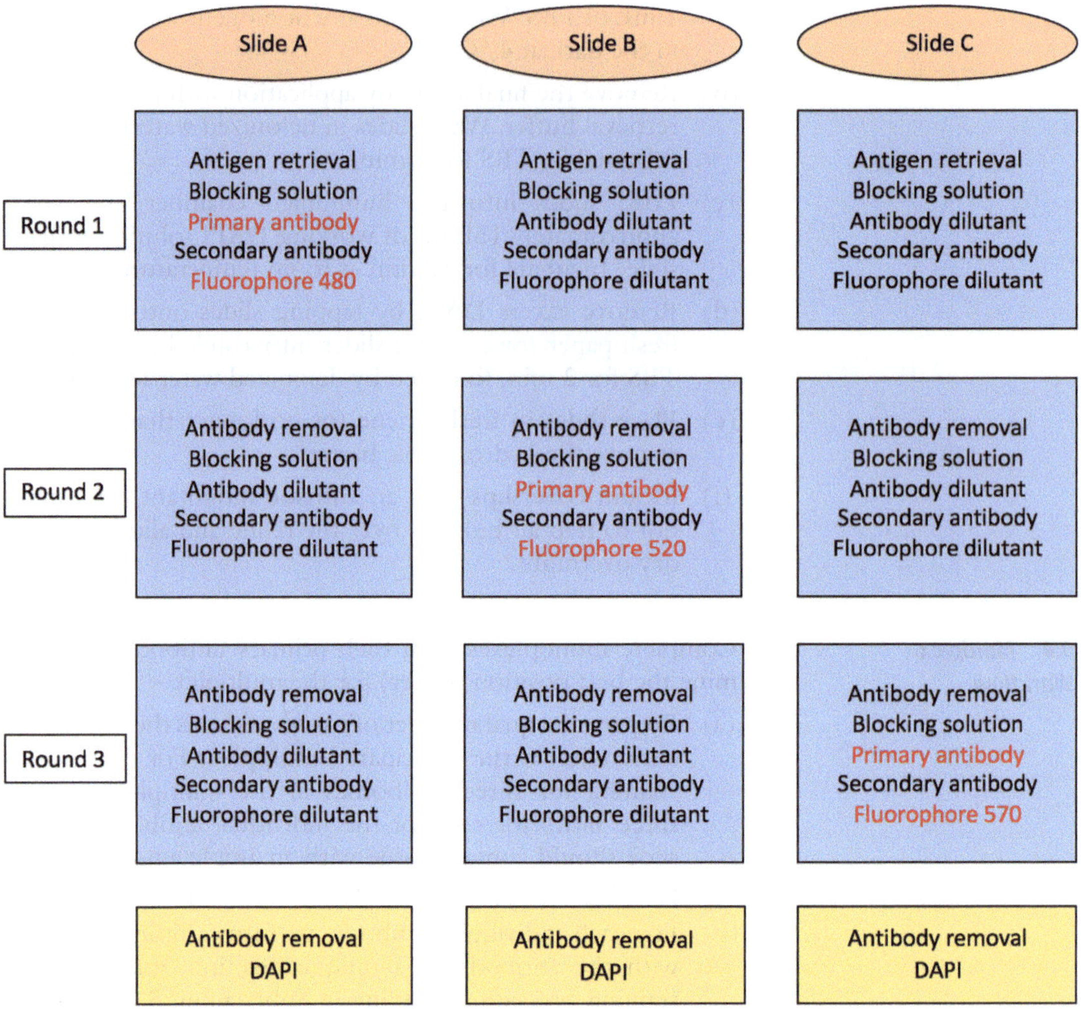


**Fig. 1** Workflow of a position test. In this example of a position test of a primary antibody for a multiplex examining three antibodies total, the primary antibody of interest is at position 1 in slide A, position 2 in slide B, and position 3 in slide C. (Adapted from Lazarus J et al (2019) J Vis Exp 149:10.3791/59915. doi: 10.3791/59915)

5–10 times higher than the background can be considered for inclusion in the final monoplex.

(d) Refer to Fig. 2 for examples of completed monoplex position tests.

3. Perform the multiplex stain (Fig. 3)

(a) Acquire slides of interest and one additional slide to serve as a blank for autofluorescence subtraction after imaging (*see* **Note 11**).

(b) Chose the appropriate order for each antibody based on monoplex staining as described in Subheading 3.4, **step 2**.

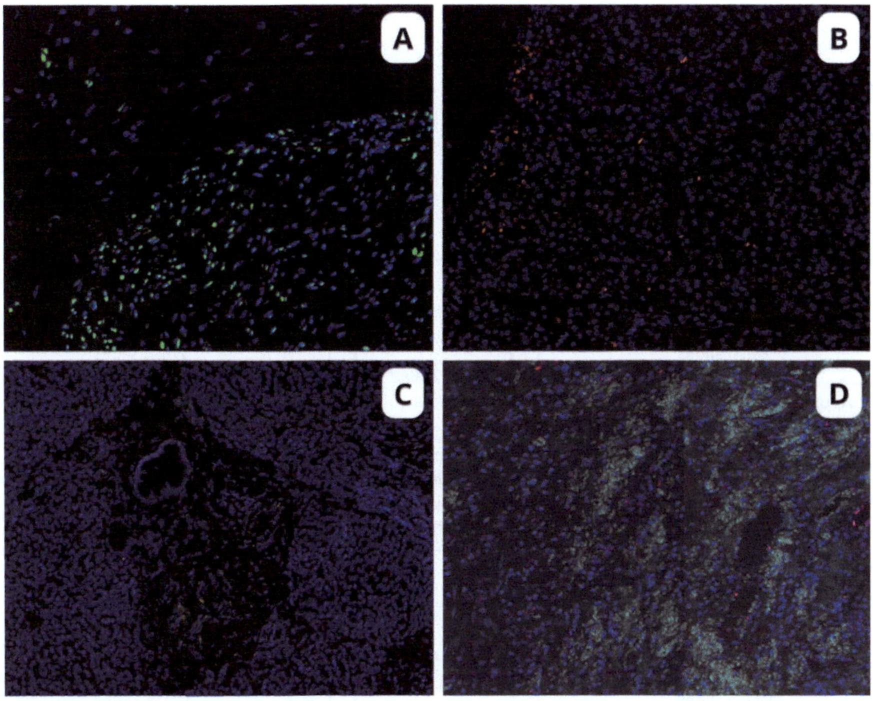


**Fig. 2** STAT3 monoplex position test on normal pancreatic tissue in four unique positions. (**a**) Staining at position 1 with Opal 520 fluorophore (green). Optimal position for multiplex with strong, specific cell staining. (**b**) Staining at position 3 with Opal 620 fluorophore (red). Intermediate quality stain with suboptimal cell specificity and signal strength. (**c**) Staining at position 2 with Opal 570 fluorophore (yellow). Poor staining quality evidenced by non-specific staining and low signal intensity. (**d**) Staining at position 4 with Opal 690 fluorophore (red). Poor staining quality evidenced by strong background autofluorescence and low signal intensity

Assign a fluorophore to each primary antibody (*see* **Note 12**).

(c) Stain the multiplex as described in Subheading 3.3, **steps 1–6**. The blank slide does not receive any primary antibody, fluorophore, or DAPI. Instead, use antibody dilutant, fluorophore dilutant, and TBS, respectively.

(d) Refer to Fig. 4 for an example of a completed multiplex stain.

## 4 Notes

1. Fresh buffer should be made prior to each net project.
2. Tissue dislodgement from slides during the multiple rounds of heating is a frequent problem so methods to ensure slide adherence should be utilized.

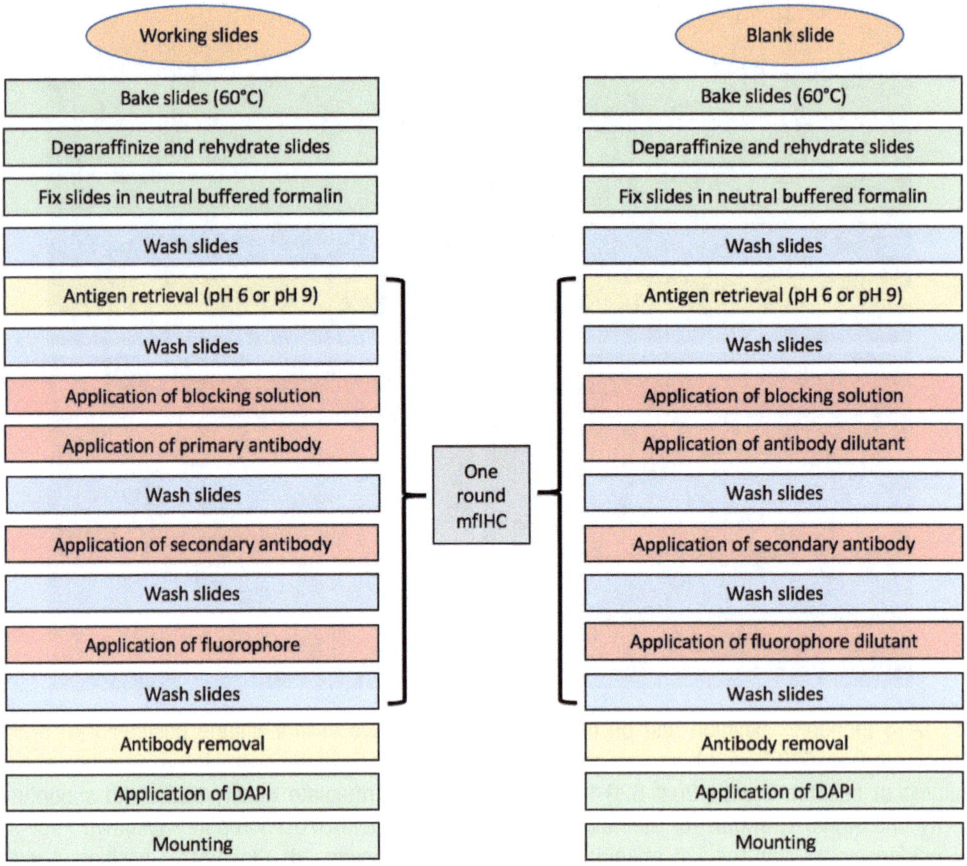


**Fig. 3** Multiplex fluorescent immunohistochemistry (mfIHC) workflow. Overview of the staining workflow for an mfIHC experiment after determining the appropriate order and position of primary antibodies. (Adapted from Lazarus J et al (2019) J Vis Exp 149:10.3791/59915. doi: 10.3791/59915)

3. For proper adherence and mounting of tissue, it is important to use distilled water.
4. The choice of pH 6 or pH 9 antigen retrieval may vary with epitope. However, we recommend starting with pH 9 for nuclear epitopes and pH 6 for membrane and cytoplasmic epitopes.
5. Microwave treatment may need optimization depending on the microwave being used. Moreover, the microwave must be equipped with inverter technology for even heating.
6. After adding the first fluorophore, it is best to limit ambient light exposure to minimize autofluorescence. Venting allows the slides to both cool down and maintain shielding from the light.
7. The amount of working solution per slide can vary depending on the tissue area, generally ranging from 100 to 200 μL. Be

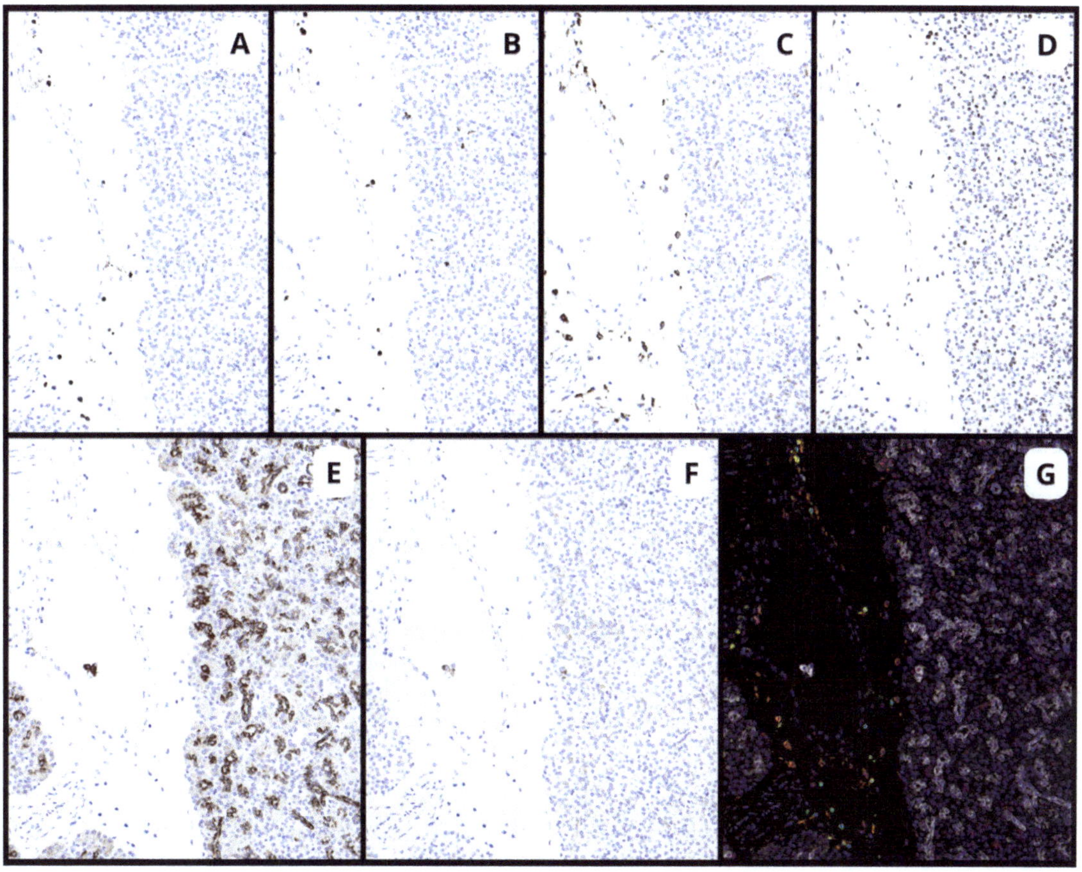


**Fig. 4** Example multiplex with six antibodies in normal pancreatic tissue. (**a**) Pathview image of CD3 stain in position 1 with Opal 480 fluorophore. (**b**) Pathview image of CD8 stain in position 2 with Opal 520 fluorophore. (**c**) Pathview image of CD163 stain in position 3 with Opal 570 fluorophore. (**d**) Pathview image of Arg1 stain in position 4 with Opal 620 fluorophore. (**e**) Pathview image of pancytokeratin stain in position 5 with Opal 690 fluorophore. (**f**) Pathview image of FoxP3 stain in position 6 with Opal 780 fluorophore. (**g**) Full multiplex composite image (CD3 = green, CD8 = yellow, CD163 = orange, Arg1 = red, pancytokeratin = white, FoxP3 = red)

sure to conserve the recommended concentrations when preparing working solutions.

8. Fluorophore concentrations may need to be optimized depending on the antibody. Additionally, it is recommended to prepare the working fluorophore solution the day it will be used. Fluorophore solutions should be stored in the dark and are stable for 24 to 48 h when stored at 4 °C.
9. Fresh TBST will help ensure a cleaner image. Be sure to prepare a large amount of fresh TBST in advance. Furthermore, gently agitate slides at the start of each TBST submersion to ensure thorough washing.
10. Care must be taken to ensure that secondary antibodies are specific for the species of the primary antibody. When possible,

HRP conjugated secondary antibodies should be used and incubated at RT for 10 min. Optimization of incubation time may be needed for the secondary. Use of unconjugated secondary antibodies will add a step and may require unique optimization.

11. It is recommended to use the same tissue type as the slides of interest to subtract autofluorescence.
12. Fluorophore assignments may require optimization. In general, use brighter fluorophores for less abundant antibody markers. Similarly, epitopes that colocalize should be matched with fluorophores at far spectrums to avoid spectral overlap.

## References


1. Gout S, Huot J (2008) Role of cancer microenvironment in metastasis: focus on colon cancer. Cancer Microenviron 1(1):69–83
2. Chen F, Zhuang X, Lin L et al (2015) New horizons in tumor microenvironment biology: challenges and opportunities. BMC Med 13: 45
3. Paluskievicz CM, Cao X, Abdi R, Zheng P, Liu Y, Bromberg JS (2019) T Regulatory Cells and priming the suppressive tumor microenvironment. Front Immunol 10:2453
4. Schulz M, Salamero-Boix A, Niesel K, Alekseeva T, Sevenich L (2019) Microenvironmental regulation of tumor progression and therapeutic response in brain metastasis. Front Immunol 10:1713
5. Katikireddy KR, O'Sullivan F (2011) Immunohistochemical and immunofluorescence procedures for protein analysis. Methods Mol Biol 784:155–167
6. Tan WCC, Nerurkar SN, Cai HY et al (2020) Overview of multiplex immunohistochemistry/immunofluorescence techniques in the era of cancer immunotherapy. Cancer Commun (Lond) 40(4):135–153
7. Stack EC, Wang C, Roman KA, Hoyt CC (2014) Multiplexed immunohistochemistry, imaging, and quantitation: a review, with an assessment of tyramide signal amplification, multispectral imaging and multiplex analysis. Methods 70(1):46–58
8. Taylor CR, Levenson RM (2006) Quantification of immunohistochemistry--issues concerning methods, utility and semiquantitative assessment II. Histopathology 49(4): 411–424
9. Lichtman JW, Conchello JA (2005) Fluorescence microscopy. Nat Methods 2(12): 910–919. https://doi.org/10.1038/nmeth817
10. Adan A, Alizada G, Kiraz Y, Baran Y, Nalbant A (2017) Flow cytometry: basic principles and applications. Crit Rev Biotechnol 37(2): 163–176
11. Zaritskaya L, Shurin MR, Sayers TJ, Malyguine AM (2010) New flow cytometric assays for monitoring cell-mediated cytotoxicity. Expert Rev Vaccines 9(6):601–616
12. Boisson A, Noel G, Saiselet M et al (2021) Fluorescent multiplex immunohistochemistry coupled with other state-of-the-art techniques to systematically characterize the tumor immune microenvironment. Front Mol Biosci 8:673042
13. Zhang W, Hubbard A, Jones T et al (2017) Fully automated 5-plex fluorescent immunohistochemistry with tyramide signal amplification and same species antibodies. Lab Investig 97(7):873–885
14. Lazarus J, Maj T, Smith JJ et al (2018) Spatial and phenotypic immune profiling of metastatic colon cancer. JCI Insight 3(22):e121932
15. Carstens JL, Correa de Sampaio P, Yang D et al (2017) Spatial computation of intratumoral T cells correlates with survival of patients with pancreatic cancer. Nat Commun 8:15095
16. Laberiano-Fernandez C, Hernandez-Ruiz S, Rojas F, Parra ER (2021) Best practices for technical reproducibility assessment of multiplex immunofluorescence. Front Mol Biosci 8:660202

# Chapter 17

# Circle Damage Sequencing for Whole-Genome Analysis of DNA Damage

**Seung-Gi Jin, Jennifer Johnson, and Gerd P. Pfeifer**

## Abstract

There are many sources of endogenous and exogenous DNA damage. Damaged bases represent a threat to genome integrity and may interfere with normal cellular processes such as replication and transcription. To understand the specificity and biological consequences of DNA damage, it is essential to employ methods that are sensitive enough to detect damaged DNA bases at the level of single nucleotide resolution and genome-wide. Here we describe in detail a method we developed for this purpose, circle damage sequencing (CD-seq). This method is based on the circularization of genomic DNA that contains damaged bases and conversion of the damaged sites into double-strand breaks using specific DNA repair enzymes. Library sequencing of the opened circles yields the precise positions of the DNA lesions that are present. CD-seq can be adopted to various types of DNA damage as long as a specific cleavage scheme can be designed.

**Key words** Circle damage sequencing (CD-seq), CD-seq sequencing library, Whole-genome, Circularization of genomic DNA, DNA damage, Double strand break

## 1 Introduction

One important goal in cancer research is to identify DNA lesions and the associated mutational processes that may explain mutational signatures of cancer genomes [1, 2]. Many DNA damaging agents are characterized by a unique spectrum of DNA lesions they produce in exposed cells or in tissues. These exposures can sometimes be linked to a cancer mutational spectrum. Mapping of DNA damage at the nucleotide level of resolution has been possible within specific genes using a method based on ligation-mediated PCR [3–5]. Today, it is of great interest to develop methodology that can be used to map DNA damage at single base resolution and genome-wide. Several such methods exist but all have certain limitations. At a lower level of resolution, antibodies against DNA lesions can be used in immunoprecipitation reactions followed by

Usha N. Kasid and Robert Clarke (eds.), *Cancer Systems and Integrative Biology*, Methods in Molecular Biology, vol. 2660, https://doi.org/10.1007/978-1-0716-3163-8_17,

sequencing of the collected DNA fragments [6, 7]. However, these methods do not achieve single-base resolution, as it would be required for interpreting mutational spectra.

One method is based on the release of small oligonucleotides surrounding a damaged DNA site by the nucleotide excision repair (NER) process. These oligonucleotides are collected and sequenced using high-throughput DNA sequencing (XR-seq) [8]. This method is very useful but is limited to DNA lesions that are processed by NER and cannot detect smaller base lesions. Other methods rely on the ligation of oligonucleotides into a DNA strand break, either a single-strand or a double-strand break [9, 10]. A somewhat related method is called CPD-seq [11, 12], in which ligation of oligonucleotides first to sonicated DNA ends and then to endonuclease-generated breaks is used for signal detection. Another method called HS-damage-seq uses ligation to sonicated ends, immunoprecipitation of the damaged molecules, and blocked primer extension at the lesion sites followed by ligation of a second primer before PCR [13].

As an alternative approach, we have been developing a new method, initially taking clues from an approach used for CRISPR/Cas9 off target identification [14]. We named this method circle-damage-sequencing (or CD-seq).

In step 1 of the procedure (Fig. 1), DNA is isolated from cells that either contain endogenous base modifications (e.g., N6-methyladenine), endogenous DNA damage, or have been treated with a DNA damaging agent.

In step 2, the DNA is then cleaved to produce DNA molecules with average fragment sizes of 300 to 400 base pairs. Initially, we used a four-base cutter restriction enzyme, such as NlaIII (5′ CATG), which leaves a 4-base overhang. The ends that are produced by these enzymes are easy to ligate. However, we found that such enzyme treatment produces a rather substantial genome bias so that genomic regions in which the cleavage sites are underrepresented will drop out from the analysis. To have a less biased distribution of fragments, we now use sonication.

After sonication and end cleaning to produce blunt ends, we then proceed to ligate the DNA into circles using diluted DNA solutions (step 3, Fig. 1). Remaining non-circularized DNA is removed with an exonuclease cocktail (Plasmid-Safe DNase). Circularized DNA is purified, but only a fraction of these circles will contain the DNA modification of interest.

During step 4 of the procedure, the modified base within the circles is then cleaved with specific enzymes to create a double-strand break at the lesion sites. Depending on the type of lesion, different repair enzymes or combinations of repair enzymes will be used. The rings are now opened and contain phosphate groups at both 5′ends.

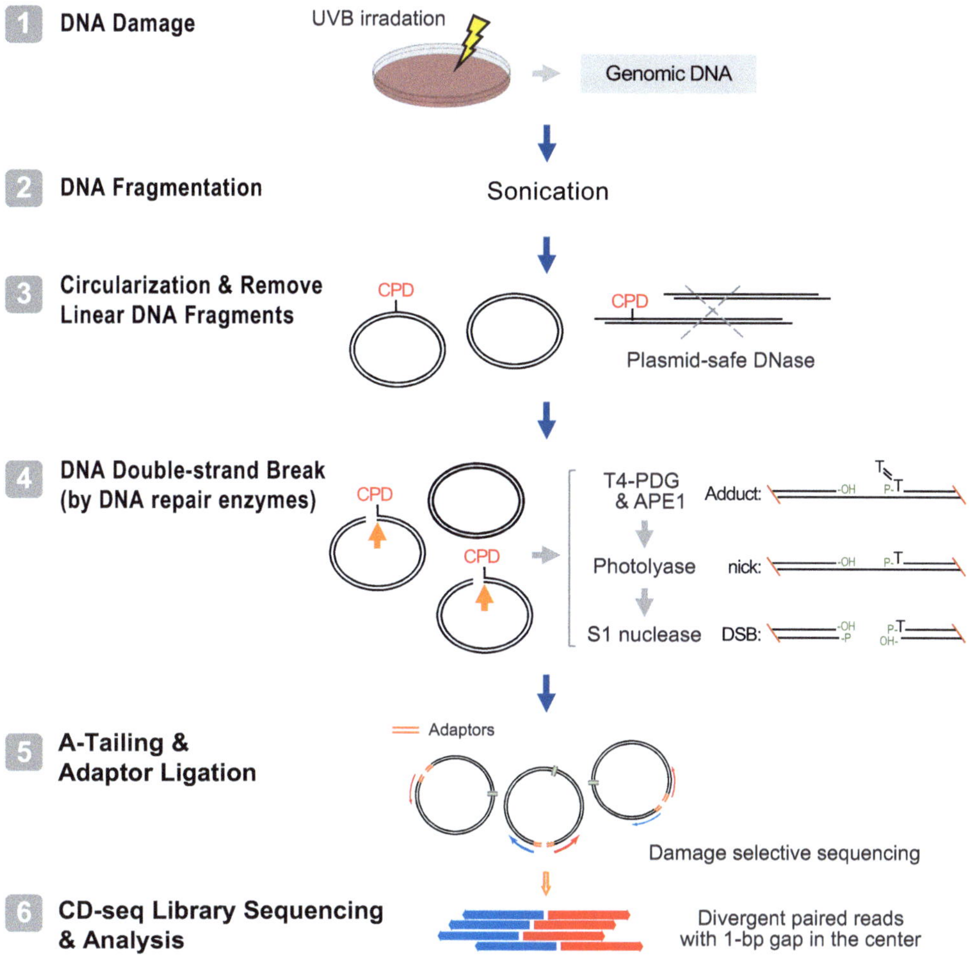


**Fig. 1** Outline of the circle-damage-sequencing method

In the subsequent step 5, we ligate Illumina sequencing linkers to the ends using standard procedures and proceed to library preparation and DNA sequencing using a high-throughput system (*see* step 6). We conduct paired-end 150 bp DNA sequencing of the library and align the reads to the genome.

When the nature of the damaged DNA base(s) is known, for example, a dipyrimidine sequence after UV irradiation or an oxidized guanine after exposure of cells to oxidative stress, we can directly infer the DNA strand on which the damage has occurred. Please note that according to this procedure, theoretically only cleaved and initially DNA damage-containing fragments will be sequenced, which dramatically reduces the complexity of the library being sequenced.

We have initially tested the CD-seq method for detecting and mapping UV-induced cyclobutane pyrimidine dimers (CPDs) in human cells [15]. These dimers occur at the positions of any two pyrimidines and may contain thymine, cytosine, or 5-methylcytosine [16]. To provide an example of the method used in this protocol, we exposed mouse embryo fibroblasts (MEFs) to a dose of 1000 J/m$^2$ of UVB light. We then proceeded to analyze CPDs in these UVB-irradiated cells and followed the scheme shown in Fig. 1.

These pyrimidine dimer DNA adducts are recognized and cleaved by the UV damage-specific DNA glycosylase T4 endonuclease V (T4-PDG). The dimerized pyrimidine bases are still present after T4-PDG cleavage and need to be reverted to monomers with *E. coli* photolyase under UVA light to create clean, ligatable DNA ends (Fig. 1). Mono-functional DNA glycosylases remove only the damaged or mismatched base leaving an apurinic/apyrimidinic (AP) site. Bifunctional DNA glycosylases (such as T4 endonuclease V) contain an additional AP lyase activity that also cleaves the sugar-phosphate backbone. AP endonuclease 1 (APE1) is employed to achieve further cleavage at DNA glycosylases-AP-lyase-incised AP sites. After processing of the 3′ends containing in some cases a sugar-derived aldehyde residue using APE1, we can then cleave the opposite strand of the nicked single base gap containing DNA with single-strand specific S1 nuclease (*see* step 4; Fig. 1). These combined treatments create a double-strand break at the sites of the lesion whereby the one base representing the damaged nucleotide will be missing.

After sequencing and alignment to the mouse mm10 genome, we expect to see divergent DNA reads from the paired-end DNA sequencing because they emanate from the same lesion-dependent DNA strand break and ligation processes. Figure 2 shows examples of genome browser snapshots with such divergent reads. From the known specificity of CPD formation, we expect that the first base of the read (after adapter trimming) would be a pyrimidine, and after genome alignment, the next base in the 5′ direction should also be a pyrimidine. A bioinformatics pipeline is used to visualize and quantitate the damaged bases genome-wide. Further downstream analysis is also possible and depends on the questions being addressed. For example, one can derive consensus sequences for the damaged DNA bases, as we have done for UVB-induced CPDs and their deaminated derivatives [15] and for oxidized guanines and thymines [17]. When appropriate, these sequence patterns can then be used for comparison with mutational spectra found in specific types of cancer [15, 17]. In addition, the general genomic patterns of the DNA damage can be established to show whether certain genome compartments are highly susceptible or less sensitive to the DNA damaging agents. As a cautionary note, while CD-seq is in principle suitable to study DNA repair, it can only be used to aggregate DNA repair rates over defined genomic features, for example, promoters or gene bodies. The method could also aggregate repair rates over individual specific genes by summing up

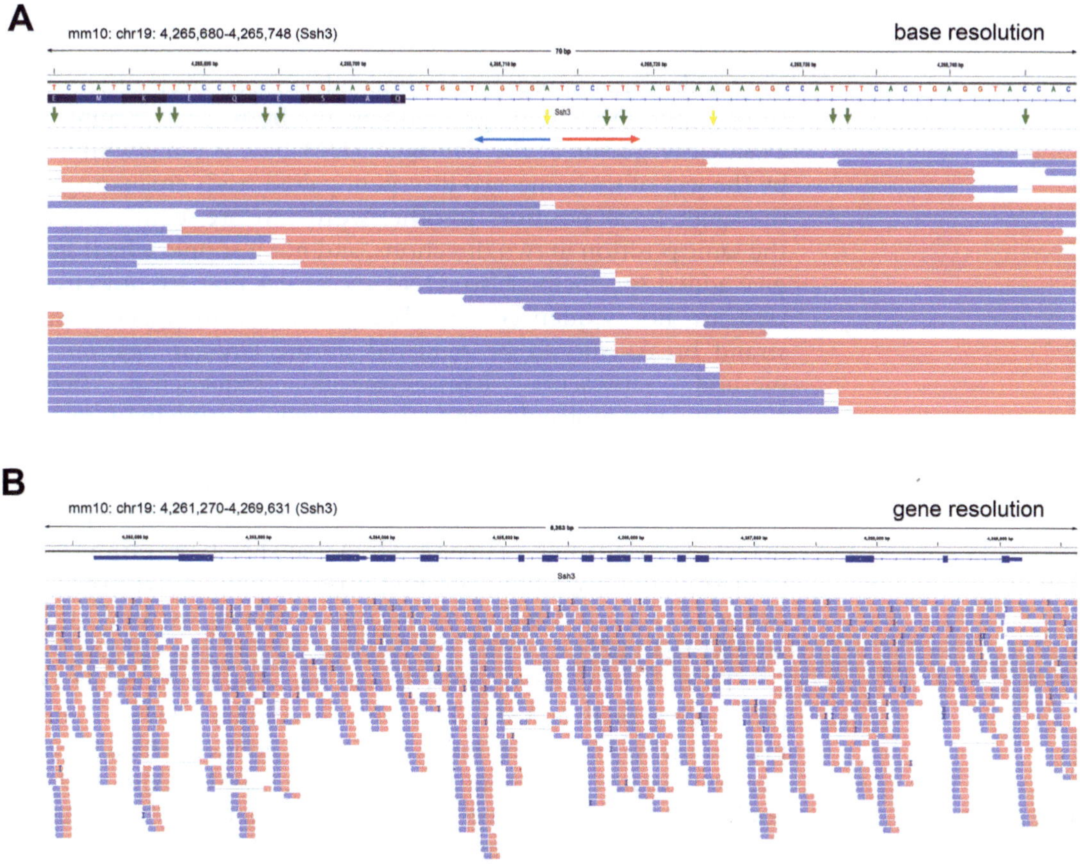


**Fig. 2** Examples of CD-seq results after genome alignment. CPD lesions at dipyrimidine sequences induced into the MEF genome by irradiation with 1000 J/m$^2$ of UVB light were visualized after filtering out divergent paired-end reads from ~100 million paired-end sequencing reads aligned to the mm10 mouse reference genome. (**a**) Snapshot of the IGV viewer displays the characteristic divergent paired reads (shown in lavender and purple colors for the forward and reverse reads, respectively) with single nucleotide gaps. The gaps indicate the positions of UVB-induced CPDs at single-base resolution (chr19: 4,265,680–4,265,748). Green arrows indicate the damage sites (T or C) on the (+) strand, and yellow arrows indicate the damage sites occurring on the (−) strand. (**b**) Snapshot of the IGV viewer shows CPD sites mapped by CD-seq for an 8.4 kb window (chr19:4,261,270-4,269,631; Ssh3) of the mouse genome that includes the *Ssh3* gene

all DNA damage sites over a time course along that gene. In such experiments, spike in DNA samples derived from a heterogeneous source (e.g., UV-irradiated lambda phage DNA) would need to be included in each time course sample. However, what the method cannot do at the present time, at least for any reasonably sized laboratory budget, is to quantify DNA damage and repair at every single possible position of a mammalian genome because extremely high read depth would need to be applied to achieve sufficient coverage.

Circle-damage-seq has several advantages over other high-throughput DNA damage mapping methods. For example, the exact cleavage site at the damaged DNA base is directly indicated by the gap of the same two divergent paired reads emanating from the single break position. In other words, we are sequencing from the same position in two directions, and not just one, which should give added specificity. Also, DNA molecules that are not circularized are not sequenced, and background from ligatable single strand breaks should be removed during circular ligation. The method represents damage-specific sequencing because non-damaged molecules are not sequenced resulting in lower background and lower sequencing cost.

CD-seq is compatible with different types of DNA damage. We have so far used DNA glycosylases which recognize base lesions. To detect 8-oxoguanine, a major DNA damage type produced by oxidative stress, we used the Fpg protein, but OGG1 could also be used [17]. For mapping of oxidized thymines, we used EndoIII and NEIL1 [17].

To detect alkylated DNA bases, one could use repair DNA glycosylases such as AAG [18], which recognize alkylated purines preferentially. For larger, bulky-type DNA adducts, the nucleotide excision repair complex in the form of *E. coli* UvrABC [19, 20] could be used. We believe that the CD-seq method has considerable versatility so that many different types of modified DNA bases should be detectable.

## 2 Materials

Buffer components and other reagents used in the CD-seq procedure must be molecular biology grade fine chemicals and be prepared fresh before use. Avoid multiple freezing and thawing cycles.

### 2.1 Genomic DNA Preparation and Formation of DNA Double-Strand Breaks

1. Quick DNA Miniprep Plus kit (Zymo research, D4074).
2. Sonicator (e.g., Covaris E220 evolution sonicator).
3. Covaris 8 microTUBE Strip V1 (130 μL volume) (Covaris, 520053).
4. DNA Clean & Concentrator-5 (Zymo research, D4013).
5. DNase/RNase free, DNA LoBind 2.0 mL microcentrifuge tubes (e.g., Eppendorf).
6. DNase/RNase free, 0.2 mL 8-strip PCR-tubes.
7. 10 mM Tris–HCl, pH 8.0.
8. Nuclease-free water.

9. PCR thermocycler.
10. PCR-grade dNTPs (10 mM).
11. T4 DNA polymerase (3 U/μL) (NEB, M0203S).
12. T4 polynucleotide kinase (T4 PNK) (10 U/μL) (NEB, M0201S).
13. RNase H (5 U/μL) (NEB, M0297S).
14. NEB T4 DNA ligase reaction buffer (NEB, B0202S): (1×) 50 mM Tris–HCl, pH 7.5, 10 mM $MgCl_2$, 1 mM ATP, 10 mM DTT.
15. NEB T4 DNA ligase (400 U/μL) (NEB, M0202S).
16. NEBuffer 2 buffer (NEB; B7002): (1×) 10 mM Tris–HCl, pH 7.9, 50 mM NaCl, 10 mM $MgCl_2$, and 1 mM DTT.
17. Plasmid-Safe ATP-Dependent DNase (10 U/μL) (Lucigen, E3101K) (*see* **Note 1**).
18. Plasmid-Safe reaction buffer: (1×) 33 mM Tris–HCl, pH 7.5, 66 mM potassium acetate, 10 mM magnesium acetate, 0.5 mM DTT, and 1 mM ATP.
19. AMPure XP beads (Beckman Coulter, 10136224).
20. T4-PDG (T4 pyrimidine dimer glycosylase) (100 U/μL) (NEB, M0308S).
21. APE1 (100 U/μL) (NEB, M0282S).
22. NEBuffer 4 buffer (NEB; B7004): (1×) 20 mM Tris–HCl, pH 7.9, 50 mM potassium acetate, 10 mM magnesium acetate, 0.5 mM DTT, supplemented with 100 μg/mL BSA.
23. PhrB photolyase (0.25 mg/mL) (Novus Biologicals, NBP2-22657).
24. PhrB photolyase reaction buffer: (1×) 50 mM Tris–HCl, pH 7.0, 50 mM NaCl, 10 mM DTT.
25. UVP 3UV lamp (Thermo Fisher Scientific) for UVA or UVB irradiation.
26. S1 nuclease (100 U/μL) (Thermo Fisher Scientific).
27. S1 nuclease reaction buffer: (1×) 40 mM sodium acetate, pH 4.5, 300 mM NaCl, 2 mM $ZnSO_4$.
28. Benchtop thermal mixer.
29. Magnetic separation rack for 0.2 mL tubes.
30. Qubit fluorometer and Qubit dsDNA HS assay kit (Thermo Fisher Scientific).

### 2.2 CD-seq Library Preparation

1. Klenow fragment (3′ → 5′ exo-) (5 U/μL) (NEB, M0212S).
2. NEBNext dA-tailing reaction buffer (NEB, B6059S): (1×) 10 mM Tris–HCl, pH 7.9, 10 mM $MgCl_2$, 50 mM NaCl, 1 mM DTT, 0.2 mM dATP.
3. NEBNext Ultra II Ligation Module (NEB, E7595). This module includes NEBNext Ultra II Ligation Master Mix and NEBNext Ligation Enhancer.
4. Hairpin adaptors for Illumina sequencing: 5′-/5Phos/GATC GGAAGAGCACACGTCTGAACTCCAGTC /ideoxyU/ ACACTCTTT CCTACACGACGCTCTTCCGATC*N-3′ (* represents phosphorothioate linkage) (*see* **Note 2**).
5. USER (Uracil-Specific Excision Reagent) Enzyme (1 U/μL) (NEB, M5505).
6. 6. Index primers for library amplification: i5 index primer; 5′--AATGATACGGCGACCACCGAGATCTACAC-i5 index (8 bp)-ACACTCTTTCCCTACACGACGCTCTTCCGAT*C-3′, i7 index primer; 5′-CAAGCAGAAGACGGCATACGAGAT-i7 index (8 bp)-GTGACTGGAGTTCAGACGTGTGCTCT TCCGAT*C-3′ (* represents phosphorothioate linkage). These index primers were modified from NEBNext Dual Index Primers (NEB, E7600) and can be synthesized by IDT.
7. NEBNext Ultra II Q5 Master Mix (NEB, M0544).
8. Agilent 2100 Bioanalyzer High Sensitivity DNA Kit (Agilent Technologies).
9. 2100 Bioanalyzer instrument (Agilent Technologies).
10. KAPA Library Quantification kit (Kapa Biosystems, KK4824).

### 2.3 Data Analysis

The following software is required:

1. FastQC https://www.bioinformatics.babraham.ac.uk/projects/fastqc/.
2. Trim Galore https://www.bioinformatics.babraham.ac.uk/projects/trim_galore/.
3. BWA MEM https://sourceforge.net/projects/bio-bwa/files/ .
4. SAMtools https://sourceforge.net/projects/samtools/.
5. IGV https://software.broadinstitute.org/software/igv/home.

## 3 Methods

The following methods have been optimized for 1 μg of input DNA and are described for mapping of UVB-induced cyclobutane pyrimidine dimers (CPDs) in mouse embryonic fibroblast (MEF) cells by using the CD-seq procedure. Please use the same amount of intact genomic DNA (gDNA) (non-treated sample) as a negative control.

### 3.1 Preparation of Circularized Genomic DNA

#### 3.1.1 Introduction of DNA CPD Adducts

1. Seed mouse embryonic fibroblast (MEF) cells on 10 cm culture plates.
2. Wash the cells with 1× phosphate buffered saline (PBS) and add 5 mL of 1× PBS to the plates.
3. Irradiate the MEF cells with UVB (e.g., at a dose of 1000 J/$m^2$) when the cells are at 80 to 90% confluence using a UVB lamp that has a peak spectral emission at or near 302 nm (*see* **Note 3**).

#### 3.1.2 Genomic DNA (gDNA) Isolation and Fragmentation

1. Harvest control (non-treated) cells and UVB-302 nm-irradiated MEF cells immediately after irradiation. Pelleted cells can be stored at −80 °C after freezing them using dry ice or liquid nitrogen until further use.
2. Isolate gDNAs from the pelleted cells using Quick DNA Miniprep Plus kit according to the manufacturer's instruction manual (*see* **Note 4**).
3. Quantify the eluted gDNA concentrations using a NanoDrop spectrophotometer.
4. Prepare 1 μg of gDNA per sample in TE buffer, pH 8.0, in a total volume of 130 μL and transfer the liquid to a Covaris microTUBE.
5. Shear the gDNAs using a E220 Covaris sonicator (or equivalent device) with the following settings, for obtaining a 300 bp average fragment size: Peak incident power (W); 140, duty factor; 10%, cycles per burst; 200, time; 80 s (*see* **Note 5**).
6. Clean up the sheared DNAs with DNA Clean & Concentrator-5 according to the manufacturer's instructions.
7. Elute the sheared DNAs in 42 μL of 10 mM Tris–HCl, pH 8.0.
8. To prepare end-repaired (blunt-ended) DNA fragments, prepare the following components in 0.2 mL nuclease-free PCR tubes (final volume 50 μL): 41 μL of sheared genomic DNA, 5 μL of T4 DNA ligase buffer (10×), 1 μL of 10 mM dNTPs, 1 μL of T4 DNA polymerase (3 U/μL), 1 μL of T4 PNK (10 U/μL), and 1 μL of RNase H and mix by pipetting followed by a quick spin.
9. Incubate the solution at 24 °C for 30 min.
10. Clean up the end-repaired DNAs with DNA Clean & Concentrator-5 according to the manufacturer's protocol (*see* **Note 6**).
11. Elute the DNAs in 50 μL of 10 mM Tris–HCl, pH 8.0.

*3.1.3 Intramolecular Circularization of DNA Fragments*

1. Prepare the following components in 0.2 mL nuclease-free PCR tubes (final volume 200 μL): 40 μL (0.5–1 μg) of sheared and blunt-ended genomic DNA, 20 μL of T4 DNA ligase buffer (10×), 4 μL of T4 DNA ligase (400 U/μL), and 136 μL of nuclease-free water and mix gently by pipetting.
2. Incubate at 16 °C in a thermocycler for overnight (15 h).
3. Clean up the DNAs with DNA Clean & Concentrator-5 according to the manufacturer's protocol.
4. Elute the DNAs in 42 μL of 10 mM Tris–HCl, pH 8.0.
5. To remove non-circularized linear DNAs from the circularized DNA pool, add 1 μL (10 U/μL) of Plasmid-Safe ATP-Dependent DNase, 5 μL of Plasmid-Safe Reaction Buffer (10×), and 2 μL of ATP (25 mM) to the ligase-treated DNA in a final volume of 50 μL.
6. Transfer the mixture to 0.2 mL PCR tubes and incubate at 37 °C in a thermocycler for 30 min.
7. Clean up the circularized DNA with 90 μL of AMPure XP beads (1.8:1 ratio of beads to sample) according to the manufacturer's instruction manual.
8. Elute the DNAs in 42 μL of 10 mM Tris–HCl, pH 8.0.

### 3.2 Introduction of Double-Strand Breaks at CPD Sites

*3.2.1 Cleavage of DNA at CPD Sites*

1. To specifically incise DNAs at CPD positions and to generate nicked DNAs, prepare the following components in 0.2 mL nuclease-free PCR tubes (final volume 50 μL): 41.5 μL of the circularized DNAs, 5 μL of NEBuffer 4 reaction buffer (10×), 0.5 μL of BSA (20 mg/mL), 1 μL of T4-PDG (10 U/μL), and 1 μL of APE1 (10 U/μL) (*see* **Note 7**).
2. Mix gently by pipetting and incubate at 37 °C in a thermocycler for 30 min.
3. Clean up the DNA with 90 μL of AMPure XP beads (1.8:1 ratio of beads to sample) and elute in 23 μL of 10 mM Tris–HCl, pH 8.0.

*3.2.2 Reversion of the Dimerized Pyrimidines*

1. To revert the dimerized pyrimidine bases remaining at the 5′ ends after T4-PDG incision, prepare the following components in 0.2 mL nuclease-free PCR tubes (final volume 50 μL), 23 μL of the T4-PDG treated DNA samples, 25 μL of phrB photolyase reaction buffer (2×), and 2 μL of *E. coli* phrB photolyase (0.25 μg/μL), and mix gently by pipetting.
2. Place the reaction tubes under a glass plate at a distance of 15 cm from a UVA lamp (UVP 3UV lamp) that has a peak spectral emission at 365 nm.
3. Incubate the samples under UVA illumination for 60 min at room temperature to induce photolyase activity.

4. Clean up the DNAs with 90 μL of AMPure XP beads (1.8:1 ratio of beads to sample) and elute the DNA in 32 μL of 10 mM Tris–HCl, pH 8.0.

*3.2.3 Generation of Double-Strand Breaks*

1. To cleave the opposite strand of the nicked DNA and to generate ligatable DNA breaks within the circularized DNA, prepare the following components in 0.2 mL nuclease-free PCR tubes (final volume 40 μL): 31 μL of the nicked DNAs, 8 μL of S1 nuclease reaction buffer (5×), and 1 μL of S1 nuclease (5 U/μL) diluted in 1× S1 nuclease reaction buffer.
2. Mix gently by pipetting and incubate for 4 min at room temperature (*see* **Note 8**).
3. Stop the reactions by adding 2 μL of 0.5 M EDTA and 1 μL of 1 M Tris–HCl, pH 8.0 to the reaction tubes followed by further incubation at 70 °C in a thermocycler for 10 min.
4. Clean up the DNAs with 72 μL of AMPure XP beads (1.8:1 ratio of beads to sample) and elute in 48 μL of 10 mM Tris–HCl, pH 8.0.
5. Quantify the concentrations of the eluted DNAs using Qubit dsDNA HS Assay kit. Typically, we obtained 50 ng to 200 ng of double-strand cleaved DNA to be used for the downstream procedures of CD-seq library preparation and sequencing.

### 3.3 Preparation of the CD-seq Sequencing Library

*3.3.1 End-Preparation and Adaptor Ligation*

1. Prepare the following components in 0.2 mL nuclease-free PCR tubes (final volume 55 μL): 47.5 μL of the S1 nuclease treated DNAs, 5.5 μL of NEBNext dA-tailing reaction buffer (10×), and 2 μL (5 U/μL) of Klenow fragment ($3' \rightarrow 5'$ exo-) and mix gently by pipetting.
2. Incubate the solution at 37 °C in a thermocycler for 30 min, and then heat-inactivate the enzymes by further incubation at 70 °C for 30 min and then hold the samples at 4 °C.
3. To perform adaptor ligation, add the following components to the end-prep reaction tubes: 30 μL of NEBNext Ultra II Ligation Master Mix, 1 μL of NEBNext Ligation Enhancer, 3 μL of 1.5 μM T-overhang hairpin adaptor [or NEBNext adaptors for Illumina sequencing (NEB, E7601A)], and 3 μL of 1.5 μM C-overhang hairpin adaptor (*see* **Note 9**).
4. Mix gently by pipetting and incubate the samples at 20 °C in a thermocycler for 15 min, and then hold at 4 °C.
5. Add 3 μL of USER enzyme (1 U/μL) to the reaction mixtures and mix gently by pipetting.
6. Incubate at 37 °C in a thermocycler for 20 min.

7. Clean up the libraries with 95 μL of AMPure XP beads (1:1 ratio of beads to sample) and elute the DNA in 30 μL of 10 mM Tris–HCl, pH 8.0.
8. Quantify the concentration of the eluted libraries using Qubit dsDNA HS Assay kit.

*3.3.2 CD-seq Library Amplification by PCR*

1. Prepare the following components in 0.2 mL nuclease-free PCR tubes (final volume 50 μL), 20 μL of the CD-seq library, 25 μL of NEBNext Ultra II Q5 Master mix (2×), 2.5 μL of 10 μM i5 index primer, and 2.5 μL of 10 μM i7 index primer (NEB), and mix gently by pipetting (*see* **Note 10**).
2. Perform PCR amplification reactions under the following conditions in a thermocycler: incubation at 98 °C for 30 s, and then 11 cycles of PCR at 98 °C for 10 s and 65 °C for 75 s, followed by a final extension step at 65 °C for 5 min (*see* **Note 11**).
3. Purify the amplified libraries with 50 μL of AMPure XP beads (1:1 ratio of beads to sample) and elute the DNA samples in 30 μL of 10 mM Tris–HCl, pH 8.0

*3.3.3 Quantification of the CD-seq Library*

1. Quantify the concentration of the purified libraries using Qubit dsDNA HS Assay kit.
2. Determine the size distributions of the CD-seq libraries using Bioanalyzer (Agilent) and estimate the average fragment sizes of the libraries (*see* **Note 12**).
3. Repeat cleanup of the libraries with 30 μL of AMPure XP beads (1:1 ratio of beads to sample) and elute in 25 μL of 10 mM Tris–HCl, pH 8.0 (*see* **Note 13**).
4. *Optional:* The CD-seq libraries can be further quantified using KAPA Library Quantification kit (Kapa Biosystems) according to the manufacturer's instruction.
5. Calculate the concentration of libraries and pool the libraries using the Pooling Calculator from the Illumina website (https://support.illumina.com/help/pooling-calculator/pooling-calculator.htm).
6. Store the library at −20 °C until sequenced on an Illumina sequencer instrument (*see* **Note 14**). We aim to obtain in the order of 100 million total reads for a standard DNA damage mapping experiment of which about 20% will be single base gapped divergent reads for UVB damage after alignment, removal of duplicates, and trimming processes. For pilot experiments, lower numbers in the range of 10–20 million total reads are sufficient. If very deep coverage is desired, we recommend one to two billion reads per sample.

### 3.4 Data Analysis

Raw data resulting from the Illumina sequencing is generally fastq formatted with 2 × 50 to 2 × 150 nucleotide read length depending on the sequencing platform used.

In paired-end sequencing, the index sequences from each index primer allows the reads to be de-multiplexed into paired fastq files with Read 1 and Read 2 for each library generated from CD-seq.

1. Trim low-quality reads and adaptor sequences with default parameters, and to generate FastQC reports on the trimmed reads, use the –fastqc option. The quality reports can be checked using MultiQC (https://github.com/ewels/MultiQC).

```
$ trim_galore --illumina --paired --phred33 --fastqc -o $OUT_-
DIR $FQ1 $FQ2
```

2. If the length of raw reads (fastq) is more than 100 bp, hard-clip from the 3′ end to 2 × 80 nucleotide length using *Trim_Galore* with default parameters as follows (e.g., with 2 × 150 bp read length):

```
$ trim_galore --illumina --paired --phred33 --fastqc --three_-
prime_clip_R1 70 --three_prime_clip_R2 70 -o $OUT_DIR $FQ1
$FQ2
```

3. Align the trimmed fastq files to the mm10 mouse reference genome using *BWA-MEM* version 0.7.17 with default parameters and remove duplicate alignments with the 'removeDups' *SAMBLASTER* version 0.1.26 option. Then, remove reads with low mapping quality (MAPQ <20) and sort reads by genomic coordinates using *SAMtools* version 1.11.

```
$ bwa mem -t 10 $mm10_ref $R1 $R2 | $ samblaster -r -d
$DISC_OUT -u $UNMAP_OUT -s $SPLIT_OUT | $ samtools sort -O
BAM -o $BAM_OUT -@ 4 | $ samtools index $BAM_OUT
```

4. To identify divergently aligned read pairs (pairs facing away from each other) with a single nucleotide gap between reads representing the positions of DNA adduct bases in CD-seq, select SAM records with TLEN of 3 and save as bam files for downstream analyses using samtools view (*see* **Note 15**).

5. To visualize the reads, use the Integrative Genomics Viewer (IGV) version 2.8.9 with bam files produced by selecting SAM records with TLEN of 3. Display the tracks across the mm10 mouse genome with a "view as pairs" option to show pairs together with a line joining the ends (Fig. 2).

## 4 Notes

1. Exonuclease V is functionally equivalent to Plasmid-Safe ATP--dependent DNase and can be used to selectively remove liner double-stranded DNAs.
2. The T-overhang hairpin adaptor is available from the NEBNext Multiplex Oligos for Illumina (Dual Index Primers) (NEB, E7600) as NEBNext Adaptor for Illumina (NEB, E7601A), and A-, G-, C-overhang hairpin adaptors can be synthesized by Integrated DNA Technologies (IDT).
3. The UVB dose can be determined using a UVX radiometer with a UVB probe.
4. To increase yield of gDNA at the final step from the gDNA prep procedures, we used 100 μL of elution buffer prewarmed to 55 °C and incubated at least 10 min at room temperature, and then eluted.
5. If a Covaris sonicator is not used, shear the gDNAs to an average length of 300 to 400 bp fragments using another appropriate DNA shearing system. We obtained best circularization results from that size range of DNA fragments.
6. Alternatively, after incubation, the reaction mixture can be heat-inactivated at 70 °C for 20 min and be directly used for downstream reaction (circularization) without further purification.
7. T4-PDG and APE1 have 100% activity in NEBuffer 4 (NEB) and 75% activity in NEBuffers 1–3 (NEB).
8. We observed increased 3′ to 5′ exonuclease events from the nicked bases with longer incubation of the S1 nuclease reaction mixture (more than 5 min), causing the increased appearance of multi-base gaps at the DNA adduct sites.
9. We observed that S1 nuclease prefers to cleave at the 5′ side of the single-stranded base gaps in **steps 1** and **2** from Subheading 3.2.3 and may leave a 3′ overhang at an adducted position that is not converted to blunt-ends by Klenow fragment (3′ → 5′ exo-). T4-PDG and S1 nuclease cleavage can generate A- or G-overhangs at the 5′ pyrimidine of CPD dimer sites within the broken circularized DNAs, and here, T- and C-overhang hairpin adaptors are used to ligate adaptors to the DNAs.

10. To add barcode sequences to each library and to pool libraries for Illumina sequencing, the combinatorial indexing of CD-seq libraries is prepared by using i5 and i7 index primer sets.
11. We generally perform 10–13 cycles of PCR amplification to avoid over-amplification and to obtain a high enough amount of library for quality measurement and sequencing.
12. We typically observed an average range of fragments between 550 and 575 bp from Bioanalyzer runs.
13. If there is no excessive adapter dimer peak at 130–150 bp in the sample run, **step 3** from Subheading 3.3.3 can be skipped.
14. Paired-end Illumina sequencing is required to create the divergently aligned read pairs with a single nucleotide gap between mated reads representing the positions of DNA adduct bases on each circularized DNA fragment.
15. TLEN field in SAM file equals the distance between the mapped end of the template and the mapped start of the template. Records with TLEN = 3 represent divergently aligned reads with one nucleotide between mates.

## Acknowledgements

We thank Dean Pettinga for help with setting up the bioinformatics pipeline. This work was supported by NIH grants CA228089 and CA276031 to GPP.

### References

1. Alexandrov LB, Kim J, Haradhvala NJ et al (2020) The repertoire of mutational signatures in human cancer. Nature 578:94–101
2. Pfeifer GP (2015) How the environment shapes cancer genomes. Curr Opin Oncol 27: 71–77
3. Denissenko MF, Pao A, Tang M, Pfeifer GP (1996) Preferential formation of benzo[a]pyrene adducts at lung cancer mutational hotspots in P53. Science 274:430–432
4. Pfeifer GP, Drouin R, Riggs AD, Holmquist GP (1991) In vivo mapping of a DNA adduct at nucleotide resolution: detection of pyrimidine (6-4) pyrimidone photoproducts by ligation-mediated polymerase chain reaction. Proc Natl Acad Sci U S A 88:1374–1378
5. Pfeifer GP, Drouin R, Riggs AD, Holmquist GP (1992) Binding of transcription factors creates hot spots for UV photoproducts in vivo. Mol Cell Biol 12:1798–1804
6. Garcia-Nieto PE, Schwartz EK, King DA, Paulsen J, Collas P, Herrera RE, Morrison AJ (2017) Carcinogen susceptibility is regulated by genome architecture and predicts cancer mutagenesis. EMBO J 36:2829–2843
7. Zavala AG, Morris RT, Wyrick JJ, Smerdon MJ (2014) High-resolution characterization of CPD hotspot formation in human fibroblasts. Nucleic Acids Res 42:893–905
8. Hu J, Adar S, Selby CP, Lieb JD, Sancar A (2015) Genome-wide analysis of human global and transcription-coupled excision repair of UV damage at single-nucleotide resolution. Genes Dev 29:948–960
9. Canela A, Sridharan S, Sciascia N, Tubbs A, Meltzer P, Sleckman BP, Nussenzweig A (2016) DNA breaks and end resection measured genome-wide by end sequencing. Mol Cell 63:898–911

10. Crosetto N, Mitra A, Silva MJ et al (2013) Nucleotide-resolution DNA double-strand break mapping by next-generation sequencing. Nat Methods 10:361–365
11. Mao P, Brown AJ, Esaki S, Lockwood S, Poon GMK, Smerdon MJ, Roberts SA, Wyrick JJ (2018) ETS transcription factors induce a unique UV damage signature that drives recurrent mutagenesis in melanoma. Nat Commun 9:2626
12. Mao P, Smerdon MJ, Roberts SA, Wyrick JJ (2016) Chromosomal landscape of UV damage formation and repair at single-nucleotide resolution. Proc Natl Acad Sci U S A 113: 9057–9062
13. Hu J, Adebali O, Adar S, Sancar A (2017) Dynamic maps of UV damage formation and repair for the human genome. Proc Natl Acad Sci U S A 114:6758–6763
14. Tsai SQ, Nguyen NT, Malagon-Lopez J, Topkar VV, Aryee MJ, Joung JK (2017) CIRCLE-seq: a highly sensitive in vitro screen for genome-wide CRISPR-Cas9 nuclease off-targets. Nat Methods 14:607–614
15. Jin SG, Pettinga D, Johnson J, Li P, Pfeifer GP (2021) The major mechanism of melanoma mutations is based on deamination of cytosine in pyrimidine dimers as determined by circle-damage-sequencing. Sci Adv 7:eabi6508
16. Pfeifer GP (2020) Mechanisms of UV-induced mutations and skin cancer. Genome Instab Dis 1:99–113
17. Jin SG, Meng Y, Johnson J, Szabo PE, Pfeifer GP (2022) Concordance of hydrogen peroxide-induced 8-oxo-guanine patterns with two cancer mutation signatures of upper GI tract tumors. Sci Adv 8:eabn3815
18. Engelward BP, Boosalis MS, Chen BJ, Deng Z, Siciliano MJ, Samson LD (1993) Cloning and characterization of a mouse 3-methyladenine/7-methyl-guanine/3-methylguanine DNA glycosylase cDNA whose gene maps to chromosome 11. Carcinogenesis 14:175–181
19. Sancar A, Rupp WD (1983) A novel repair enzyme: UVRABC excision nuclease of Escherichia coli cuts a DNA strand on both sides of the damaged region. Cell 33:249–260
20. Tang MS, Zheng JB, Denissenko MF, Pfeifer GP, Zheng Y (1999) Use of UvrABC nuclease to quantify benzo[a]pyrene diol epoxide-DNA adduct formation at methylated versus unmethylated CpG sites in the p53 gene. Carcinogenesis 20:1085–1089

# Chapter 18

# CRISPR-Directed Gene Editing as a Method to Reduce Chemoresistance in Lung Cancer Cells

**Natalia Rivera-Torres, Pawel Bialk, and Eric B. Kmiec**

## Abstract

We are advancing a novel strategy for the treatment of solid tumors by employing CRISPR-directed gene editing to reduce levels of standard of care required to halt or reverse the progression of tumor growth. We intend to do this by utilizing a combinatorial approach in which CRISPR-directed gene editing is used to eliminate or significantly reduce the acquired resistance emerging from chemotherapy, radiation therapy, or immunotherapy. We will utilize CRISPR/Cas as a biomolecular tool to disable specific genes involved in the sustainability of resistance to cancer therapy. We have also developed a CRISPR/Cas molecule that can distinguish between the genome of a tumor cell in the genome of a normal cell, thereby conferring target selectivity onto this therapeutic approach. We envision delivering these molecules by direct injection into solid tumors for the treatment of squamous cell carcinomas of the lung, esophageal cancer, and head and neck cancer. We provide experimental details and methodology for utilizing CRISPR/Cas as a supplement to chemotherapy to destroy lung cancer cells.

**Key words** Gene editing, CRISPR/Cas9, Chemoresistance, Squamous cell carcinoma of the lung

## 1 Introduction

The American Cancer Society estimates there will be almost 600,000 deaths due to cancer in the United States this year, as well as a total of almost two million new cancer diagnoses. These new cases will be treated with a regimen consisting of a combination of chemotherapy, immunotherapy, radiation, and surgery. As with any medical treatment, all these therapies have their own risks and complications. In fact, many patients are not able to tolerate their full chemotherapy course due to side effects. Immunotherapy is the newest tool in the box and can have less frequency of side effects than chemotherapy, but when they do occur they have been reported to be more severe. Even with all these treatment options, there are many types of common cancers that still exhibit poor

Usha N. Kasid and Robert Clarke (eds.), *Cancer Systems and Integrative Biology*, Methods in Molecular Biology, vol. 2660, https://doi.org/10.1007/978-1-0716-3163-8_18,

survival. The overall 5-year survival rate for some of the more common cancers (lung, liver, pancreas, and esophageal) is still less than 20%, with minimal improvement from the survival in the 1970s.

We intend to develop a CRISPR/Cas [1, 2] gene editing therapeutic to disable genes functioning to promote or enable resistance to standard of care therapy, most prominently chemotherapy, for solid tumors [3, 4]. Our initial targets are based on foundational molecular biology information describing the activity of critical genes that promote chemoresistance, etc. [5]. Our first genetic target is the gene NRF2, but other targets are emerging including EGFR and KRAS [6]. Critical endpoints include halting the progression of tumor growth, inducing tumor cell death, and reducing tumor size to facilitate surgical resection.

### *1.1 Significance and Therapeutic Methodological Goals*

In developing CRISPR/Cas technology as a treatment for solid tumors, we aim to add another tool to the therapeutic armament. While a primary endpoint for cancer trials is overall survival or progression-free survival, we also envision including an endpoint to measure a reduction in the amount of chemotherapy needed for treatment. In other words, if we can help to reduce side effects from the primary treatment, then in addition to prolonging survival and enhancing the potential to complete the treatment regimen, we may also help their quality of life. A separate goal may be to allow more patients to become surgical candidates due to the reduction in tumor size by CRISPR-directed gene editing. We do not plan initially to use CRISPR as a stand-alone drug but are hoping that we can incorporate it to optimize the standard therapy.

### *1.2 The NRF2 Gene as the Target*

NRF2 is a global transcriptional regulator that functions to activate genes in response to cellular stress [6]. Under normal growth conditions, stress emerges during metabolic processes including cell division and expansion. But, when heat, radiation, or any type of oxidative stress are present in the environment, the Nrf2 protein functions to hyper-activate stress response pathways. This response appears to be evolutionary in nature, and under normal conditions, it ensures continual well-being and survival of the cell. If the cell is transformed, however, and oncogenesis progresses, the NRF2 gene exhibits its dark side and enables the cell to resist standard of care strategies designed precisely to destroy it. These strategies include chemotherapy, radiation therapy, and now, even immunotherapy. Since this gene engages a naturally occurring pathway, a brief knockdown by molecular tools such as RNAi, siRNA, or even antisense therapy will simply delay the inevitable tumor progression. Our objective is to use CRISPR-directed gene editing to disable NRF2 functions so that the cancer cell becomes sensitive to the standard of care therapy. Our preliminary work both in cell culture and in xenograft animal models confirms our belief that this gene is a viable target for a gene editing tool to be used in therapeutic approaches to solid tumors [7]. In the protocol detailed below, we outline the experimental strategies we have used to determine the preliminary efficacy of CRISPR-directed gene editing as a supplemental approach to the killing of lung cancer cells. We describe the use of a nonviral (CRISPRMax) delivery mode to introduce the CRISPR/Cas payload into mammalian cells. While a great deal of activity has surrounded the use of viral vectors for therapeutic purposes, it is now believed that non-viral vectors including the use of lipo-nanoparticles (LNPs) may provide a more safe and more cost-effective way to introduce biologically relevant molecules into human cells. The pioneering work of the biotechnology company, Intellia [8], has helped guide our thinking, and we continue to explore possibilities of utilizing nonviral delivery strategies.

## 2 Methods and Protocols

### 2.1 Cell Line and Culture Conditions

Human lung squamous cell carcinoma NCI-H1703 [H1703] cells (#CRL-5889, passage number 1–12) were purchased from ATCC (Manassas, VA, USA). NCI-H1703 cells were grown in RPMI 1640 medium (ATCC) supplemented with 10% fetal bovine serum (FBS) (ATCC). Cells were grown at 37 °C in 5% CO2. Cells were tested for *Mycoplasma* upon thawing and before use in experiments using the MycoScope PCR Mycoplasma Detection Kit (Genlantis, Cat. MY01100).

### 2.2 NFR2 R34G Cell Line Engineering

The NRF2 gene-coding sequence was analyzed on SnapGene and the following gRNAs were selected for targeting exon 2: (1) 5′- TGGAGGCAAGATATAGATCT-3′, (2) 5′- GATATAG ATCTTGCAGTATC-3′. Synthetic single gRNAs and recombinant spCas9 protein were ordered from Synthego (Menlo Park, California, USA). The sgRNA and SpCas9 protein were mixed at a 5:1 ratio (250:50 pmol) and set to incubate at room temperature for 20 min before transfections. Alt-R repair templates were ordered from Integrated DNA Technologies (Coralville, Iowa, USA): (1) 5′- TTAAAAAACATGAGCTCTCTCCTTCCTTTTTTTGTCT TAAACATAGGACATGGATTTGATTGACATACTTTGGAGGC AAGATATAGATCTTGcAGTAtcTgGAGAAGTATTTGACTTCA GTCAGCGACGGAAAGAGTATGAGCTGGAAAAACAGAAAA AACTTGAAAAGGAAAGACAAGAACAACTCCAAAAGGAGC AAG-3′, (2) 5′- TTAAAAAACATGAGCTCTCTCCTTCCTTTT TTTGTCTTAAACATAGGACATGGATTTGATTGACATACTT TGGAGGCAAGATATAGATCTTGGAGTAAGTgGAGAAGTAT TTGACTTCAGTCAGCGACGGAAAGAGTATGAGCTGGAAA AACAGAAAAAACTTGAAAAGGAAAGACAAGAACAACTCC AAAAGGAGCAAG-3′.

To develop the R34G cell line, we developed a two-step approach to achieve the mutation of interest through CRISPR/Cas9 gene editing. H1703 cells were seeded 24 h before transfection and allowed to reach 60–80% confluency. On the day of the transfection, cells were harvested by trypsinization and neutralized with media. Cells were resuspended at a concentration of 1 × 106 cells/100 μL in SF/supplement solution. 5 μL of RNP complex 1 and 6 uL of repair template 1 were added. Lonza program C M-130 was used, and after 15 min of rest, cells were transferred to a T-25 flask for 48 h before sorting. Transfected cells were sorted in 96-well plates with the Namocell Hana (Namocell, Mountain View, CA, USA). Clones were expanded and transferred to larger plates as the individual clones reached confluence. At the 24-well stage, clonal cells were screened by genomic DNA isolation to identify the mutation of interest and expanded for the second step of transfection. Clone 44 (homozygous knock-in of repair template 1) was later transfected and sorted similarly, but with 5 uL of RNP complex 2 and 3 uL of repair template 2 included in the transfection solution. The appropriate clones from this transfection (homozygous knock-in of repair template 2 – Clones 44–25 and 44–8) were banked for use.

### 2.3 NRF2 R34G Targeting with LNP

2.3.1. Seed $0.4 \times 10^5$ adherent cells in each well of a 24-well plate in complete media (RPMI 1640 + 10% FBS).

2.3.2. Label two tubes: Tube 1 and Tube 2.

2.3.3. In Tube 1, add the following components in the given volumes (mix Cas9 nuclease/sgRNA solution with Cas9 Plus reagent). Mix well.

| Component (Tube 1) | Volume/concentration |
|---|---|
| Opti-MEM I medium | 25 μL |
| spCas9 | 10 pmol |
| sgRNA | 10 pmol |
| Cas9 plus reagent | 2.5 μL |

2.3.4. In Tube 2, add the following components in the given volumes (dilute CRISPRMAX Reagent in Opti-MEM I medium). Mix well.

| Component (Tube 2) | Volume/concentration |
|---|---|
| Opti-MEM I medium | 25 μL |
| CRISPRMAX reagent | 1.5 μL |

2.3.5. Immediately add solution from Tube 1 and Tube 2, and then mix well. This will allow the formation of the Cas9 nuclease/gRNA/transfection reagent complex.

2.3.6. Incubate complex for 10 min at room temperature. (**Note**: *Do not incubate for >30 min).*

2.3.7. Aspirate the cell media to the well of the 24-well plate. Replace with 500 μL of complete media (RPMI 1640 + 10% FBS).

2.3.8. Add the Cas9 nuclease/gRNA/transfection reagent complex (~50 μL) to the cells in the well.

2.3.9. Incubate cells with the Cas9 nuclease/gRNA/transfection reagent complex for 72 h.

2.3.10. After incubation, remove the culture medium and rinse cells with 50–500 μL PBS. Cells can now be used for downstream analysis.

### 2.4 Gene Editing Analysis

Cellular genomic DNA was isolated from each clonal cell line using the DNeasy Blood and Tissue Kit (Qiagen, Cat. 69506). The region surrounding the CRISPR target site was PCR amplified using the Q5 High-Fidelity 2X Master Mix (New England BioLabs, Cat. M0492) (Exon 2 – FWD primer 5′--CACCATCAACAGTGGCATAATGTGAA-3′, REV primer 5′-AACTCAGGTTAGGTACTGAACTCATCA-3′). The PCR reaction was purified using the QIAquick PCR Purification Kit (Qiagen, Cat. 28106) and Big Dye Terminator PCR was performed using Big Dye Terminator v3.1 (Thermofisher). PCR products were purified once more using the Big Dye Xterminator kit (Thermo Fisher) and then sequenced using the SeqStudio Genetic Analyzer (Applied Biosystems). Clonal allelic analysis was conducted using the software program, DECODR, available at https://decodr.org/analyze.

### 2.5 Chemosensitivity Testing

Cell viability after drug exposure was evaluated using the CellTiter 96 Aqueous Non-Radioactive Cell Proliferation Assay (Promega, Madison, WI). R34G cell lines were plated in quadruplicate at $4 \times 10^3$ cells per well and allowed to culture for 24 h. The cells were then treated with cisplatin at increasing concentrations for 3 days. The cell media was aspirated, the cells washed with PBS, and then exposed to the MTS reagent for 3 h. After 3 h of MTS bio-reduction by proliferating cells, the formazan product's absorbance was measured using a 450 nm filter on an Infinite 2000 PRO microplate reader (Tecan, Mannadorf, Switzerland). Each experiment was conducted twice, and the data was normalized and averaged for each clone at each concentration. The standard error of the mean was calculated using the normalized value of all eight data points for each clone at each concentration.

## 3 Results

To provide further guidance on the protocol and to outline our workflow and predicted results, we have provided three figures performed as described above. In Fig. 1, our routine and standardized experimental flow is described. H1703 cells containing the NRF2 R34G mutation, propagated in culture, were freshly plated 24 h prior to the introduction of the CRISPR/Cas payload. To deliver the bioactive molecules, we employed the nonviral liposomal formulation CRISPRMax at the optimized ratios (see above) and allowed the cells to incubate for 72 h. It is important to note that the transformation of established cell lines by any bioactive molecule requires significant levels of optimization. We have found that the length of time that cells are cultured and passaged in a continual line can affect this efficiency. It is also important to establish the optimal density for the transformation and transfection of any molecule into cultured cells. Serum levels in media constituents are also determining factors in the efficiency of uptake, especially in the case of viral delivery modes. Importantly, optimization of successful transfection must include an analysis of the level of active molecules introduced into the nucleus, not just the cell. We also monitor the health of the cell during this period to ensure that at least 80–90% of the cells are remaining intact and viable. Transfection under any conditions is a harsh cellular process, and the cells need a period to recover prior to executing the intended molecular outcome. After 72 h, the level of gene editing is assessed by using a sample from the treated cell culture. We recommend this quick look-in at the progress of the gene editing activity prior to the addition of anticancer therapy, in this case, chemotherapeutic drugs. This allows for an early analysis of the efficiency of the gene editing activity which can help troubleshoot field experiments obtained after cell viability measurements. Once completed, the

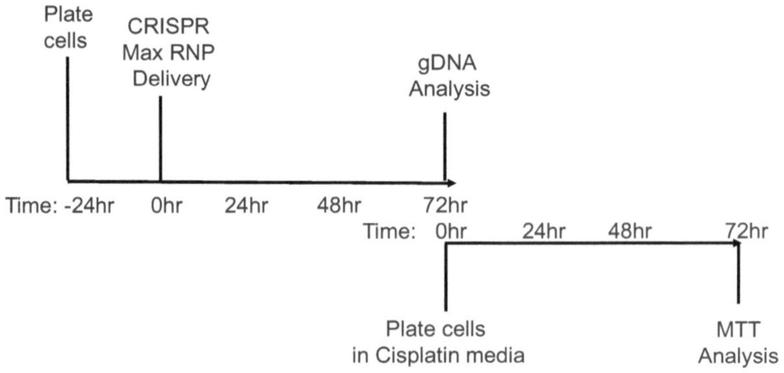


**Fig. 1** NRF2 R34G targeting and cisplatin treatment. Experimental CRISPR/Cas9 RNP delivery workflow for targeting NRF2 R34G cell lines (see text for details)

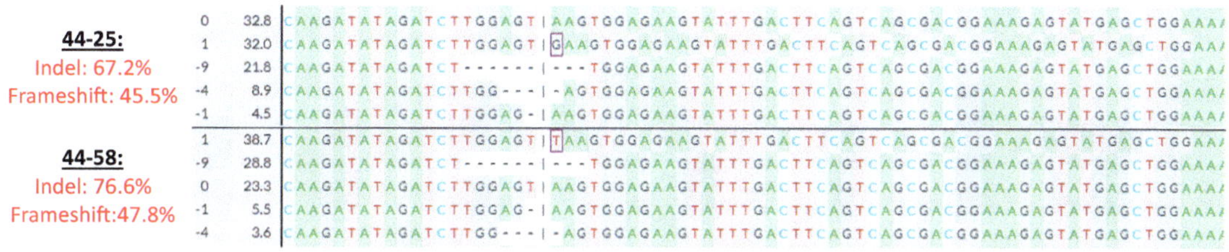


**Fig. 2** Genomic analyses of R34G cell lines after CRISPR targeting. Genomic DNA from R34G cell lines, 44–25 and 44–58, was isolated and amplified across exon 2 of the *NRF2* gene. Amplicons were Sanger sequenced and analyzed for indels at the CRISPR target site. Raw sequence files were aligned using the software program, DECODR, to display the *NRF2* allele-specific indel pattern (listed as INDEL and %)

targeted cells are then plated in the presence of the desired chemotherapeutic reagent, in this case, cisplatin. Incubation continues for another 72 h, and viability testing completes the experimental protocol.

Figure 2 illustrates a standard analysis and profile of an experiment carried out using the protocol described above. Here the sequence files are aligned to demonstrate the efficiency of genetic disruption caused by the CRISPR/Cas complex as well as the revelation of the categories of indels created in each allele. The software program DECODR [9] is the best analytical tool for deciphering the outcome of CRISPR-directed gene editing using Sanger sequencing. Here, two clonal cell lines were examined, and each cell line shows to have gene editing executed at a frequency of over 60%. Overall indel formation above 60–70% is an excellent outcome but more importantly, frameshift mutations at approximately 50% ensure the generation of clonal populations that have a disrupted target gene. We consider the frameshift mutation frequency to be the standard by which we judge genetic knockout activity. In some cases, deletions or point mutation changes that do not lead to frameshift mutations can often result in the generation of truncated proteins or enable genetic rearrangements such as exon skipping.

Figure 3 depicts the corresponding phenotypic readout of gene editing by measuring the increased sensitivity of targeted cells to cisplatin. A standard viability assay is used to assess the overall effect of genetic disruption at several concentrations of cisplatin. It is important to assess the sensitivity range of cells to chemotherapeutic drugs independent of genetic manipulation prior to the beginning of this assay. In our case, we are utilizing concentrations of cisplatin that are known to enable cell viability at lower levels. Several important observations are instructive in this figure. First, CRISPR-directed gene editing is effective in reducing the growth of tumor cells in response to cisplatin. Second, clonal isolates from the same cell line bearing similar genetic disruption patterns exhibit slightly different sensitivity to the chemotherapeutic agent. The second observation is a key to developing effective anticancer strategies since solid tumors are composed of a mosaic of genetic

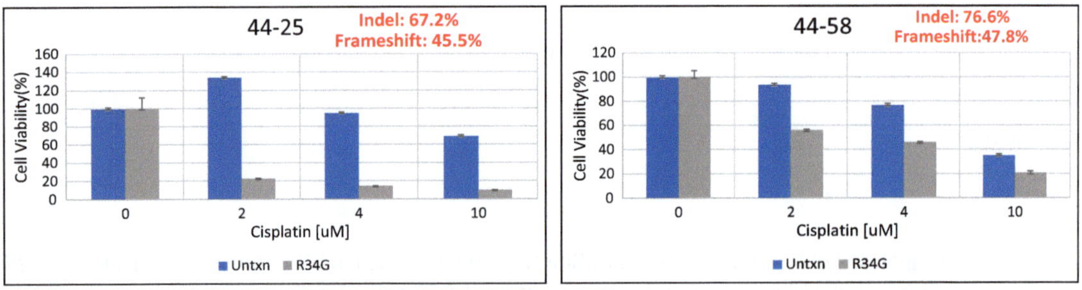


**Fig. 3** Chemosensitivity testing in response to NRF2 R34G targeting. Chemosensitivity was measured via bio-reduction of MTS to a formazan product. Cells were treated with increasing concentrations of cisplatin for 72 h and then evaluated for cell proliferation. The average relative proliferation of cells in response to cisplatin is graphed above

signatures, and thus the correct range of chemotherapeutic treatment must be assessed carefully. And, as the mutated phenotype within tumor cells expands, changes in response to varying doses of chemotherapy can change rapidly. Our data suggest that the combinatorial treatment of CRISPR-directed gene editing with chemotherapy can overcome a significant degree of chemoresistance in tumor cells. Furthermore, more effective cell killing is observed at lower concentrations of chemotherapy, an important clinical outcome for the development of safer and more tolerable anticancer therapies.

## 4 Discussion and Conclusion

Chemotherapy remains an important option in the treatment of lung cancer, but issues involving chemoresistance and toxicity are often problematic with extended care. A clear demonstration that the genetic knockout of NRF2 improves the efficacy of chemotherapy would provide foundational information to the field at large and potentially introduce a new weapon in the anticancer treatment armament [3, 4, 6]. Because we are embedded in a community cancer center, we embrace the development of innovative therapies for lung cancer as the research focus of our laboratory. This method provides foundational information for workers in the gene editing field to establish a reliable and robust assay system to analyze the effect of genetic knockout of key genes involved in the development of resistance to standard of care. For example, oncogenic KRAS mutations are found in approximately 25% of non-small cell lung cancers and are associated with resistance to standard platinum-based chemotherapy, associated with a poor overall prognosis. This unique population of patients with KRAS-mutated non-small cell lung cancers with chemotherapy-resistant tumors is the ideal initial target population for this type of innovative combinatorial therapy. They have limited options and have shown to be refractory to most forms of currently practiced treatments.

Although multiple breakthroughs in cancer management have been clinically approved and have significantly improved the quality of life in patients with cancers, the sad fact is that only 10–20% of cancer patients in almost every type of cancer remain or develop resistance to any type of treatment [10, 11]. Therefore, the development of novel therapeutics is critical. With the advancement of next-generation sequencing, there now is a growing list of tumor driver mutations in NRF2, EGFR, KRAS, TP53, ERBB2 (HER2/neu), and BRAF that have provided information detailing genetic signatures expanding its clinical applications in cancer diagnosis, prognosis, and therapeutic decisions by becoming more readily accessible to clinicians and translational scientists. The mutations of these genes often lead to the accumulation of functionally altered proteins within the tumor cells and constitutively enable resistance to treatment. At present, there are few effective therapeutics capable of removing the mutated proteins within the tumor cells; therefore, CRISPR/Cas9-mediated precision targeting of tumor driver mutations would be an ideal intervention through the disruption of genes that code for these proteins. Ultimately, integrating effective chemotherapy, targeted drugs, immunotherapy, and now gene editing should give these patients the best chance at longer disease control and improve quality of life.

## References


1. Doudna JA, Charpentier E (2014) The new frontier of genome engineering with CRISPR-Cas9. Science 346. https://doi.org/10.1126/science.1258096
2. Hsu PD, Lander ES, Zhang F (2014) Development and applications of CRISPR-Cas9 for genome engineering. Cell 157:1262–1278
3. Chen J, Solomides C, Simpkins F, Simpkins H (2017) The role of Nrf2 and ATF2 in resistance to platinum-based chemotherapy. Cancer Chemother Pharmacol 79:369–380
4. Wang XJ, Sun Z, Villeneuve NF et al (2008) Nrf2 enhances resistance of cancer cells to chemotherapeutic drugs, the dark side of Nrf2. Carcinogenesis 29:1235–1243
5. Hayden A, Douglas J, Sommerlad M et al (2014) The Nrf2 transcription factor contributes to resistance to cisplatin in bladder cancer. Urol Oncol 32:806–814
6. Bialk P, Wang Y, Banas K, Kmiec EB (2018) Functional gene knockout of NRF2 increases chemosensitivity of human lung cancer A549 cells in vitro and in a xenograft mouse model. Mol Ther Oncolytics 11:75–89
7. Banas K, Modarai S, Rivera-Torres N, Yoo B-C, Bialk P, Barrett C, Batish M, Kmiec EB (2022) Exon skipping induced by CRISPR-directed gene editing regulates the response to chemotherapy in non-small cell lung carcinoma cells. Gene Ther 29:357–367
8. Gillmore JD, Gane E, Taubel J et al (2021) Crispr-cas9 in vivo gene editing for transthyretin amyloidosis. N Engl J Med 385:493–502. https://doi.org/10.1056/NEJMoa2107454
9. Bloh K, Kanchana R, Bialk P, Banas K, Zhang Z, Yoo BC, Kmiec EB (2021) Deconvolution of complex DNA repair (DECODR): establishing a novel deconvolution algorithm for comprehensive analysis of CRISPR-edited Sanger sequencing data. CRISPR J 4(1): 120–131. https://doi.org/10.1089/crispr.2020.0022
10. Zhang D, Rennhack J, Andrechek ER, Rockwell CE, Liby KT (2018) Identification of an unfavorable immune signature in advanced lung tumors from Nrf2-deficient mice. Antioxid Redox Signal 29:1535–1552
11. Homma S, Ishii Y, Morishima Y et al (2009) Nrf2 enhances cell proliferation and resistance to anticancer drugs in human lung cancer. Clin Cancer Res 15:3423–3432

# Chapter 19

# Integrated In Silico Analysis of Proteogenomic and Drug Targets for Pancreatic Cancer Survival

**Alakesh Bera, Digonto Chatterjee, Jack Hester, and Meera Srivastava**

## Abstract

Pancreatic cancer remains a major health concern, being among the deadliest forms of cancer with over 80% of the patients presenting with metastatic disease. According to the American Cancer Society, for all stages of pancreatic cancer combined, the 5-year survival rate is less than 10%. Genetic research on pancreatic cancer has generally been focused on familial pancreatic cancer, which is only 10% of all pancreatic cancer patients. This study focuses on finding genes that impact the survival of pancreatic cancer patients which can be used as biomarkers and potential targets to develop personalized treatment options. We used cBioPortal platform using NCI-initiated The Cancer Genome Atlas (TCGA) dataset to find genes that were altered differently in different ethnic groups which can serve as potential biomarkers and analyzed the genes' impact on patient survival. MD Anderson Cell Lines Project (MCLP) and genecards.org were also utilized to identify potential drug candidates that can target the proteins encoded by the genes. The results showed that there are unique genes that are associated with each race category which may influence the survival outcomes of patients, and their potential drug candidates were identified.

**Key words** Proteogenomics, Proteomics, Genomics, Survival outcomes, Drug targets, cBioPortal, The Cancer Genome Atlas (TCGA), MD Anderson Cell Lines Project (MCLP), Copy number alterations (CNAs), Socioeconomic status (SES), Disease-free survival (DFS), Overall survival (OS), Pancreatic cancer

## 1 Introduction

Pancreatic cancer is among the deadliest forms of cancer. It is the seventh most common cancer, yet it is the second leading cause of cancer deaths in the United States [1]. It is estimated that in 2021, 48,220 patients will die from pancreatic cancer. Risk factors for pancreatic cancer include smoking, diabetes, obesity, chronic pancreatitis, and family history [2]. Over 80% of the patients present with metastatic disease. Despite advances in chemotherapy, the average survival remains less than 5 years even after surgery [3]. Genetic testing has been primarily focused on familial pancreatic cancer, which only accounts for ten percent (10%) of all pancreatic

Usha N. Kasid and Robert Clarke (eds.), *Cancer Systems and Integrative Biology*, Methods in Molecular Biology, vol. 2660, https://doi.org/10.1007/978-1-0716-3163-8_19,

cancers. A study on non-familial pancreatic cancer patients revealed six genes, namely, CDKN2A, TP53, MLH1, BRCA2, ATM, and BRCA1, with significant associations between pancreatic cancer and mutations in these genes [4]. A previous study has shown that patient survival was associated with mutations in KRAS, CDKN2A, SMAD4, and TP53 [5].

Similar to other common malignancies, pancreatic cancer is associated with disparities by socioeconomic status (SES), ethnic minority status, and insurance [6, 7]. In contrast to other types of cancer (breast, colon) where screening can detect early-stage disease, no screening modality exists for pancreatic cancer. Thus, disparities in outcomes for pancreatic cancer do not result from lack of screening [8]. There are currently limited data on the genetic susceptibility of pancreatic cancer survival based on race or ancestral origin. The association of driver gene alterations based on racial category and their association with patient outcomes has not been clearly established. Therefore, we performed a quantitative genomic analysis to find genes associated with patient survival for the pancreatic cancer patients based on their racial categories. In this current study, we tried to establish a method to identify the genetic basis of racial disparities in cancer survival. We tried to determine whether White, African American/Black, and Asian patients have different gene alterations and whether or not they can be diagnosed and managed based on their specific signature profile. Identification, prevention, and management of factors based on race may help find effective strategies for clinical management of pancreatic cancer.

## 2 Clinical Data from cBioPortal

cBioPortal was utilized to access and analyze the public database on cancer tissues generated by the TCGA project (https://www.cancer.gov/tcga). The Cancer Genome Atlas (TCGA) is an NCI-initiated and established cancer genomic database which is publicly available. There are over thirty thousand tumor samples that have already been sequenced from over twenty different types of cancer. The cBioPortal allows users to question datasets across data types including genes and clinical samples, providing an opportunity to investigate several different biologically and/or clinically relevant hypotheses. For this study, all datasets for pancreatic cancer were selected particularly for patient's survival with genetic signature. Data also included the specific patient's number and their contributions from different studies. Customized datasets were created from Pancreatic Cancer Studies in cBioPortal along with age, race, and sex distributions of patients (Fig. 1). Datasets that did not provide race information were not included in this analysis.

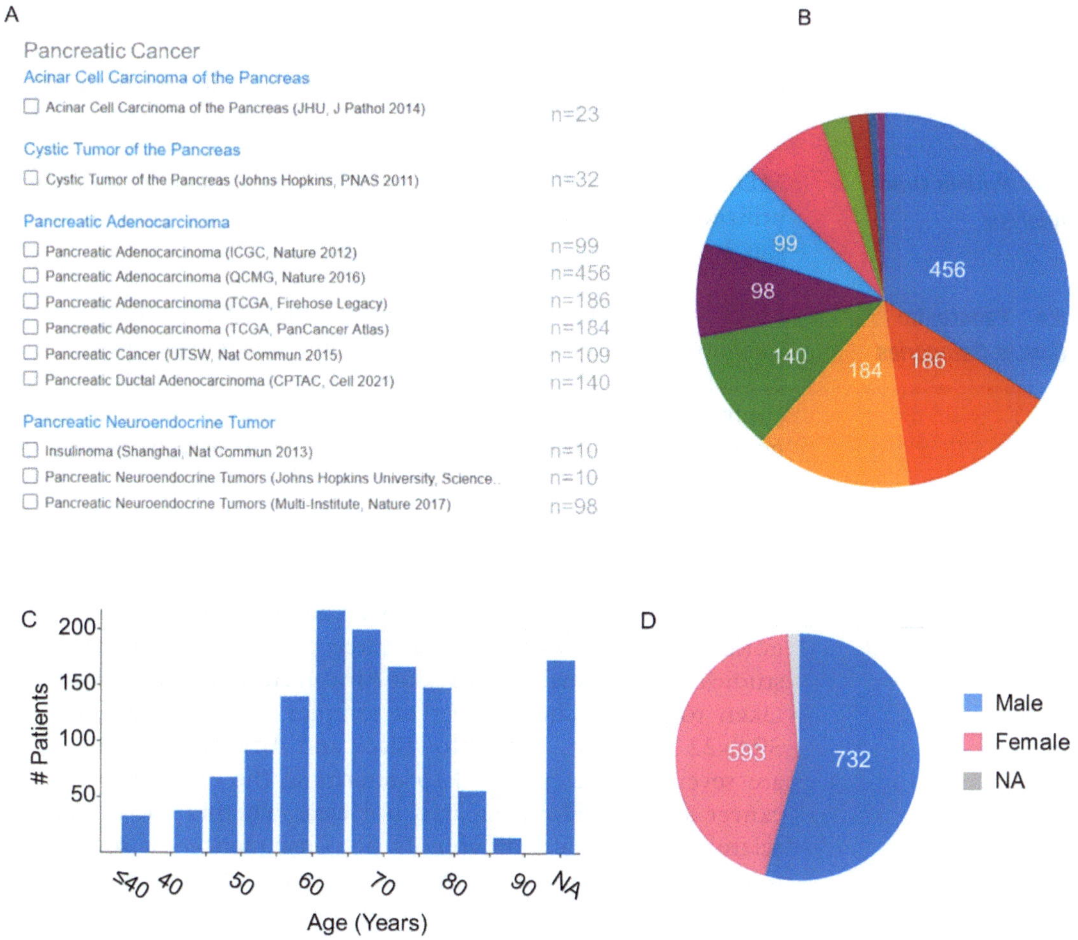


**Fig. 1** Different studies included in this analysis to study the pancreatic cancer patient's survival with genetic signature in particular. (**a**, **b**) Data also included the specific patient's number and their contributions from different studies. (**c**, **d**) Customized dataset created from Pancreatic Cancer Studies in cBioPortal. (**c**) Data indicated the age distributions of patients. (**d**) Male vs female ratio

## 3 Methods

### 3.1 Copy Number Alterations

Copy number alteration data was analyzed in each of three racial categories, (1) White, (2) African American and Black, and (3) Asian with customized datasets. Within those separate virtual studies, the genes that were amplified in the most patients in terms of CNA (copy number alterations) were recorded and compared among the different virtual datasets.

### 3.2 Survival Curves

Genes that were altered differently in the racial groups were selected to analyze the genes' impact on patient survival. Both overall survival (OS) and disease-free survival (DFS) curves were computed. This was done by utilizing the inbuilt statistical analysis tools available within the cBioPortal platform. The data was tested for significance using a log-rank test to compare the survival distributions between two samples. This was done automatically using cBioPortal [9, 10].

### 3.3 Protein Drug Analysis

MD Anderson Cell Lines Project (MCLP) and genecards.org were utilized to find protein drug interactions using the genes found in this study as drug targets [11, 12].

### 3.4 Pancreatic Cancer Categories

More than 1200 patient data spanning ten [10] different pancreatic cancer clinical studies were analyzed. The disease types were (1) acinar cell carcinoma of the pancreas, (2) cystic tumor of the pancreas, (3) five pancreatic adenocarcinoma studies, and (4) three pancreatic neuroendocrine tumor studies (Fig. 1). The data were stratified according to three race categories self-identified by patients: (1) White, (2) African American or Black, and (3) Asian. There were only two samples that were identified as Hispanic; therefore the Hispanic category was not included in our study. A total of 436 patient samples were customized based on three (3) race categories. The samples were from three pancreatic adenocarcinoma studies. For our analysis, Black or African American and Black were taken together. The number of samples for White patients was 394, 22 for Asian, and 19 for Black and African American. There are several studies pursued to determine the racial disparities in cancer survival. However, it is not clear whether the disparity is related to biological differences or consequences of social, economic, or cultural environments. Given the challenges associated with race and ethnicity definition and disease-linked biological observations in this context, results from our study need to be independently validated in larger patient population with ancestry information.

### 3.5 Data Analysis

#### 3.5.1 Copy Number Alterations Based on Race Categories

The analysis began with an investigation of copy number alterations (CNA) at the genomic DNA level for pancreatic cancer samples. Copy number alterations (CNAs) refer to changes in the number of copies of a genetic region caused by deletion or duplication events in the genome. Recent technological advances have enabled the identification of CNAs across the entire genome, associating cancer with the alteration rates of various genes in cancer [8].

CBioPortal was utilized, which contains comprehensive genomic and transcriptomic data from various cancer studies [13]. Focusing on the customized pancreatic cancer datasets created based on race categories as mentioned in the above section, genes with the highest frequency of copy number alterations (CNA) in each race category were determined. A few unique genes specific to each race category that had copy number amplifications were found (Table 1). The altered genes' locations are in different cytobands for each race category. As previously reported,

**Table 1**
**Genes with copy number alterations unique to each race category**

| Race | Genes with copy number alterations (AMP) | Cytoband | Alteration frequency (%) |
|---|---|---|---|
| White | GATA6, RECQL4, MIB1 | 18q11.2 (GATA5, MIB1)<br>8q24.3 (RECQL4) | 14.8 (GATA6)<br>14.0(RECQL4)<br>12.8%(MIB1) |
| Black | PKDL1, GARS1, NEUROD3 | 7p12.3(PKDL1)<br>7p14.3(GA RS1, NEUROD3) | 28.6 |
| Asian | FGFR3, ABCA11P, UVSSA (etc.) | 4p16.3 | 27.3 |

CDKN2A was deleted in 30% of patients in every race category [5], confirming our data analysis approach. Most genes that had deep deletions in copy number were common among the different race categories and therefore were not included in further investigation and analysis.

### 3.5.2 Association of CNA Amplification with Survival Outcome

An evaluation was carried out between the association of the gene alterations unique to each category and most frequently amplified in each race category with survival outcome. The most frequently altered genes in each category were found to have a direct association with patient survival. Kaplan–Meyer curves for overall survival (OS) and disease-free survival (DFS) were generated using the cBioPortal platform for each gene in the specific race categories. P-values were generated by cBioPortal using the log-rank test for survival between 2 samples. This was compared to a significance level of 0.05. The survival was calculated in median months, and the 95% confidence interval was given as well.

Figure 2 shows representative Kaplan–Meier curves for disease-free survival (DFS) outcomes associated with alterations of genes in patients from different race categories. The P-values are 0.0031 for White patients, 0.0516 for Asian patients, and 0.8680 for African American and Black patients.

The results showed that there are unique genes that are associated with each race category which have an effect on the survival outcomes of patients. For patients that had expression of these genes, the median month of survival was less compared to those who did not have amplified expression.

***White Patients*** For the genes amplified in the White race category, the effect on OS was minimal while the DFS was significantly worse than in patients where there was no amplification with a p-value of 0.0031. The genes amplified in the White race category were from the cytoband 18p11.2 (GATA6, MIB1) and 8q24.3 (RECQL4). It is notable that the genes that were from the same cytoband showed nearly identical data.

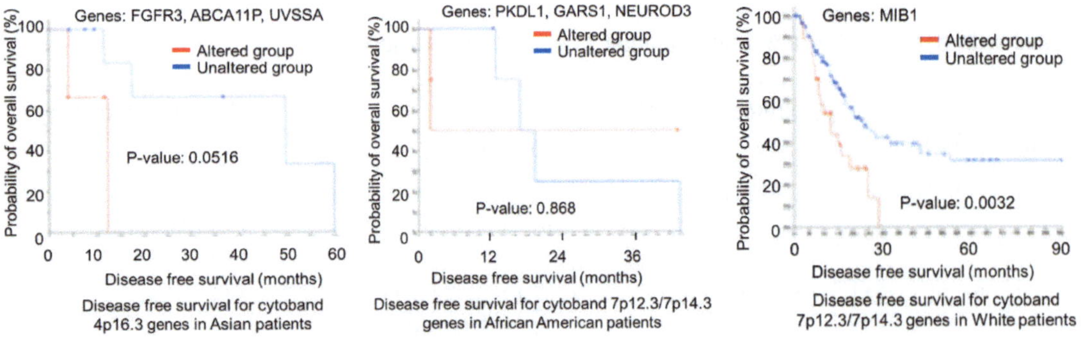


**Fig. 2** Kaplan–Meyer curves for disease-free survival in each race category. Statistical values (P-value) are included in the figure

**Table 2**
**Survival outcomes (*in months*) associated with altered genes in each race category**

| Race category | Genes altered | Overall Survival (OS) unaltered | Overall Survival (OS) altered | Disease Free Survival (DFS)-unaltered | Disease Free Survival (DFS)-altered |
|---|---|---|---|---|---|
| | GATA6 | 20.17 | 20.35 | 20.4 | 12.43 |
| White | RECQL4 | 20.35 | 15.11 | 20.37 | 9.57 |
| | MIB1 | 20.19 | 20.34 | 23.52 | 12.42 |
| African American & black | PKDIL1. GARS1, NEUROD6 | 17.03 | 2.01 | NA | NA |
| Asian | FGFR3. ABCA11P, UVSSA. (etc.) | 66.89 | 11.61 | 12.32 | 49.68 |

***Asian Patients*** For genes amplified in the Asian race category, the survival outcomes for patients with amplified genes was significantly worse for both overall survival and disease-free survival compared to patients who did not have amplified genes with a p-value of 0.0516. The genes in the Asian race category were all from the same cytoband (4p16.3), and while only 3 were listed in Table 2, there were 39 genes from the cytoband which were all amplified and showed the same survival outcomes.

***African American and Black Patients*** For genes amplified in the African American and Black race category, the effect on overall survival for patients that had amplified genes in the 7p cytoband had significantly poor outcomes (2 months) with alterations in the genes PKD1L1, GARS1, and NEUROD6 compared to patients with no amplification (17 months). There was no data for disease-

**Table 3**
**Role of the genes in cancers and as potential drug candidates**

| Genes/race categories | Role in pancreatic cancer | Role in other cancers (prognostic value) | Drug candidates |
|---|---|---|---|
| GATA6 (White) | No | Renal (unfavorable) (detected in many) | Spautin.l; parbendazole |
| RECQL4 (White) | No | Liver (unfavorable) (detected in all) | Rec 15/2615 dihydrochloride |
| MIB1 (White) | No | None (detected in all) | SIB 1893 |
| PKD1L1 (Black) | No | None (detected in many) | JKC 363; HS 014 |
| GARS1 (Black) | NA | NA | MRS 1220; RS 100329 hydrochloride |
| NEUROD6 (Black) | No | None (not detected) | SB 224289 hydrochloride; Pancuronium dibromide |
| FGFR3 (Asian) | No | Endometrial (unfavorable) (detected in many) | PD 173074 |
| ABCA11P (Asian) | NA | NA | Bobcat339, TBCA |
| UVSSA (Asian) | No | Renal (unfavorable) urothelial (favorable) detected in all | Iressa, ASB 14780 |

free survival due to the limited number of samples. However, it is notable to mention that data for the African American and Black categories were limited (African American/Black, $n = 19$) and were not statistically significant.

*3.5.3 Protein Expression and Protein–Drug Interaction*

Using the Protein Atlas, protein–drug interaction from MD Anderson Cancer-Cell Lines Project (MCLP) dataset [11, 12] and genecards.org data, an analysis of the role of the genes in cancers that are the focus of our study was done, and drug candidates that can target the proteins encoded by the genes were determined. Potential drug candidates that can be further tested in clinical trials to provide gene and race specific therapeutic options for pancreatic cancer patients were proposed. Table 3 shows the role of the genes in cancers, their role as prognostic markers, and drugs that can be used to target these genes. The top-ranked 2 drugs targeting each gene were included.

## 4 Notes

1. This study demonstrates that there are unique sets of genes commonly associated with self-identified race.
2. Alterations in these genes are associated with patient outcomes in patients with pancreatic cancer.
3. These data if validated in large population with ancestry information may serve as potential biomarkers for specific race groups to predict patient outcomes.
4. Additionally, several potential drug candidates proposed in our study targeting the specific proteins encoded by the genes specific to the individual race categories can be further studied in clinical settings to develop more personalized race-based therapeutic options.
5. Besides, understanding the molecular events and mechanisms that determine patient outcomes has the potential to develop new and improved treatment approaches for patients with pancreatic cancer.
6. These results are correlated by other studies, as GATA6 is now a known oncogene [14], meaning it has the potential to cause cancer. Similarly, studies have shown copy number alterations in the cytoband 8q24.3 having a role in cancer [15–17].
7. The role of genes and their respective cytobands from the Black and African American as well as Asian race category was more novel. Of the genes found to be amplified in these two race categories, only FGFR3 is known to be an oncogene [18].
8. It is notable that the cytoband 4p16.3 which was amplified frequently in the Asian race category, showing 39 genes all with frequent alteration, causes Wolf–Hirschhorn syndrome when deleted [19]; however, there has been little investigation on its amplification and no study on its unique amplification in the Asian race category or pancreatic cancer.
9. The biggest shortcoming was that the sample size was low, especially for the Black and African American and Asian race categories. Despite this, the results can be used as a good starting point for further, more robust investigation.
10. Future goals include developing new and improved treatment for patients with pancreatic cancer, potentially by targeting the genes that were noted in this study.

11. Another potential goal would be to do statistical modeling and larger population study to determine if these genes can be used for diagnosis of pancreatic cancer, which would be especially noteworthy as pancreatic cancer is known to be hard to diagnose early.
12. Furthermore, the specific role of each cytoband that was largely amplified and its association with pancreatic cancer can be investigated.

## Acknowledgements

Alakesh Bera and Digonto Chatterjee contributed as co-equal first authors on this study.

**Disclaimer** The opinions or assertions contained herein are the private ones of the authors and are not to be construed as official or reflecting the views of the Department of Defense, the Uniformed Services University of the Health Sciences, or any other agency of the U.S. Government.

### References

1. Siegel RL, Miller KD (2019) Jemal A. Cancer statistics. CA Cancer J Clin 69(1):7–34. https://doi.org/10.3322/caac.21551
2. Ryan DP, Hong TS, Bardeesy N (2014) Pancreatic adenocarcinoma. N Engl J Med 371(22):2140–2141. https://doi.org/10.1056/NEJMc1412266
3. Noel M, Fiscella K (2019) Disparities in pancreatic cancer treatment and outcomes. Health Equity 3(1):532–540. https://doi.org/10.1089/heq.2019.0057
4. Hu C, Hart SN, Polley EC et al (2018) Association between inherited germline mutations in cancer predisposition genes and risk of pancreatic cancer. JAMA 319(23):2401–2409. https://doi.org/10.1001/jama.2018.6228
5. Qian ZR, Rubinson DA, Nowak JA et al (2018) Association of alterations in main driver genes with outcomes of patients with resected pancreatic ductal adenocarcinoma. JAMA Oncol 4(3):e173420. https://doi.org/10.1001/jamaoncol.2017.3420
6. Khawja SN, Mohammed S, Silberfein EJ et al (2015) Pancreatic cancer disparities in African Americans. Pancreas 44(4):522–527. https://doi.org/10.1097/MPA.0000000000000323
7. Shapiro M, Chen Q, Huang Q et al (2016) Associations of socioeconomic variables with resection, stage, and survival in patients with early-stage pancreatic cancer. JAMA Surg 151(4):338–345. https://doi.org/10.1001/jamasurg.2015.4239.
8. Zhang Q, Zeng L, Chen Y et al (2016) Pancreatic cancer epidemiology, detection, and management. Gastroenterol Res Pract 8962321. https://doi.org/10.1155/2016/8962321
9. Cerami E, Gao J, Dogrusoz U et al (2012) The cBio cancer genomics portal: an open platform for exploring multidimensional cancer genomics data. Cancer Discov 2(5):401–404. https://doi.org/10.1158/2159-8290.CD-12-0095
10. Gao J, Aksoy BA, Dogrusoz U et al (2013) Integrative analysis of complex cancer genomics and clinical profiles using the cBioPortal. Sci Signal 6(269):11. https://doi.org/10.1126/scisignal.2004088
11. Chen MM, Li J, Mills GB, Liang H (2020) Predicting cancer cell line dependencies from the protein expression data of reverse-phase protein arrays. JCO Clin Cancer Inform 4:357–366. https://doi.org/10.1200/CCI.19.00144
12. Li J, Zhao W, Akbani R et al (2017) Characterization of human cancer cell lines by reverse-phase protein arrays. Cancer Cell 31(2):

225–239. https://doi.org/10.1016/j.ccell.2017.01.005

13. Bera A, Subramanian M, Karaian J et al (2020) Functional role of vitronectin in breast cancer. PLoS One 15(11):e0242141. https://doi.org/10.1371/journal.pone.0242141
14. Deng X, Jiang P, Chen J et al (2020) GATA6 promotes epithelial-mesenchymal transition and metastasis through MUC1/beta-catenin pathway in cholangiocarcinoma. Cell Death Dis 11(10):860. https://doi.org/10.1038/s41419-020-03070-z
15. Bera A, Radhakrishnan S, Russ E et al (2022) Functional role of long non-coding RNA YTHDF3-AS1 in breast cancer. Cancer Res 82(12_Suppl):5828
16. Bera A, Karaian J, Subramanian M et al (2020) EXOSC4, a novel gene at chromosome 8q24 loci is linked with breast cancer progression and is a prognostic marker for breast cancer survival. Cancer Res 80(16_Supplement):4322
17. Brusselaers N, Ekwall K, Durand-Dubief M (2019) Copy number of 8q24.3 drives HSF1 expression and patient outcome in cancer: an individual patient data meta-analysis. Hum Genomics 13(1):54. https://doi.org/10.1186/s40246-019-0241-3
18. Zingg D, Bhin J, Yemelyanenko J et al (2022) Truncated FGFR2 is a clinically actionable oncogene in multiple cancers. Nature 608 (7923):609–617
19. Zollino M, Orteschi D, Ruiter M et al (2014) Unusual 4p16.3 deletions suggest an additional chromosome region for the Wolf-Hirschhorn syndrome-associated seizures disorder. Epilepsia 55(6):849–857

# Chapter 20

# Advanced Computational Methods to Evaluate Vascular Heterogeneity in Tumor Tissue Based on Single Plane Illumination Microscopy

**Felix T. Kurz and Artur Hahn**

## Abstract

During tumor growth, the complex composition of vasculature is prone to dynamic changes due to mechanic and biochemical challenges. Perivascular invasion of tumor cells to co-opt existing vasculature, but also formation of de-novo vasculature and other effects on the vascular network, may lead to altered geometric vessel properties as well as changes in vascular network topology, which is defined by vascular multifurcations and connections between vessel segments. The intricate organization and heterogeneity of the vascular network can be analyzed with advanced computational methods to uncover vascular network signatures that may allow differentiating between pathological and physiological vessel regions. Herein, we present a protocol to evaluate vascular heterogeneity in whole vascular networks, using morphological and topological measures. The protocol was developed for single plane illumination microscopy images of mice brain vasculature but can be applied to any vascular network.

**Key words** Single plane illumination microscopy (SPIM), Perivascular invasion, Vascular heterogeneity, Vascular network analysis, Vascular network topology, Vasculotome, Vasculature-alterating therapies, Glioblastoma

## 1 Introduction

Vascular networks are highly complex entities that not only ensure nutrient and oxygen supply to biological tissue but also show a remarkable adaptability to environmental challenges such as changes in vascular flow, e.g., due to atherosclerosis, or to secure tissue homeostasis in regulating ion transit across vessel walls [1]. These changes can occur on largely different time scales: while tissue homeostasis regulation may be immediate and vascular-based thermoregulation fast, adaptation to an increasing infiltration of tumor cells that co-opt existing and build new vessels to supply tumor tissue typically occurs in weeks to months [2]. Microvascular proliferation is especially important in the

Usha N. Kasid and Robert Clarke (eds.), *Cancer Systems and Integrative Biology*, Methods in Molecular Biology, vol. 2660, https://doi.org/10.1007/978-1-0716-3163-8_20,

pathophysiology of the aggressive and common brain tumor glioblastoma [3] and its inhibition in the form of anti-angiogenic therapy with vascular endothelial growth factor inhibitor bevacizumab, to prolong overall survival [4]. However, anti-angiogenic therapy may also impact supply of conventional, cytostatic therapy agents, leading to a decreased clinical efficacy [5, 6]. It is therefore important to find measures of vascular architecture and vascular changes that may allow differentiating between therapy responders and non-responders, e.g., using magnetic resonance imaging in a clinical setting [7, 8], or using tissue clearing and selective single plane illumination microscopy (SPIM) in an experimental setting [9].

While changes in geometric properties between tumor vasculature and healthy vasculature such as radii or vessel densities may be less pronounced [10], other descriptors of vasculature become more important, including functional measures such as vascular topology, which considers the distribution of intervascular connections [11, 12]. Specifically, glioblastoma vascular topology was found to significantly differ to that of healthy vasculature and more so than simple geometric properties averaged over voxels the size of an MR imaging voxel [11, 13]. This functional vascular organization may result from scaling of glioblastoma microenvironment interactions; however, it is mainly these vascular architecture characteristics, together with changes in local tumor cell density, that potentially impact the quantification of diagnostic biomarkers in routine diagnostic imaging, including the machine-learning based detection of pathological image signatures [14, 15].

The protocol described herein applies to the quantification of vascular heterogeneity in tumor tissue, based on a procedure that was developed to describe differences between whole vascular networks of glioblastoma and healthy tissue [11, 16]. The procedure was used on SPIM images of tissue-cleared mice brain, where vessels were marked with fluorescent lectin [9] (*see also* Fig. 1), but it can be applied to any vascular network. We provide a detailed step-by-step procedure from vascular segmentation to vascular network topology analysis, to obtain a range of vascular parameters that may aid in differentiating pathological vasculature from healthy vasculature.

## 2 Materials

### 2.1 Experimental Agents

#### 2.1.1 Selective Plane Illumination Microscopy on Healthy Mice Brain and Mice Injected with Glioblastoma/Glioma Tumor Cells

1. U-87MG glioblastoma or GL261 glioma cells are available from ATCC® and the National Cancer Institute NCI, Bethesda, MD, USA.
2. Nine-week-old, male NOD Scid Gamma mice (for U87-MG glioblastoma cell implantation), or 6–8-week-old, female C57 BL/6J mice (for GL261 glioma cell implantation; Charles River Laboratories, Sulzfeld, Germany).

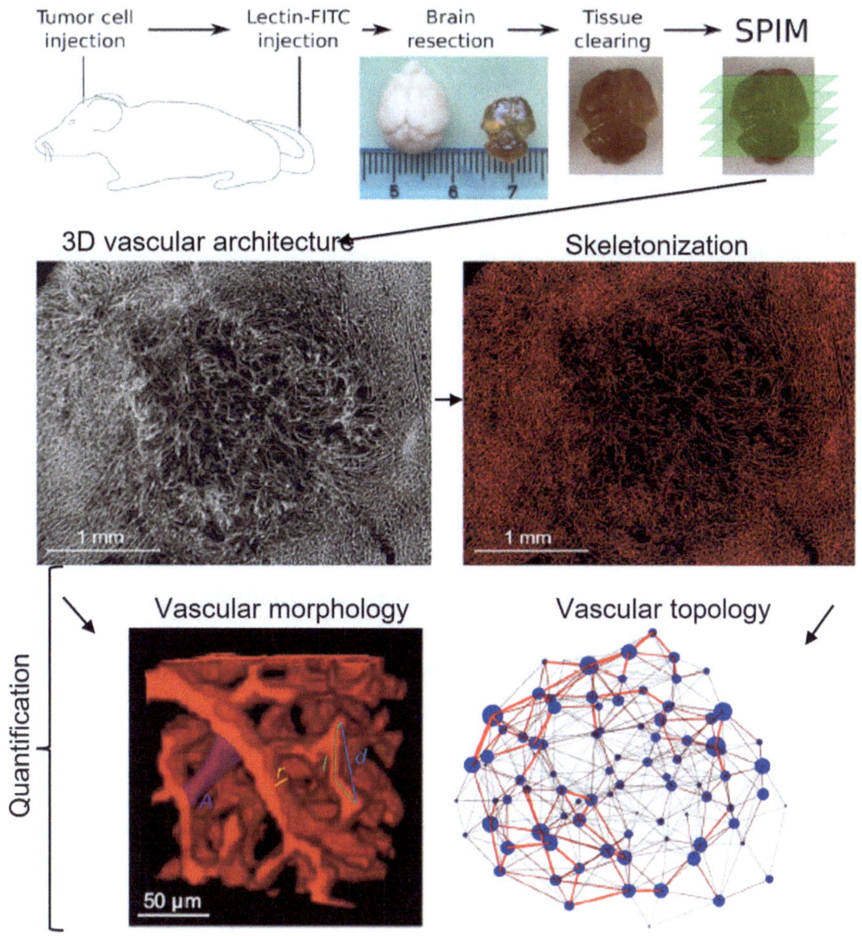


**Fig. 1** Vascular architecture acquisition and processing. Upper panel: Illustration of the experimental procedure to image whole vascular networks. From left to right: mice brain are injected with tumor cells; after sufficient tumor growth, mice are injected with a fluorescent vessel marker (lectin-FITC). The subsequently resected brain is cleared using the FluoClearBaBB protocol and imaged using selective single plane illumination microscopy (SPIM) with an in-plane resolution of approximately 3 µm × 3 µm. Middle panel: The resulting high-resolution stacks (left) are skeletonized (right). Lower panel: This allows obtaining quantifiable measures on vessel geometry and topology, such as mean vessel segment radii or vessel segment surface (left), and vascular communities in a vascular network representation (right). *(Adapted from Fig. 1 in Ref.* [11]*, with permission from Ref.* [11]*. Copyright 2019)*

3. Lectin-FITC (Sigma-Aldrich, St. Louis, MO, USA) to provide selective vessel fluorescence.
4. Ultramicroscope II setup (LaVision Biotec, Bielefeld, Germany).

### *2.2 Image Processing and Analysis Software*

1. Brain microvasculature can be processed and analyzed using the open-source software interactive learning and segmentation toolkit ilastik [17], the open-source software Fiji 2.0.0-rc-43/1.51r or higher [18], and Matlab R2016b (MathWorks, Natick, MA, USA) or higher.

## 3 Methods

### 3.1 Experimental Methods

#### 3.1.1 Tumor Cell Injection (Example for Glioblastoma and Glioma Cells)

1. Use $7.5 \cdot 10^4$ U-87MG cells or $10^5$ GL261 glioma cells.
2. Check biweekly for mycoplasma contamination.
3. Dilute cells in 5 μL sterile phosphate and buffer in saline (PBS, Sigma-Aldrich Chemie GmbH, Taufkirchen, Germany).
4. Inject cells in the striatum of the right hemisphere of 9-week-old, male NOD Scid Gamma mice (U-87MG cells) or in 6–8-week-old, female C57 BL/6J mice (GL261 cells). Inject 2 mm lateral and 2 mm ventral of the bregma.

#### 3.1.2 Lectin Injection, Mouse Sacrifice, and Clearing Protocol for Selective Plane Illumination Microscopy

1. After 21 days post tumor cell implantation for mice injected with U87 cells, and 28 days post tumor cell implantation for mice injected with GL261 cells: inject intravenously 300 μL lectin-FITC (Sigma-Aldrich, St. Louis, MO, USA) with concentration 1 mg/mL.
2. Sacrifice mice after 3 min of incubation by a ketamine/xylazine overdose.
3. Perfuse sacrificed mice transcardially with 20 mL PBS and 20 mL 4% paraformaldehyde solution.
4. Explant and optically clear mice brain with the FluoClearBaBB protocol [19].
5. Use SPIM to image the microvasculature of the entire brain, using lectin fluorescence. Resolution: 3.25 × 3.25 μm in-plane and 5 μm inter-plane.

Image with the following parameters on the Ultramicroscope II: Andor camera exposure time of 686.35 ms, 100% laser power, 5 μm stepsize, dynamic focus on 5–10 steps, 16-bit low noise gain, left and right light sheet together.

### 3.2 Vascular Segmentation and Image Post-Processing

#### 3.2.1 Tumor Volume Segmentation

1. Draw binary mask manually over each imaging data set to identify tumor volume. Use only regions with sufficiently suppressed blurring. Exclude the ventricular system (that may exhibit false lectin fluorescence).
2. Binarize tumor cores based on the transition from a dense outer vasculature to a less dense vascularized center within the tumor segmentation.

#### 3.2.2 Vessel Segmentation Using Ilastik

1. Before loading the image data set into ilastik, convert your raw image data to hdf5/h5 format (e.g., using the Fiji plugin "HDF5").
2. Load image file/stack in ilastik and use the "Feature Selection" tab to select vessels with different standard deviations of the Gaussian smoothing kernel.

3. Use the "Training" tab to add about 4–5 classes of objects (e.g., background, foreground, specific vessel type, etc.) to begin multi-class segmentation, making also use of the zoom function to zoom into imaging data. The ilastik guidelines recommend labeling individual pixels or short lines instead of large areas. Label classes in different areas of the image.
4. Use the "Live Update" functionality to overlay your image with class probability maps based on your current labeling. Segmentation layers can be turned on and off during the process.
5. Correct labels in places where the segmentation failed, always controlling with the "Live Update" function.

   Tipp: use the "Uncertainty" map to show areas where the algorithm has problems making a good prediction; this can be very helpful for refinement.
6. Once an adequate vessel segmentation is accomplished, export the segmentation data using all labels.
7. Use the "Transpose to Axis Order" command with the "xyzc" option (alternatively, the "xyc" option), to export images with the same orientation as the imported raw image. The export imaging data can be controlled with the "Prediction Export" tab.
8. Save the ilastik project and use the "Batch processing" tab to process new imaging data sets with the same segmentation routine. Several image data sets can also be trained simultaneously.

#### 3.2.3 Vascular Segmentation Post-Processing

1. Use the "Gaussian Blur 3D" plugin in Fiji to reduce noise with a 3D-Gaussian filter with isotropic standard deviation of 1 voxel unit.
2. Binarize the segmentation volume again with an intensity threshold at half of the maximum voxel value.
3. Remove noise using Matlab's *bwconncomp* function: remove all 6-connected voxel structures whose volume is smaller than spheres with a radius of 6 μm.
4. Use Matlab *imfill* function to fill holes in the vascular structures (with *imfill*(binary image, 'holes'))—this concerns holes as a set of background voxels that cannot be reached by filling in the background from the edge of the image.
5. Fill big holes with Matlab's *bwareaopen* function, using 20.000 voxels as a cutoff.
6. Check if big holes are left with Matlab's *bwconncomp* function.

#### 3.2.4 Vascular Skeletonization

1. Use Fiji to skeletonize image data sets using the "Skeletonize (2D/3D)" plugin (without the pruning option).
2. Subsequently run the "Analyze Skeleton (2D/3D)" plugin in Fiji to obtain a tagged skeleton on the segmented vascular structure. This assigns a tag to each skeleton voxel that identifies vascular endpoints (skeleton voxels with only one neighboring skeleton voxel), slab or segment voxels (skeleton voxels with exactly two neighboring voxels), and junctions (skeleton voxels with more than two neighboring voxels). The skeletons can then be used for labeling of vessels and nodes and determination of vascular connectivity.

#### 3.2.5 Vascular Thinning Due to Lectin Fluorescence Overexposure

1. To compensate for effects of fluorescent overexposure in the segmentation of vascular structures, where vessels tend to appear thicker since background voxels next to vascular structures are illuminated as well, vessels need to undergo a thinning procedure. We use Matlab *bwperim* function to thin vessels around their respective skeleton segment by a 1 voxel layer.

### 3.3 Vascular Geometry Analysis

1. Determine the fraction of blood-filled vascular structures in a tiling-box approach that quantifies a fractional vessel volume on cubic subvolumes with side length of 500 μm as the quotient of the number of vascular voxels versus voxels in the cubic subvolume.
2. Determine the microvascular density as the quotient of the number of individual vessel segments per cubic subvolume versus $(500\ \mu m)^3$.
3. Determine the vessel length density as the total vessel length, i.e., the sum of all vessel skeleton segment lengths, per cubic subvolume.
4. Determine the vascular surface density as the lumen surface area per cubic subvolume. Lumen surface is determined for each vessel segment using Matlab's *null* function to create an orthonormal basis with each vascular segments' start and end voxel. Then, calculate the intersection of that plane with the respective cubic subvolume to determine the binarized voxels in that plane. Finally, for each vessel segment voxel and its respective orthogonal plane, use Matlab's *bwperim* function to determine the vascular perimeter voxels within the orthogonal plane. The sum of all perimeter voxels for one vascular segment yields the vessel segment lumen surface per segment. The lumen surface per cubic subvolume is then determined as the sum of all lumen surfaces within one cubic subvolume.

5. In the same orthogonal planes as above, for each vessel segment voxel, determine the average voxel-specific vascular diameter with Matlab's *regionprops* function and option 'EquivDiameter'. Define the voxel-specific radius as half of the voxel-specific vascular segment diameter. Determine the vascular segment radius as the weighted sum of all voxel-specific vascular radii along one vascular segment.

### 3.4 Vascular Network Topology

The vascular network topology is modeled as an undirected network with nodes being vessel branchings or multifurcations that are interconnected by vessel segments as network edges. Connectivity properties of the vascular topology can then be extracted with graph theory.

Using the vascular skeleton, vascular nodes are assigned to junction and end-point voxels, see Subheading 3.2.4, while vascular edges are assigned to skeleton segment voxels.

#### 3.4.1 Scale-Free and Basic Topological Characteristics of the Vascular Network

If a vascular network is scale-free, the occurrence of a vascular hub with a lot of branchings increases the probability of another hub with many branchings in its vicinity. Such behavior has been observed in healthy and pathological vascular networks [11, 20], but also in neuronal networks [21] or in mitochondrial networks [22, 23]. Specifically, in GL261 tumors, the peripheral vascular network architecture was found to be highly scale-free, when compared to U87 tumors or healthy brain vasculature [11].

1. To evaluate scale-free properties of the vascular network, determine the degree $k$ of every node as the number of attached vessel branches, i.e., the number of directly neighboring skeleton voxels at junction voxels.
2. Using logarithmic scales, visualize relative frequencies of nodes with degree $k$ versus degree $k$, see also Fig. 2a in [11]. Model the resulting curve with a power law $P(k) \propto k^{-\gamma}$, with degree exponent $\gamma$, using Matlab's *polyfit* function for linear fits in log space. Use $k \in [3, 20]$ to fit power law.
3. Determine the average clustering coefficient of the vascular network, $C$, as the average of the clustering coefficients $C_i$ of each vascular node $i$ with $n_i$ neighboring (i.e., directly to node $i$ connected) nodes and $E_i$ connections between these neighboring nodes as $C_i = \frac{2E_i}{n_i(n_i-1)}$. Then, $C = \Sigma_i C_i / \Sigma_i$. Clustering coefficients measure the network propensities toward nodes that tend to be surrounded by well-interconnected nodes. Increased clustering has been observed in tumor vasculature compared to healthy brain vasculature, especially in the tumor core [11].

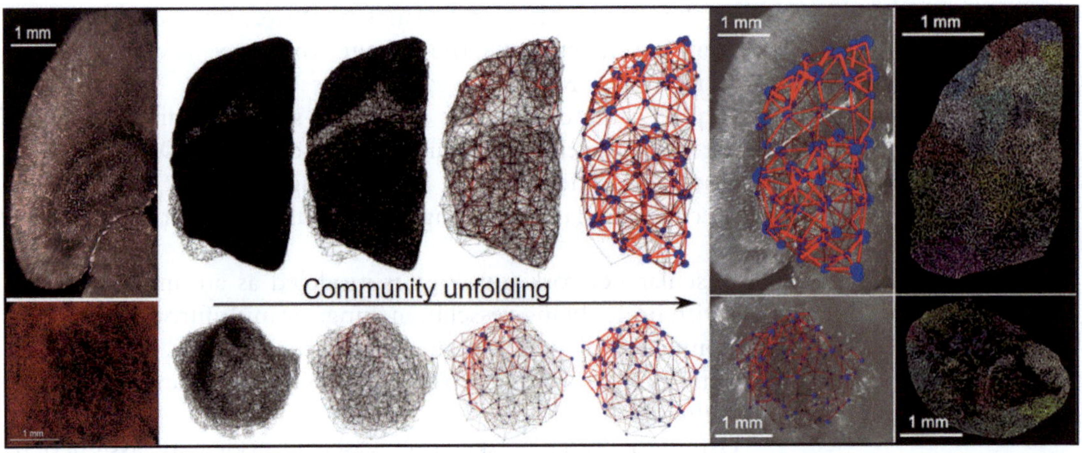


**Fig. 2** Community unfolding in a modular network structure. Healthy mouse brain hemisphere with segmented vasculature (upper panel on left) and glioblastoma vasculature in the same species (lower panel on left), based on SPIM imaging. The community unfolding process is visualized on both networks, whereby each level of partitioning represents a local maximum in network modularity (for details see text) for increasing vascular network communities. The clustering scheme (second-to-last on the right-hand-side) is obtained at global maximum modularity in a slice of the original SPIM image: communities are shown in blue spheres, where the size of the sphere is proportional to the respective cluster size, while the weight of a connection between clusters (or the number of intercommunity vessel segments) is shown in red with varying edge thickness and brightness according to weight value. Right-hand side: Community affiliations at global maximum modularity. *(Adapted from Figs. 1 and 3 in Ref.* [11], *with permission from Ref.* [11]. *Copyright 2019)*

4. The characteristic vascular network path length is determined as the average number of edges on a geodesic that links any two nodes that are connected by a path on the graph. To determine the characteristic path length, use Johnson's algorithm for the shortest paths problem with MatlabBGL library version 4.0 from function *johnson_all_sp* [24].
5. Likewise, determine the network diameter as the greatest separation between a pair of nodes, given by the maximum of all shortest paths. Such a diameter is a measure of the linear size of the graph that characterizes the network.

#### *3.4.2 Community Measures on the Vascular Network*

Communities within a network represent sets of densely interconnected nodes that are potentially overlapping. This translates into vessels that tend to be more interconnected, when they are in the same vascular community. For example, for tumor vascular networks in brain tumors, communities tend to be much smaller than in healthy vasculature, especially in GL261 tumors [11], giving rise to the assumption that tumor cell invasion leads to a re-organization of the formerly healthy vasculature that lead to smaller and interspersed vascular communities.

The propensity of networks to have nodes organized in such communities or modules can be measured with the network modularity, $Q$. The modularity takes values between −0.5 and 1 to measure the density of vascular connections inside a vascular community compared to vascular connections between communities.

To uncover an optimal community structure, one needs to apply the Louvain community unfolding algorithm [25, 26].

1. First, assume that each node represents its own community $c_i$.
2. Determine the adjacency matrix $A$ that holds the number of edges between nodes $i$ and $j$ at $A(i,j)$.
3. Determine $m = \frac{1}{2}\sum_{ij} A(i,j)$ as the sum of all edges within the original network graph.
4. For every node $i$, determine the number $n_i$ of edges attached to it.
5. Determine the network modularity as $Q = \frac{1}{2m}\sum_{ij}\left[A_{ij} - \frac{n_i n_j}{2m}\right]\delta(c_i, c_j)$.
6. Then, assign a node $i$ to a neighboring node $j$ while removing it from its own position.
7. Calculate the new number of edges between nodes, considering that nodes may now have more than one (undirected) edge between each other.
8. Determine the modularity of the new network with **steps 2–5**.
9. Determine the change in modularity toward the original network.
10. Repeat **steps 6–9** for each neighboring node $j$ to find the greatest change in modularity. If no increase in modularity is found, node $i$ remains in its own community. Otherwise, place node $i$ into the community of node $j$ with the greatest modularity change.
11. Repeat **steps 6–10** for all nodes until modularity does not increase anymore.
12. Define new graph with new nodes as the resulting node communities from the previous steps. Links between nodes of the previous network, which are in the same community, are then self-loops on the new community node. Repeat **steps 2–11**.
13. Repeat **steps 2–12** until no further reassignments can increase the network modularity.

#### 3.4.3 Community Structure and Connectivity

To study the resulting community structure from above and its relevance for vascular architecture, one can assign spatial coordinates to each community as the centroid from all spatial coordinates of its inherent nodes (Fig. 2).

1. Determine the size of each community as the mean distance between a community node's spatial position and the community centroid's spatial position.
2. Determine the number of vessel segments within each vascular community or cluster to obtain another measure of community size.
3. Determine the community's topological perimeter as the number of connecting vessel segments of a community to other communities, which reflects a connectivity measure or the vascular community's supply situation.
4. Determine the degree of a community as the sum of twice the number of vessel segments within each community and the topological perimeter. This measure conveys the relevance of a community as a supply entity within the vascular network.
5. Determine network properties from **steps 1** to **5** in Subheading 3.4.1 for each community and determine averages of these measures across communities.
6. Determine the shortest path between communities by the physical separation of their centroids.

## 4 Notes

1. The program ilastik does not have an "Undo" button, meaning that labels cannot simply be undone if new labels make the segmentations worse. Then, labels must be erased manually with the "Eraser" function [27].
2. Too many labels in ilastik lead to worse segmentation results.
3. The SPIM imaging procedure only images perfused vessels, which may impact the results of the vascular geometry analysis, but not those of the vascular topology analysis [11].
4. Irregular surfaces and vessel segment boundaries may cause voxel stubs in the skeletonization process that have a degree of 2. These nodes do not contribute to the vascular topology; however, they may impact topology measures involving node degrees. Pruning may help to eliminate such nodes [28].
5. Consecutive branching points in a vessel segment may combine to vascular nodes of a high degree if imaging resolution is not sufficiently high [11].

## 5 Conclusion

SPIM imaging of tissue-cleared brains allows a morphological and topological analysis of whole vascular networks, using advanced vascular segmentation and network theoretical analysis methods. These tools may uncover collective changes within the "vasculotome" that are otherwise concealed in local analyses. They provide the means to study functional vascular relations regarding vascular supply properties, to extract and grade pathological vascular network segments, or to develop and evaluate vasculature-alterating therapies.

## Acknowledgements

Experiments that form the basis of this protocol were performed in the laboratory of B. Tews (German Cancer Research Center) and F. Winkler (Heidelberg University Hospital and German Cancer Research Center).

**Author Contributions** F.T.K. and A.H. conceived and wrote the manuscript.

### References

1. Huwiler A, Pfeilschifter J (2021) Recuperation of vascular homeostasis. Circ Res 129:237–239. https://doi.org/10.1161/CIRCRESAHA.121.319558
2. Hardee ME, Zagzag D (2012) Mechanisms of glioma-associated neovascularization. Am J Pathol 181:1126–1141. https://doi.org/10.1016/j.ajpath.2012.06.030
3. Das S, Marsden PA (2013) Angiogenesis in glioblastoma. N Engl J Med 369:1561–1563
4. Wang N, Jain RK, Batchelor TT (2017) New directions in anti-angiogenic therapy for glioblastoma. Neurotherapeutics 14:321–332. https://doi.org/10.1007/s13311-016-0510-y
5. Carmeliet P, Jain RK (2011) Principles and mechanisms of vessel normalization for cancer and other angiogenic diseases. Nat Rev Drug Discov 10:417–427. https://doi.org/10.1038/nrd3455
6. Lopes-Coelho F, Martins F, Pereira SA, Serpa J (2021) Anti-angiogenic therapy: current challenges and future perspectives. Int J Mol Sci 22:3765. https://doi.org/10.3390/ijms22073765
7. Emblem K, Mouridsen K, Bjornerud A et al (2013) Vessel architectural imaging identifies cancer patient responders to anti-angiogenic therapy. Nat Med 19:1178–1183. https://doi.org/10.1038/nm.3289
8. Buschle LR, Ziener CH, Zhang K et al (2018) Vessel radius mapping in an extended model of transverse relaxation. MAGMA 31(4):531–551. https://doi.org/10.1007/s10334-018-0677-9
9. Breckwoldt MO, Bode J, Kurz FT et al (2016) Correlated magnetic resonance imaging and ultramicroscopy (MR-UM) is a tool kit to assess the dynamics of glioma angiogenesis. elife 5:e11712–e11712. https://doi.org/10.7554/eLife.11712
10. Forster JC, Harriss-Phillips WM, Douglass MJJ, Bezak E (2017) A review of the development of tumor vasculature and its effects on the tumor microenvironment. Hypoxia 5:21–32. https://doi.org/10.2147/HP.S133231
11. Hahn A, Bode J, Krüwel T et al (2019) Glioblastoma multiforme restructures the topological connectivity of cerebrovascular networks.

Sci Rep 9:11757. https://doi.org/10.1038/s41598-019-47567-w

12. Ji X, Ferreira T, Friedman B et al (2021) Brain microvasculature has a common topology with local differences in geometry that match metabolic load. Neuron 109:1168–1187.e13. https://doi.org/10.1016/j.neuron.2021.02.006
13. Hahn A, Bode J, Krüwel T et al (2020) Gibbs point field model quantifies disorder in microvasculature of U87-glioblastoma. J Theor Biol 494:110230–110230. https://doi.org/10.1016/j.jtbi.2020.110230
14. Tomaszewski MR, Gillies RJ (2021) The biological meaning of radiomic features. Radiology 298:505–516. https://doi.org/10.1148/radiol.2021202553
15. Hahn A, Bode J, Schuhegger S et al (2022) Brain tumor classification of virtual NMR voxels based on realistic blood vessel-induced spin dephasing using support vector machines. NMR Biomed 35(4):e4307. https://doi.org/10.1002/nbm.4307
16. Breckwoldt MO, Bode J, Sahm F et al (2019) Correlated MRI and ultramicroscopy (MR-UM) of brain tumors reveals vast heterogeneity of tumor infiltration and neoangiogenesis in preclinical models and human disease. Front Neurosci 2:1004. https://doi.org/10.3389/fnins.2018.01004
17. Berg S, Kutra D, Kroeger T et al (2019) Ilastik: interactive machine learning for (bio)image analysis. Nat Methods 16:1226–1232. https://doi.org/10.1038/s41592-019-0582-9
18. Schindelin J, Arganda-Carreras I, Frise E et al (2012) Fiji: an open-source platform for biological-image analysis. Nat Methods 9:676–682. https://doi.org/10.1038/nmeth.2019
19. Schwarz MK, Scherbarth A, Sprengel J et al (2015) Fluorescent-protein stabilization and high-resolution imaging of cleared, intact mouse brains. PLoS One 10:e0124650. https://doi.org/10.1371/journal.pone.0124650
20. Albert R, Barabási AL (2002) Statistical mechanics of complex networks. Rev Mod Phys 74:47–97. https://doi.org/10.1103/RevModPhys.74.47
21. Eguíluz VM, Chialvo DR, Cecchi GA et al (2005) Scale-free brain functional networks. Phys Rev Lett 94:018102
22. Kurz FT, Aon MA, O'Rourke B, Armoundas AA (2014) Cardiac mitochondria exhibit dynamic functional clustering. Front Physiol 5:329–329
23. Kurz FT, Aon MA, O'Rourke B, Armoundas AA (2017) Functional implications of cardiac mitochondria clustering. Adv Exp Med Biol 982:1–24
24. Gleich DF (2009) Models and algorithms for PageRank sensitivity. Ph.D. thesis, Stanford University, Chapter 7 on MatlabBGL
25. Blondel VD, Guillaume JL, Lambiotte R, Lefebvre E (2008) Fast unfolding of communities in large networks. J Stat Mech 8:P10008. https://doi.org/10.1088/1742-5468/2008/10/P10008
26. Barabási AL (2016) Network science, 1st edn. Cambridge University Press, United Kingdom. Available: http://networksciencebook.com/
27. Haubold C, Schiegg M, Kreshuk A, Berg S, Koethe U, Hamprecht FA (2016) Segmenting and tracking multiple dividing targets using ilastik. Adv Anat Embryol Cell Biol 219:199–229
28. Hirsch S, Reichold J, Schneider M, Székely G, Weber B (2012) Topology and hemodynamics of the cortical cerebrovascular system. J Cereb Blood Flow Metab 32:952–967. https://doi.org/10.1038/jcbfm.2012.39

# Chapter 21

# Illuminating DEPDC1B in Multi-pronged Regulation of Tumor Progression

**Howard E. Boudreau, Jennifer Robinson, and Usha N. Kasid**

## Abstract

DEPDC1B (aliases BRCC3, XTP8, XTP1) is a DEP (Dishevelled, Egl-1, Pleckstrin) and Rho-GAP-like domains containing predominately membrane-associated protein. Earlier, we and others have reported that DEPDC1B is a downstream effector of Raf-1 and long noncoding RNA lncNB1, and an upstream positive effector of pERK. Consistently, DEPDC1B knockdown is associated with downregulation of ligand-stimulated pERK expression. We demonstrate here that DEPDC1B N-terminus binds to the p85 subunit of PI3K, and DEPDC1B overexpression results in decreased ligand-stimulated tyrosine phosphorylation of p85 and downregulation of pAKT1. Collectively, we propose that DEPDC1B is a novel cross-regulator of AKT1 and ERK, two of the prominent pathways of tumor progression. Our data showing high levels of DEPDC1B mRNA and protein during the G2/M phase have significant implications in cell entry into mitosis. Indeed, DEPDC1B accumulation during the G2/M phase has been associated with disassembly of focal adhesions and cell de-adhesion, referred to as a DEPDC1B-mediated de-adhesion mitotic checkpoint. DEPDC1B is a direct target of transcription factor SOX10, and SOX10-DEPDC1B-SCUBE3 axis has been associated with angiogenesis and metastasis. The Scansite analysis of the DEPDC1B amino acid sequence shows binding motifs for three well-established cancer therapeutic targets CDK1, DNA-PK, and aurora kinase A/B. These interactions and functionalities, if validated, may further implicate DEPDC1B in regulation of DNA damage–repair and cell cycle progression processes. Finally, a survey of the publicly available datasets indicates that high DEPDC1B expression is a viable biomarker in breast, lung, pancreatic and renal cell carcinomas, and melanoma. Currently, the systems and integrative biology of DEPDC1B is far from comprehensive. Future investigations are necessary in order to understand how DEPDC1B might impact AKT, ERK, and other networks, albeit in a context-dependent manner, and influence the actionable molecular, spatial, and temporal vulnerabilities within these networks in cancer cells.

**Key words** DEPDC1B, CRAF/Raf-1, lncNB1, ERK, p85/PI3K, AKT1, G2/M phase, Cell entry into mitosis, SOX10, SCUBE3, Multi-pronged regulation, Tumor progression, Breast cancer, Lung cancer, Pancreatic cancer, Renal cell carcinoma, Melanoma

## 1 Introduction

DEPDC1B (aliases BRCC3, XTP8, XTP1) is a DEP (Dishevelled, Egl-1, Pleckstrin) domain containing protein. *DEPDC1B* gene was discovered by mRNA expression profiling of human breast cancer

Usha N. Kasid and Robert Clarke (eds.), *Cancer Systems and Integrative Biology*, Methods in Molecular Biology, vol. 2660, https://doi.org/10.1007/978-1-0716-3163-8_21,

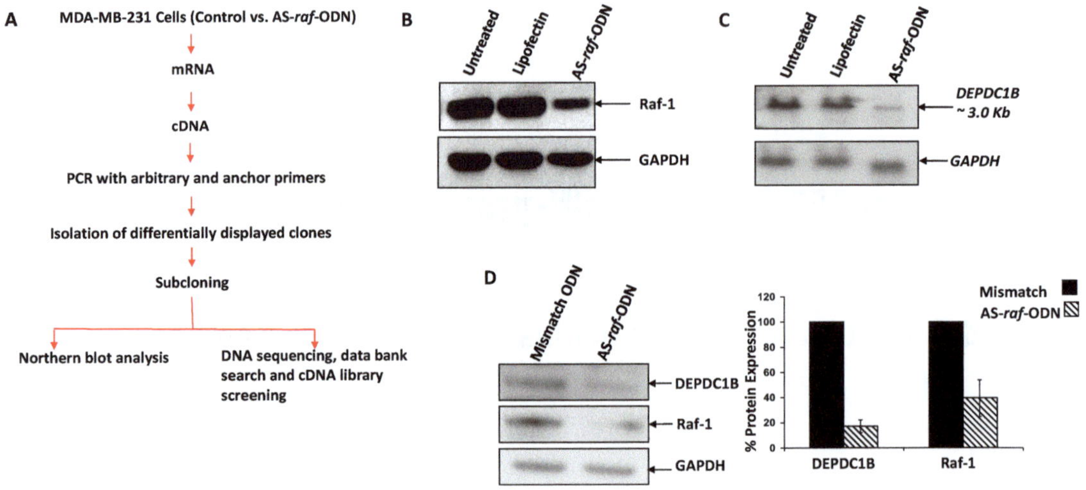


**Fig. 1** Knockdown of Raf-1 expression is associated with downregulation of DEPDC1B mRNA and protein expression. (**a**) Experimental scheme to identify mRNA expression profile of Raf-1 knockdown MDA-MB-231 breast cancer cells. (**b**) Antisense knockdown of Raf-1 protein expression. MDA-MB-231 cells were treated with 2 μM AS-*raf*-ODN (5′-TCC-CGC-CTG-TGA-CAT-GCA-TT-3′) for 48 h as described earlier [5], followed by sequential Western blotting with anti-Raf-1 and anti-GAPDH antibodies. (**c**) *DEPDC1B* mRNA expression in Raf-1 knockdown MDA-MB-231 cells. Northern blot was probed with a radiolabeled *DEPDC1B* cDNA and re-probed with a radiolabeled *GAPDH* cDNA probe. (**d**) DEPDC1B protein expression in Raf-1 knockdown cells. MDA-MB-231 cells were treated with 2 μM of mismatch oligodeoxynucleotide (Mismatch ODN) (5′-TCC-CGC-GCA-CTT-GAT-GCA-TT-3′) or AS-*raf*-ODN (5′-TCC-CGC-CTG-TGA-CAT-GCA-TT-3′) for 48 h as in panel b. Western blot analysis was performed using rabbit polyclonal antibody generated against a DEPDC1B-specific custom peptide PFQPFRTRSFRM [5]. The blot was subsequently reprobed with anti-Raf-1 and anti-GAPDH antibodies (left panel), and the signals were quantified using ImageQuant software and normalized against GAPDH (right panel). Data shown are mean ± s.d. from two independent experiments

cells. Treatment of MDA-MB-231 breast cancer cells with c-*raf*-1 antisense oligodeoxynucleotide (AS-*raf*-ODN) resulted in knockdown of Raf-1 and concomitant decrease in *DEPDC1B* mRNA and protein expression (Fig. 1) [1–5]. Genomic *DEPDC1B* is localized at the human chromosome 5q12.1. Sequence homology search of the chromosome 5 genomic DNA revealed several CCAAT-boxes and E-boxes (CANNTG) within the putative *DEPDC1B* promoter region (Fig. 2a). The CCAAT boxes are potential binding sites of transcription factor NF-Y, whereas E-boxes have been shown to bind upstream stimulatory factors (USFs) [6, 7]. Similar to *Cyclin B*, *DEPDC1B* mRNA is likely to be regulated by transcriptional enhancer CCAAT/E-box motifs in a cell cycle-dependent manner. *DEPDC1B* transcript (3.0 kb) was detected in just a few adult human normal tissues (testes, bone marrow, placenta) and in most of the human cancer cell lines tested (Fig. 2b and c). In other studies, RNA sequencing of MYCN amplified and non-amplified human neuroblastoma cell lines led to identification of a long noncoding RNA lncNB1 as an overexpressed transcript in MYCN amplified neuroblastoma cells. Here, lncNB1 was shown to

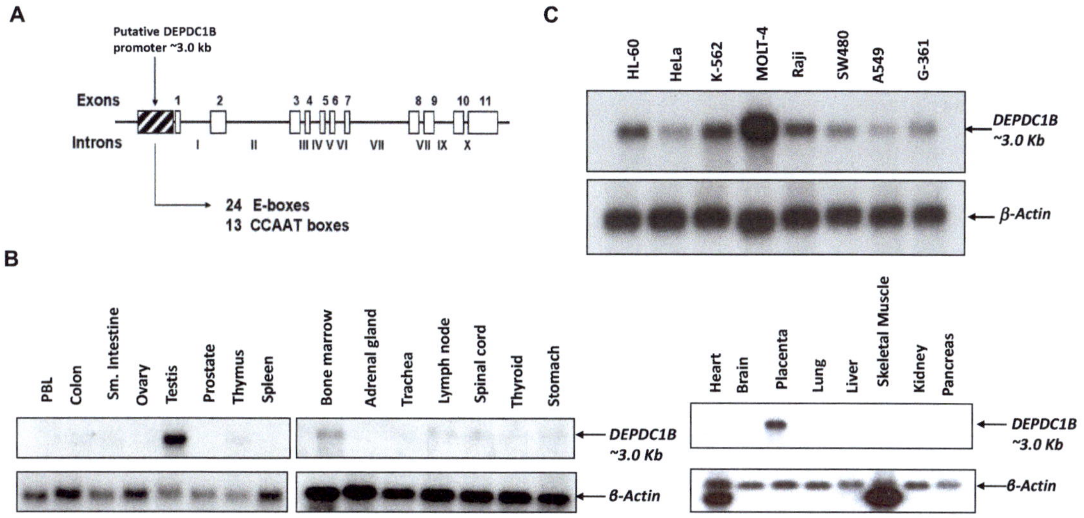


**Fig. 2** Genomic *DEPDC1B* and *DEPDC1B* mRNA expression analyses. (**a**) Schematic of *DEPDC1B* exon and intron regions and putative promoter within human chromosome 5. *DEPDC1B* cDNA sequence (GenBank accession #BC019075) was blasted against a human chromosome 5 clone (GenBank accession #AC109133). Approximately 3000 bp upstream of the 5′-end of *DEPDC1B* mRNA were selected as a putative promoter region and searched using TransFac database for putative transcription factor binding site(s). (**b**) Northern blot analysis of *DEPDC1B* mRNA in normal human adult tissues. (**c**) Northern blot analysis of *DEPDC1B* mRNA in human cancer cell lines. Northern blots shown in panels b and c were probed with a radiolabeled *DEPDC1B* cDNA and re-probed with a radiolabeled *β-Actin* as described earlier [5]

bind to the ribosomal protein RPL35, leading to enhanced synthesis of E2F1 protein which, in turn, promotes *DEPDC1B* gene transcription and tumorigenesis [8]. More recently, an in silico approach led to identification of five consensus binding motifs of a neural crest lineage transcription factor SOX10 within the *DEPDC1B* promoter region. Consistently, *DEPDC1B* is a direct downstream target of SOX10 in melanoma cells [9]. Together these data suggest context-dependent regulation of *DEPDC1B* mRNA expression in cancer cells by Raf-1, lncNB1, and SOX10.

Full-length DEPDC1B open reading frame (529 amino acids) codes for a ~ 60 kDa predominately membrane-associated protein [4, 5]. GEPIA2 tool was used to generate the isoform structure of two protein coding variants of DEPDC1B, DEPDC1B-001 (529 amino acids), and DEPDC1B-002 (467 amino acids). Both isoforms contain a DEP domain (green, amino acids 24–108) and a Rho-GAP-like domain (orange, amino acids 201–393) (Fig. 3). The Rho-GAP family of GTPase activating proteins have important implications in cancer progression [10, 11]. DEPDC1B Rho-GAP domain lacks the arginine finger residue critical for its Rho GTPase activity. However, it may bind to a target Rho protein and influence activity of cognate Rho GTPase [12]. The Scansite search showed potential interactions of DEPDC1B with the p85 regulatory subunit of membrane lipid kinase PI3K (at P111 and P116) and DNA

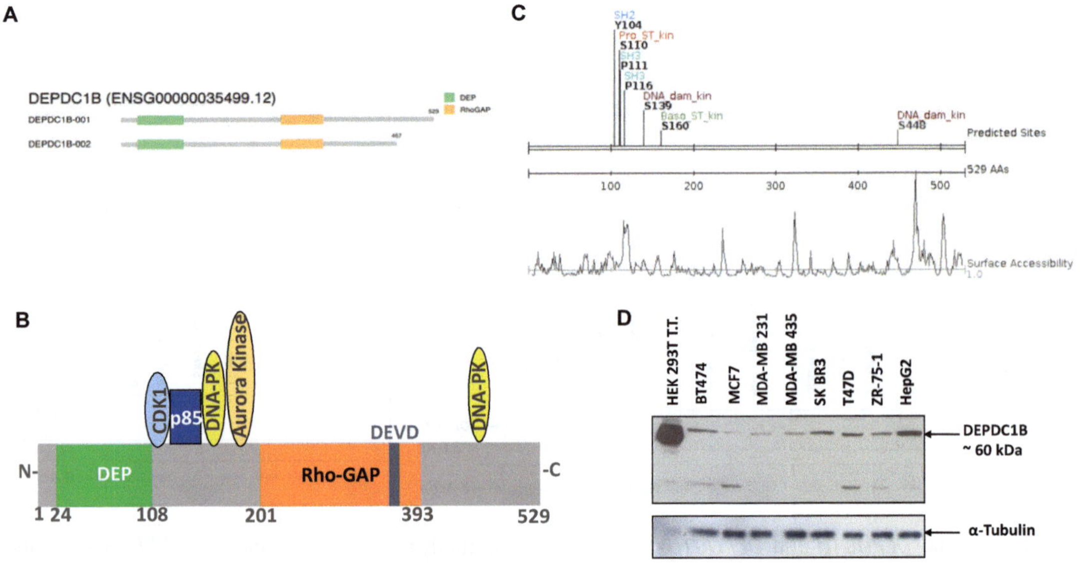


**Fig. 3** Characterization of DEPDC1B protein. (**a**) DEPDC1B isoforms (http:/gepia2.cancer-pku.cn/#isoform). (**b**) Schematic of DEPDC1B conserved domains, predicted protein–protein interaction sites, and a DEVD cleavage site. (**c**) DEPDC1B protein motifs. The amino acid sequence of DEPDC1B was scanned for protein motifs using Scansite (Scansite 4.1.02021 https:/scansite4.mit.edu/#scanProtein). (**d**) Western blot analysis of DEPDC1B expression in human cancer cell lines using custom generated anti-DEPDC1B antibody as stated in legend to Fig. 1d. HEK293T T.T., HEK293T cells transiently transfected with Myc-tagged DEPDC1B

double-strand break repair enzyme DNA protein kinase (DNA-PK) (at S139 and S448). Additional predicted binding motifs present are for CRK SH2 at Y104, CDK1 at S110, and aurora kinase A/B at S160 (Fig. 3b and c). If validated, any one or more these interactions are likely to shed new insights into DEPDC1B-mediated mechanisms of tumor progression and therapy response.

## 2 High DEPDC1B Expression During the G2/M Phase Promotes Disassembly of Focal Adhesion, De-adhesion, and Cell Entry into Mitosis

Earlier, we investigated the DEPDC1B mRNA and protein expression throughout the cell cycle. MDA-MB-231 human breast cancer cells were enriched in G1 phase by treatment with aphidicolin, followed by incubation in aphidicolin-free medium for various times. We found that DEPDC1B mRNA and protein levels were elevated in G2/M phase [4, 5]. Similar results were observed in MDA-MB-435 melanoma cells (Table 1) [5].

High expression of DEPDC1B during the G2/M phase has been shown to promote progression of cells into mitosis via disassembly of focal adhesions leading to detachment of cells (i.e., de-adhesion), referred to as a DEPDC1B-mediated de-adhesion mitotic checkpoint [13, 14]. Mechanistically, cell de-adhesion, a

**Table 1**
**Enhanced expression of DEPDC1B protein during G2/M phase of cell cycle[a]**

| Cells | Treatment | G1 (%cells) | S (%cells) | G2/M (%cells) | DEPDC1B Expression (Increase in G2/M vs untreated) |
|---|---|---|---|---|---|
| MDA-MB-435 | Untreated | 39% | 47% | 13% | |
| | Aphidicolin | 52% | 47% | 0% | |
| | 12 h post Aphidicolin | 24% | 23% | 52% | Fivefold |
| | Nocodazole | 4% | 24% | 71% | 3.4-fold |
| HeLa | Untreated | 60% | 26% | 13% | |
| | Nocodazole | 0% | 10% | 90% | Fourfold |

[a]Cells were synchronized at the G1/S phase with aphidicolin (4 μg/mL, 24 h), followed by incubation in aphidicolin-free medium for 12 h. Alternatively, cells were enriched in G2/M phase by treatment with nocodazole (100 ng/mL, 16 h) [5]

requisite for mitotic entry, is achieved by DEPDC1B accumulation during G2 phase, DEPDC1B-mediated inhibition of the small GTPase RhoA via competitive binding of DEPDC1B to protein tyrosine phosphatase receptor type, F (PTPRF), and preventing GEF-H1-mediated activation of RhoA (DEPDC1B-RhoA-PTPRF axis). DEPDC1B silencing has been associated with delayed mitotic entry and delayed activation of mitosis promoting factors. DEPDC1B amino acid sequence shows a CDK1 binding motif (S110) (Fig. 2c). Whether DEPDC1B influences CDK1 activity is unknown. In a recent report, under reduced growth factor signaling conditions, DEPDC1B seems to promote focal adhesion disassembly by the membrane lipid phosphatidylinositol-3,4-bisphosphate (PtdIns(3,4)$P_2$)-dependent inactivation of RhoA. In this scenario, a synergism between DEPDC1B and protein kinase N2 (PKN2) results in synthesis of the PtdIns(3,4)$P_2$ by class II PI3K-C2β, and PtdIns(3,4)$P_2$, in turn, inactivates RhoA through recruitment of the ARAP3 GAP [15]. In summary, DEPDC1B high expression in the G2/M phase is critical to inactivation of RhoA, disassembly of focal adhesions, and cell de-adhesion, thereby leading cell entry into mitosis.

## 3 DEPDC1B Is a Positive Upstream Effector of pERK

The MAPK canonical signaling pathway (RAS-CRAF/Raf-1-MEK-ERK) is activated in a majority of cancers, and CRAF/Raf-1 and MEK have been extensively investigated as promising druggable targets [16–21]. Raf-1-induced transcriptional changes are

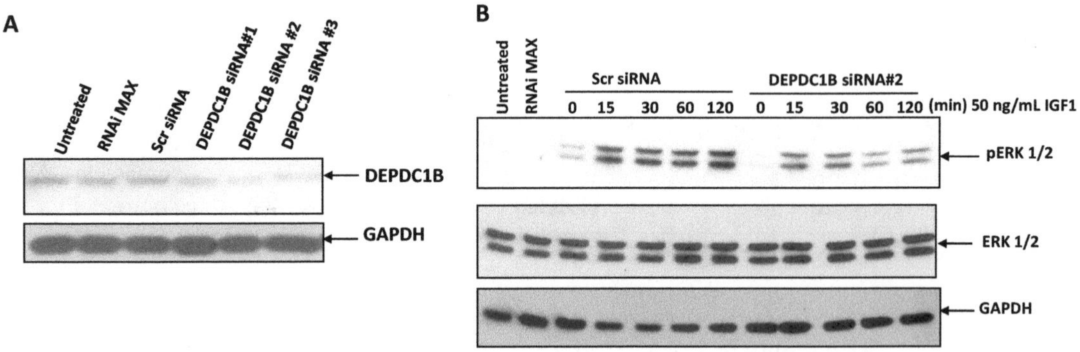


**Fig. 4** DEPDC1B knockdown is associated with decreased induction of pERK in response to IGF1 in MCF-7 cells. (**a**) Validation of siRNA knockdown of DEPDC1B. Cells were treated with 50 nM DEPDC1B siRNA (DEPDC1B Stealth siRNA #1, #2, or #3, Invitrogen) for 72 h, followed by sequential Western blot analysis as shown. (**b**) Decreased IGF1-mediated induction of pERK in DEPDC1B knockdown MCF-7 cells. Cells were treated with 50 nM DEPDC1B Stealth siRNA #2 for 72 h, followed by IGF1 treatment as shown. Whole cell lysates were analyzed by sequential Western blotting using anti-pERK(E10), anti-ERK, and anti-GAPDH antibodies

dependent, in part, on phosphorylation and activation of ERK. However, regulation of pERK via Raf-1 transcriptome is unclear. Earlier we have reported that siRNA knockdown of Raf-1 in MDA-MB-231 cells resulted in decreased expression of DEPDC1 and pERK; and siRNA knockdown of DEPDC1B in MDA-MB-231 cells was associated with decrease in pERK expression [5]. We have also demonstrated that transient or stable expression of Myc-tagged DEPDC1B in COS-1, HEK293T, and MCF-7 cells led to increased pERK levels compared to vector control [5]. In addition to the enhanced basal levels of pERK, DEPDC1B expression also seems to correlate with IGF1-stimulated expression of pERK. In this context, our data shows that siRNA knockdown of DEPDC1B is associated with reduced IGF1-induced pERK expression in MCF-7 breast cancer cells (Fig. 4). DEPDC1B expression has also been correlated with enhanced pERK expression in neuroblastoma cells [8]. Together these data suggest that DEPDC1B, a component of Raf-1 transcriptome, is a positive upstream effector of both the constitutive and ligand-stimulated pERK expression.

## 4 DEPDC1B Is a Novel Binding Partner of the p85 Subunit of PI3K, and DEPDC1B Expression Correlates with Downregulation of Insulin-Stimulated Tyrosine Phosphorylation of p85 and pAKT1 Expression

As mentioned above, DEPDC1B amino acid sequence shows putative binding motifs of the p85 subunit of PI3K at P111 and P116 (Fig. 3). Here we have verified this interaction by reciprocal co-immunoprecipitation using whole cell lysates from HEK293T

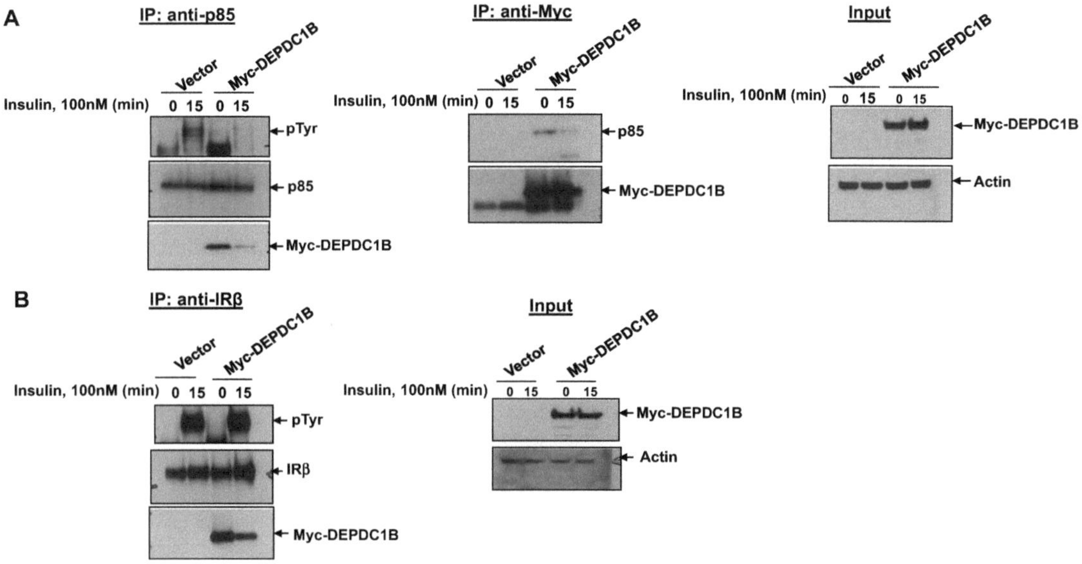


**Fig. 5** DEPDC1B constitutively interacts with the p85 subunit of PI3K, and the interaction and tyrosine phosphorylated p85 level are reduced in insulin-treated HEK293T cells. (**a**) The Myc-tagged full-length DEPDC1B cDNA was amplified by PCR using the forward primer containing BglII restriction site (bold), Kozak sequence, translation initiation codon, and the Myc epitope (underlined) 5′-G**AGATCT**GCCATGGAGCA-GAAACTCATCTCTGAAGAGGACCTGATGGAGCATCGCATCGTGGGG-3′, and the reverse primer containing Mlu1 restriction site (bold) 5′ - G**ACGCGT**CTAGCACCTGTTGCTGTGGAAG-3′, and cloned into the pCR3.1 plasmid (Myc-DEPDC1B) as detailed earlier [5]. Left panel, cells were transiently transfected with pCR3.1 vector or pCR3.1 Myc-DEPDC1B. Thirty-six hours after transfection, the cells were serum-starved for 16 h and then stimulated with insulin as shown. Whole cell lysates were pre-cleared with protein A/G beads for 1 h at 4 °C. The pre-cleared lysates, 750 μg of protein, were subjected to immunoprecipitation (IP) with anti-p85 antibody-conjugated agarose beads for 24 h. The immunoprecipitated proteins were analyzed by Western blotting as shown. Middle panel, cells were transiently transfected and treated with insulin as above. The pre-cleared whole cell lysates (750 μg of protein) were used for IP with anti-Myc antibody-conjugated agarose beads, followed by sequential immunoblotting with anti-p85 and anti-Myc antibodies. (**b**) Left panel, cells were transfected and treated as in panel a. The pre-cleared cell lysates (750 μg of protein) were used for IP with anti-Insulin Receptorβ subunit (IRβ) antibody-conjugated agarose beads, followed by Western blotting with anti-pTyr, anti-IRβ (C-19), and anti-Myc antibodies. (**a and b**) Right panels, Western blots using 1/10 (75 μg) the amount of protein from IP experiments for input references

cells transfected with Myc-DEPDC1B. Interestingly, the interaction was found to be reduced in response to insulin. In addition, insulin-stimulated tyrosine phosphorylation of p85 was negated in cells expressing Myc-DEPDC1B as compared to vector control (Fig. 5a). DEPDC1B appears to also bind to insulin receptor (IRβ), and this binding was also decreased in the presence of insulin (Fig. 5b). However, in contrast to p85, insulin-stimulated tyrosine phosphorylation of IRβ was unaffected in DEPDC1B overexpressing HEK293T cells. Transient transfection experiments using the N-terminus deletion construct (DEPDC1B-NT) or the C-terminus deletion construct (DEPDC1B-CT) of

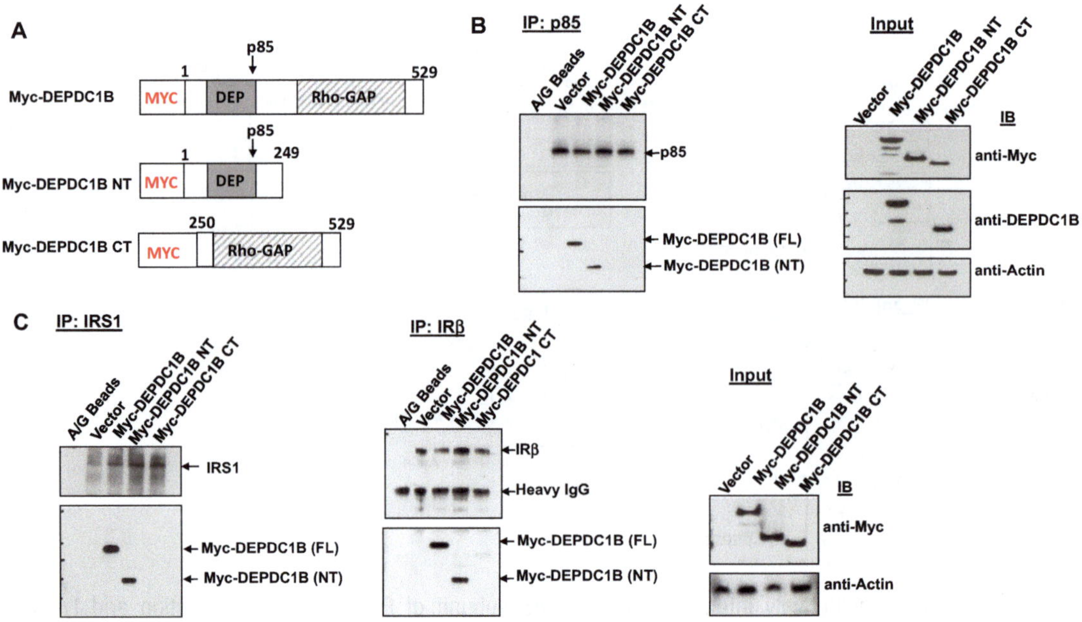


**Fig. 6** N-terminus of DEPDC1B interacts with the p85 subunit of PI3K in vivo. (**a**) Schematic of the full-length DEPDC1B, and two deletion constructs, N-terminus (NT) containing the conserved DEP domain and putative p85 binding site, and C-terminus (CT) containing the Rho-GAP domain. All constructs were Myc-epitope tagged (MYC) as shown. For the Myc-DEPDC1B N-terminus construct (Myc-DEPDC1B NT), forward primer sequence containing *BglII* restriction site (bold), Kozak sequence, translation initiation codon, and the Myc epitope (*underlined*) was 5′- G**AGATCT**GCCATGGAGCAGAAACTCATCTCTGAAGAGGACCTGATGGAGCATCG-CATCGTGGGG-3′, and the reverse primer sequence was 5′-CACGCGTTTACACTTCATAGCTGACAGC-3′. For the Myc-DEPDC1B C-terminus construct (Myc-DEPDC1B CT), forward primer sequence was 5′-G**AGATCT** GCCATGGAGCAGAAACTCATCTCTGAAGAGGACCTGTTGGCAAATTGGCCCAACTG-3′, and the reverse primer sequence containing *Mlu1* restriction site (bold) was 5′-G**ACGCGT**CTAGCACCTGTTGCTGTGGAAG-3′. The PCR conditions and vector for deletion constructs were the same as for full-length Myc-DEPDC1B [5]. (**b**) Left panel and (**c**) left and middle panels. HEK293T cells were transiently transfected with pCR3.1 empty vector, Myc-DEPDC1B, Myc-DEPDC1B NT, or Myc-DEPDC1B CT. After 48 h, whole cell lysates (1 mg of protein) were precleared and subjected to immunoprecipitation (IP) with anti-p85, anti-IRβ, or anti-IRS-1 antibody-conjugated agarose beads. The immunoprecipitated proteins were analyzed by Western blotting with anti-Myc antibody. The same blot was stripped and re-probed with the antibodies as indicated. (**b** and **c**) Right panels, immunoblotting (IB) with indicated antibodies using lysates (100 μg of protein) from IP experiments for input references

Myc-DEPDC1B and co-immunoprecipitation showed that N-terminus of DEPDC1B containing the p85 binding motifs and not the C-terminus of Myc-DEPDC1B interacts with p85, IRS-1, and insulin receptor (Fig. 6). These data demonstrate direct and constitutive interaction between DEPDC1B and the p85 subunit of PI3K and suggest that DEPDC1B may also associate with IRS1 and IRβ. We conclude that the N-terminal portion of DEPDC1B contains at least one bona fide p85 binding motif required for its binding to the p85 subunit of PI3K.

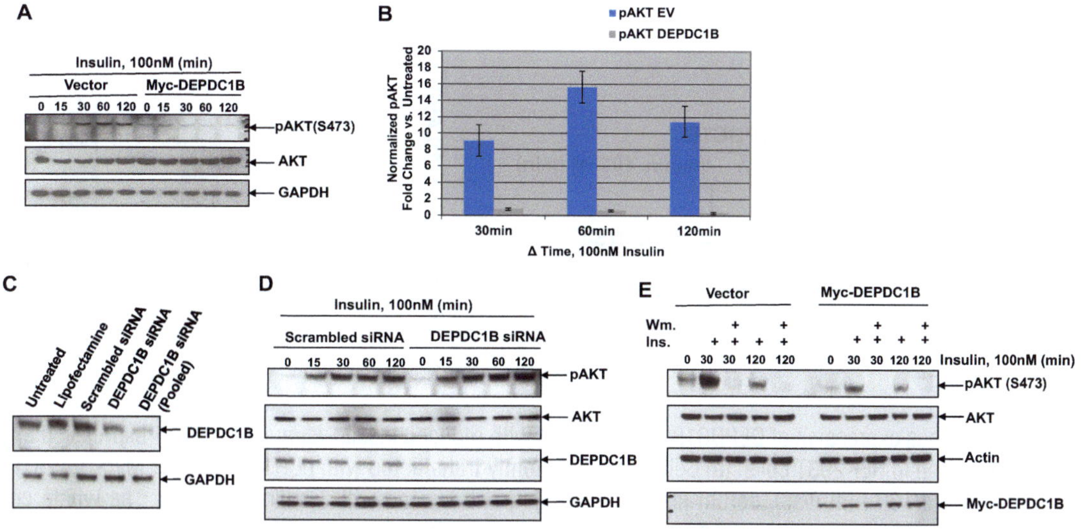


**Fig. 7** Exogenous DEPDC1B expression is associated with decreased insulin-induced p-AKT via PI3K in HEK293T cells. (**a**) Cells were transiently transfected with either pCR3.1 vector or Myc-DEPDC1B using LipofectAmine 2000. Twenty-four hours after transfection, the cells were serum-starved overnight and then treated with insulin as shown. Whole cell lysates were analyzed by sequential Western blotting with anti-pAKT (S473), anti-AKT, and anti-GAPDH antibodies. (**b**) The Western blot signals in panel a were quantified using ImageQuant software, and pAKT expression was normalized against GAPDH signal. Data shown are representative of three independent experiments. (**c**) Cells were seeded in 60mm$^2$ tissue culture dishes ($5 \times 10^5$ cells/dish) in DMEM supplemented with 10% FBS. The next day, the medium was switched to 2 mL OPTI-MEM medium containing scrambled siRNA (200 nM) or DEPDC1B siRNA (200 nM) (siGENOME, Dharmacon) for 72 h, and whole cell lysates were analyzed by sequential immunoblotting with anti-DEPDC1B and anti-GAPDH antibodies. (**d**) Cells were treated with scrambled siRNA (200 nM) or DEPDC1B siRNA (200 nM) for 48 h, followed by serum starvation for 16 h. Cells were then treated with insulin, and the cell lysates were analyzed by sequential immunoblotting as shown. (**e**) Cells were transiently transfected as in panel a, followed by serum starvation for 16 h, and treatment with wortmannin (Wm., 100 nM) for 15 min, and then with human recombinant insulin as indicated. The whole cell lysates were subjected to sequential immunoblotting with anti-pAKT(S473), anti-AKT, anti-Myc (9E10), and anti-Actin antibodies

Insulin binding to the insulin receptor tyrosine kinase (IR) induces autophosphorylation of the receptor and activation of the IR-IRS1-PI3K pathway. Activated class I PI3K phosphorylates PtdIns-4,5P$_2$ to PtdIns-3,4,5P$_3$. PtdIns-3,4,5P3 interacts with the Pleckstrin homology domain of AKT1 (PKBα), recruiting AKT1 to the membrane where it is phosphorylated at T308 and S473 by PDK1 and mTORC2, respectively [22–25]. As shown in Fig. 7a and b, Myc-DEPDC1B expression in HEK293T cells correlated with a substantial reduction of insulin-induced pAKT (S473) level compared to vector control. Conversely, siRNA knockdown of endogenous DEPDC1B was associated with relatively higher levels of insulin-stimulated pAKT versus scrambled siRNA control (Fig. 7c and d). To verify that DEPDC1B-mediated negative regulation of pAKT1 indeed occurs downstream of PI3K, HEK293T cells were treated with wortmannin, a PI3K-specific

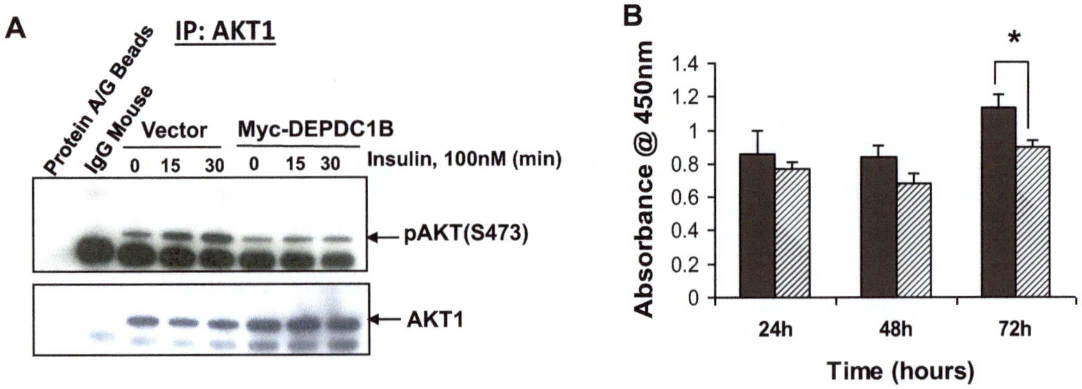


**Fig. 8** Exogenous expression of DEPDC1B is associated with reduced insulin-stimulated pAKT1 expression and cell proliferation in HEK293T cells. (**a**) Cells were transiently transfected with pCR3.1 vector or Myc-DEPDC1B (10 μg plasmid/100 mm dish), serum-starved overnight, and treated with insulin as shown. The whole cell lysates (1 mg protein) were subjected to immunoprecipitation (IP) using anti-AKT1 antibody. Immunoprecipitated proteins were analyzed by Western blotting sequentially with anti-pAKT(S473) and anti-AKT1 antibodies. (**b**) Cells ($5 \times 10^5$) were transiently transfected with 2 μg of Myc-DEPDC1B (hatched) or empty vector (black) and seeded at $1 \times 10^4$ per well in a 96-well plate for 36 h. Cells were then serum-starved and treated with 100 nM insulin for 16 h. Cell viability was assayed at 24, 48, and 72 h by WST-1 assay. Data presented are the mean +/− s.d. of two independent experiments, each performed in triplicate. *, $p < 0.001$

inhibitor, prior to insulin stimulation. As shown in Fig. 7e, insulin-stimulated pAKT(S743) levels were (1) significantly inhibited within 30 min of insulin treatment in Myc-DEPDC1B transfected cells as compared to control vector and (2) completely inhibited in the presence of wortmannin in both the Myc-DEPDC1B transfected cells and control vector transfectants versus no wortmannin control counterparts. Furthermore, using a combination of the immunoprecipitation and immunoblotting assays, overexpression of DEPDC1B was found to suppress insulin-induced phosphorylation of AKT1 (S473) (Fig. 8a). DEPDC1B overexpression in HEK293T cells also attenuated the cell viability in insulin-treated cells at 72 h after transfection compared to vector control (Fig. 8b).

Insulin receptor substrates (IRS1, IRS2) are known positive scaffolding adaptors linking the insulin and IGF1 receptors to activation of PI3K/AKT. DEPDC1B appears to function as a newly discovered negative adaptor in this pathway, further exposing AKT1 as a targetable vulnerability in cancer cells. Previously, pAKT1-mediated inhibition of invasion and migration has been reported [26]. In MCF-10A cells, knockdown of AKT1 was shown to enhance ERK activation and cell migration in response to IGFI or EGF [27]. Other reports suggest cell context-dependent cross-regulation of AKT1 and ERK signaling pathways. Specifically, activated AKT1-mediated phosphorylation of cRAF/Raf-1 (S259) has been shown to inhibit Raf-1 activity [28, 29]. It remains to be determined whether modulation of DEPDC1B influences Raf-1 activity and the outcome of Raf-1/MEK-targeted therapies. Nonetheless, we conclude that DEPDC1B is central to cross-regulation of the AKT1 and ERK pathways.

## 5 DEPDC1B Is Implicated in Other Protein–Protein Interactions and Tumor Progression

Several other cell context-dependent mechanisms of DEPDC1B-mediated oncogenesis, angiogenesis, and metastasis have been reported. For instance, while DEPDC1B expression had no effect on RHOA-GTP/RAC1-GTP levels or WNT/$\beta$-catenin signaling in melanoma cells, activation of SOX10-DEPDC1B-SCUBE3 axis was found to promote melanoma angiogenesis and metastasis [9]. In this study, DEPDC1B was identified as a direct downstream target of transcription factor SOX10 and shown to interact with ubiquitin ligase CDC16, thereby preventing degradation of SCUBE3, a member of the SCUBE family of secreted glycoproteins. In contrast, DEPDC1B seems to enhance cell migration and invasion in non-small cell lung carcinoma through activation of Wnt/β-catenin signaling [30]. In a predicted 3D model of DEPDC1B, amino acid residues interacting with Rac-1 have been identified, and DEPDC1B has been shown to promote cell migration and invasion through the Rac1/PAK1-LIMK1-Cofilin1 signaling pathway in pancreatic cancer cells [31, 32]. In prostate cancer cells, DEPDC1B was shown to bind to Rac1, activating the Rac1-PAK1 signaling pathway and inducing epithelial- mesenchymal transition [33].

STRING protein–protein interaction network analysis shows a number of predicted DEPDC1B interacting proteins (Fig. 9). These are Rho GTPase activating protein 11a (ARHGAP11A), mitotic checkpoint serine/threonine (BUB1), cyclin-dependent kinase-1 (CDK1), centrosomal protein of 55 kDa (CEP55), disk large-associated protein 5 (DLGAP5), hyaluronan-mediated motility receptor (HMMR), kinesin-like protein-15 (KIF15), kinesin-like protein-2C (KIF2C), LMBR1 domain-containing protein-2 (LMBRD2), and exportin-7 (XPO7). Many of these interactions, if validated, are likely to expand the repertoire of DEPDC1B-centric networks implicated in regulation of tumor progression and therapy response.

High expression of DEPDC1B has been demonstrated in several cancer types [9, 30, 32–34]. Moreover, a survey of publicly available datasets showed significantly high DEPDC1B expression in breast, lung, and pancreatic carcinoma and melanoma compared to matched benign specimens (Fig. 10a). Furthermore, high DEPDC1B expression has been correlated with decreased probability of survival in patients with breast cancer, lung cancer, pancreatic adenocarcinoma, and renal cell carcinoma (Fig. 10b). Thus, DEPDC1B is a viable target in a clinical setting.

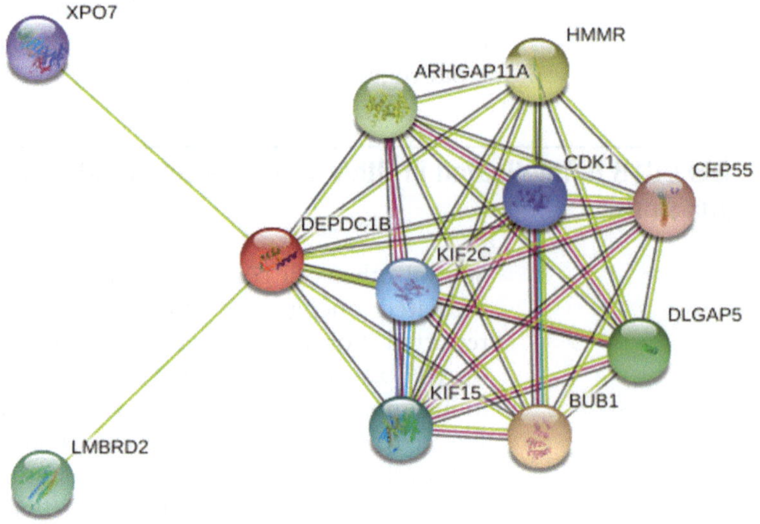


**Fig. 9** STRING protein–protein interaction network-based identification of DEPDC1B interacting proteins. The network is composed of nodes, which are the proteins, and the color-coded edges provide the association each node has with another node. The empty nodes are proteins of unknown 3D structure and filled nodes are proteins with predicted or known 3D structure. The pink and blue edges are known interactions either experimentally determined or from curated databases. The green, red, and blue edges are predicted interactions based on gene neighborhood, gene fusions, or gene co-occurrence, respectively. Proteins shown were identified as known or predicted to interact with DEPDC1B. DEPDC1B STRING (Version 11.5 August 2022) data can be found at https:/version-11-5.string-db.org/cgi/network?networkId=bFZBQ7CtAKS3(DEPDC1B protein (human) - STRING interaction network (string-db.org)

## 6 Conclusions and Future Perspectives

Molecular targets functioning at the interface of multiple signaling pathways are likely to engage a broad spectrum of downstream effectors and thereby contribute to plasticity of both the canonical pathways and genome-wide networks, two of the major challenges in tackling tumor evolution and therapy response. While our understanding of the DEPDC1B systems and integrative biology as yet is rudimentary, several reports to date and new data presented here provide an evidence of DEPDC1B as a model target with multipronged influence on the tumor systems biology (Fig. 11). Current highlights of this area of research are listed as follows (1). DEPDC1B is a new binding partner of the p85 subunit of PI3K (2). DEPDC1B cross-regulates AKT and ERK pathways through downregulating the ligand-stimulated tyrosine phosphorylation of p85 and expression of pAKT1 and promoting pERK, a hallmark of tumor progression. (3) High DEPDC1B expression during the

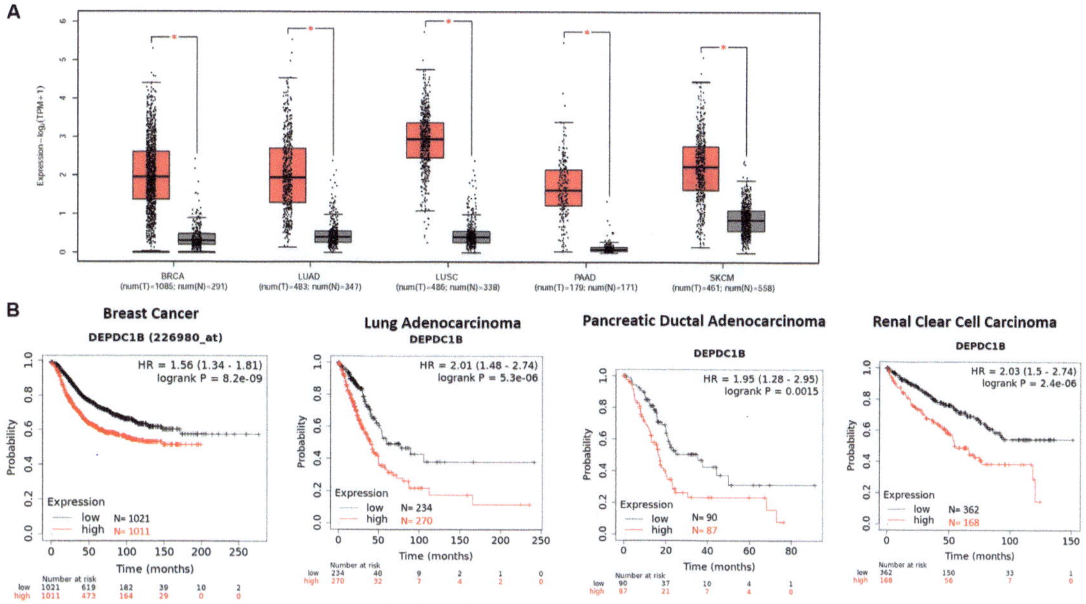


**Fig. 10** DEPDC1B is a viable cancer biomarker. (**a**) *DEPDC1B* gene expression in tumor (T) and adjacent normal (N) tissues from TCGA and the Genotype Tissue Expression (GTEx) datasets. BRCA breast invasive carcinoma, LUAD lung adenocarcinoma, LUSC lung squamous cell carcinoma, PAAD pancreatic adenocarcinoma, and SKCM skin cutaneous melanoma. Tumor, red box; normal, black box. The *p* value cutoff was 0.01. Expression shown is -$\log_2$(TPM + 1) (Transcripts Per Million), GEPIA2 (2018) box plots: http:/gepia2.cancer-pku.cn/#analysis. (**b**) Relationship between *DEPDC1B* expression and survival probability in cancer patients. Survival curves were generated with The KMplotter (July 2022). Pan-cancer online datasets and the breast cancer datasets are linked as: https:/kmplot.com/analysis/index.php?p=background

G2/M phase is a cell de-adhesion mitotic checkpoint, a prerequisite for cell entry into mitosis (4). DEPDC1B mRNA expression is upregulated via expression of Raf-1 and long noncoding RNA lncNB1, and DEPDC1B is a direct target of transcription factor SOX10. DEPDC1B prevents SCUBE3 degradation, and SOX10-DEPDC1B-SCUBE3 axis promotes angiogenesis and metastasis.

Future investigations are necessary to (a) address whether DEPDC1B functions to integrate, albeit in a context-dependent manner, the AKT and ERK networks and known feedback mechanisms and (b) identify the actionable molecular, spatial, and temporal vulnerabilities within the DEPDC1B-guided versions of the integrated AKT and ERK networks in cancer cells. In addition, DEPDC1B amino acid sequence shows binding motifs for three clinically relevant therapeutic targets CDK1, DNA-PK, and aurora kinase A/B. These partnerships and functionalities, if validated, are likely to further implicate DEPDC1B in regulation of DNA damage–repair and cell cycle progression processes critical to disease prognosis and therapy response.

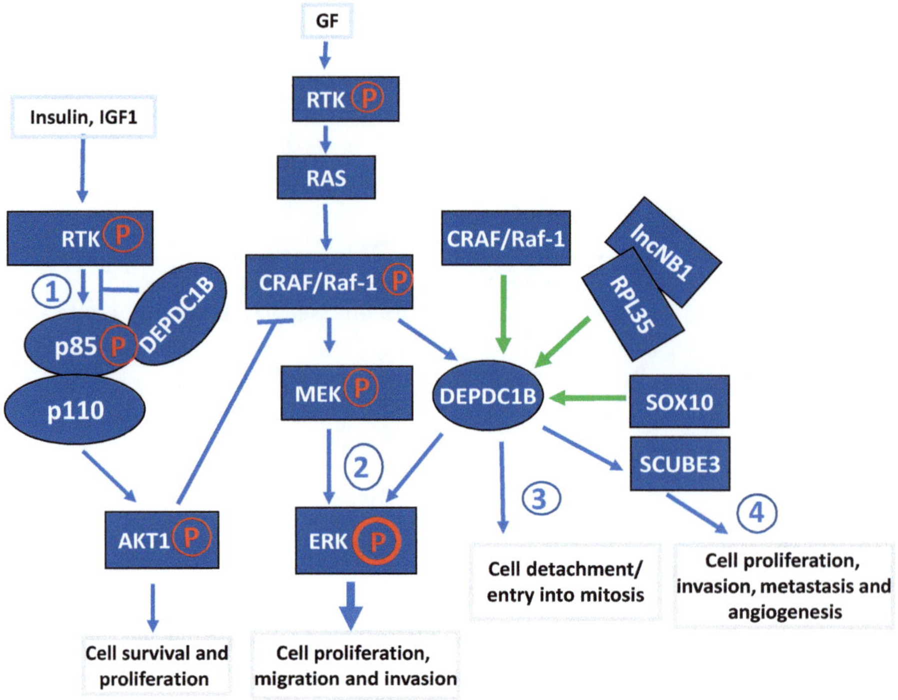


**Fig. 11** Partnership between DEPDC1B, a downstream effector of Raf-1/lncNB1/SOX10, and the p85 subunit of PI3K: A paradigm of interception of ligand-induced PI3K-mediated activation of pAKT1 and promotion of pERK activity. Green arrows indicate upregulation of *DEPDC1B* mRNA expression. GF Growth factor, RTK Receptor tyrosine kinase. (See text for details)

## Acknowledgements

The authors thank Doug Joubert, Coordinator, The National Institutes of Health (NIH) Library Bioinformatics Support Program, for assistance with bioinformatic tools and resources, and their laboratory colleagues for contributions during the initial phase of this project. This work was supported, in part, by grants from the National Institutes of Health (CA58984, CA65012, CA68322, CA74175) and NeoPharm, Inc. Portion of the work presented in this chapter was carried out by HEB toward his Ph.D. dissertation. UNK and HEB are co-inventors of the Georgetown University DEPDC1B (BRCC3) patented technology "Gene BRCC-3 and Diagnostic and Therapeutic Uses Thereof" US patent 7,442,520 B2. HEB is currently a Staff Scientist at National Institutes of Health, NIAID/LCIM/MDS, Bethesda, MD, USA. Several cell lines were obtained from the Tissue Culture Shared Resource of the Georgetown Lombardi Comprehensive Cancer Center. All shared resources were supported by the NIH grant P30-CA51008.

## References


1. Gokhale PC, Zhang C, Newsome J, Pei J, Ahmad I, Rahman A, Dritschilo A, Kasid U (2002) Pharmacokinetics, toxicity, and efficacy of ends-modified raf antisense oligodeoxyribonucleotide encapsulated in a novel cationic liposome (LErafAON). Clin Cancer Res 8: 3611–3621
2. Rudin CM, Marshall JL, Huang CH, Kindler HL, Zhang C, Kumar D, Gokhale PC, Steinberg J, Wanaski S, Kasid UN, Ratain MJ (2004) Delivery of a liposomal c-raf-1 antisense oligonucleotide by weekly bolus dosing in patients with advanced solid tumors: a phase I study. Clin Cancer Res 10(21):7244–7251
3. Dritschilo A, Huang CH, Rudin CM, Marshall J, Collins B, Dul JL, Zhang C, Kumar D, Gokhale PC, Ahmad A, Ahmad I, Sherman JW, Kasid UN (2006) Phase I study of liposome-encapsulated c-raf antisense oligodeoxyribonucleotide infusion in combination with radiation therapy in patients with advanced malignancies. Clin Cancer Res 12: 1251–1259
4. Boudreau HE, Broustas CG, Gokhale PC, Mewani RR, Ahmad I, Kasid U (2006) Inhibition of Raf-1 in human breast cancer cells is associated with decreased expression of BRCC3, a novel cell cycle regulated molecule. 97th annual American Association of Cancer Research Meeting, Washington, DC., Abstract No. 2586
5. Boudreau HE, Broustas CG, Gokhale PC, Kumar D, Mewani RR, Rone JD, Haddad BR, Kasid U (2007) Expression of BRCC3, a novel cell cycle regulated molecule, is associated with increased phospho- ERK and cell proliferation. Int J Mol Med 19:29–39
6. Roy AL, Meisterernst M, Pognonec P, Roeder RG (1991) Cooperative interaction of an initiator-binding transcription initiation factor and the helix-loop-helix activator USF. Nature 354:245–248
7. Wasner M, Haugwitz U, Reinhard W et al (2003) Three CCAAT-boxes and a single cell cycle genes homology region (CHR) are the major regulating sites for transcription from the human cyclin B2 promoter. Gene 312: 225–237
8. Liu PY, Tee AE, Milazzo G et al (2019) The long noncoding RNA lncNB1 promotes tumorigenesis by interacting with ribosomal protein RPL35. Nat Commun 5:5026
9. Hu F, Fong KO, Cheung MPL et al (2022) DEPDC1B promotes melanoma angiogenesis and metastasis through sequestration of ubiquitin ligase CDC16 to stabilize secreted SCUBE3. Adv Sci (Weinh) 9(10):e2105226
10. Kreider-Letterman G, Carr NM, Garcia-Mata R (2022) Fixing the GAP: the role of RhoGAPs in cancer. Eur J Cell Biol 101(2):151209
11. Schaefer A, Der CJ (2022) RHOA takes the RHOad less traveled to cancer. Trends Cancer 8:655–669
12. Amin E, Jaiswal M, Derewenda U et al (2016) Deciphering the molecular and functional basis of RHOGAP family proteins: a systematic approach toward selective inactivation of RHO family proteins. J Biol Chem 29: 20353–20371
13. Marchesi S, Montani F, Deflorian G et al (2014) DEPDC1B coordinates de-adhesion events and cell-cycle progression at mitosis. Dev Cell 31:420–433
14. Garcia-Mata R (2014) Arrested detachment: a DEPDC1B-mediated de-adhesion mitotic checkpoint. Dev Cell 31:387–389
15. Posor Y, Kampyli C, Bilanges B et al (2022) Local synthesis of the phosphatidylinositol-3,4-bisphosphate lipid drives focal adhesion turnover. Dev Cell 57:1694–1711
16. Karreth FA, Frese KK, DeNicola GM, Baccarini M, Tuveson DA (2011) C-Raf is required for the initiation of lung cancer by K-Ras G12D. Cancer Discov 1:128–136
17. Kasid U, Pfeifer A, Brennan T, Beckett M, Weichselbaum RR, Dritschilo A, Mark GE (1989) Effect of antisense c-raf-1 on tumorigenicity and radiation sensitivity of a human squamous carcinoma. Science 243:1354–1356
18. Klomp JE, Klomp JA, Der CJ (2021) The ERK mitogen-activated protein kinase signaling network: the final frontier in RAS signal transduction. Biochem Soc Trans 49:253–267
19. Yaeger R, Corcoran RB (2019) Targeting alterations in the RAF–MEK pathway. Cancer Discov 9:329–341
20. Karoulia Z, Gavathiotis E, Poulikakos PI (2017) New perspectives for targeting RAF kinase in human cancer. Nat Rev Cancer 17: 676–691
21. Kasid U, Suy S, Dent P, Ray S, Whiteside TL, Sturgill TW (1996) Activation of Raf by ionizing radiation. Nature 382:813–816
22. Hoxhaj G, Manning BD (2020) The PI3K-AKT network at the interface of oncogenic signalling and cancer metabolism. Nat Rev Cancer 20:74–88
23. Vanhaesebroeck B, Guillermet-Guibert J, Graupera M, Bilanges B (2010) The emerging

mechanisms of isoform-specific PI3K signaling. Nat Rev Mol Cell Biol 11:329–341

24. Vanhaesebroeck B, Perry MWD, Brown JR, André F, Okkenhaug K (2021) PI3K inhibitors are finally coming of age. Nat Rev Drug Discov 20:741–769
25. Manning BD, Toker A (2017) AKT/PKB signaling: navigating the network. Cell 169:381–405
26. Yoeli-Lerner M, Yiu GK, Rabinovitz I, Erhardt P, Jauliac S, Toker A (2005) Akt blocks breast cancer cell motility and invasion through the transcription factor NFAT. Mol Cell 20: 539–550
27. Irie HY, Pearline RV, Grueneberg D et al (2005) Distinct roles of Akt1 and Akt2 in regulating cell migration and epithelial–mesenchymal transition. J Cell Biol 171:1023–1034
28. Rommel C, Clarke BA, Zimmermann S et al (1999) Differentiation stage-specific inhibition of the Raf-MEK-ERK pathway by Akt. Science 286:1738–1741
29. Zimmermann S, Moelling K (1999) Phosphorylation and regulation of Raf by Akt (protein kinase B). Science 286:1741–1744
30. Yang Y, Liu L, Cai J et al (2014) DEPDC1B enhances migration and invasion of non-small cell lung cancer cells via activating Wnt/β-catenin signaling. Biochem Biophys Res Commun 450:899–905
31. Ahuja P, Singh K (2016) In silico approach for SAR analysis of the predicted model of DEPDC1B: a novel target for oral cancer. Adv Bioinform 3136024:1
32. Zhang S, Shi W, Hu W et al (2020) DEP domain-containing protein 1B (DEPDC1B) promotes migration and invasion in pancreatic cancer through the Rac1/PAK1-LIMK1-Cofilin1 signaling pathway. OncoTargets Ther 13: 1481–1496
33. Li Z, Wang Q, Peng S et al (2020) The metastatic promoter DEPDC1B induces epithelial-mesenchymal transition and promotes prostate cancer cell proliferation via Rac1-PAK1 signaling. Clin Transl Med 10(6):e191
34. Dang XW, Pan Q, Lin ZH et al (2021) Overexpressed DEPDC1B contributes to the progression of hepatocellular carcinoma by CDK1. Aging (Albany NY) 13(16):20094–20115

# Chapter 22

# Single-Cell and Spatial Analysis of Emergent Organoid Platforms

**Aditi Kumar, Shuangyi Cai, Mayar Allam, Samuel Henderson, Melissa Ozbeyler, Lilly Saiontz, and Ahmet F. Coskun**

## Abstract

Organoids have emerged as a promising advancement of the two-dimensional (2D) culture systems to improve studies in organogenesis, drug discovery, precision medicine, and regenerative medicine applications. Organoids can self-organize as three-dimensional (3D) tissues derived from stem cells and patient tissues to resemble organs. This chapter presents growth strategies, molecular screening methods, and emerging issues of the organoid platforms. Single-cell and spatial analysis resolve organoid heterogeneity to obtain information about the structural and molecular cellular states. Culture media diversity and varying lab-to-lab practices have resulted in organoid-to-organoid variability in morphology and cell compositions. An essential resource is an organoid atlas that can catalog protocols and standardize data analysis for different organoid types. Molecular profiling of individual cells in organoids and data organization of the organoid landscape will impact biomedical applications from basic science to translational use.

**Key words** Organoid heterogeneity, Organoid atlas, Organogenesis, Single-cell and spatial analysis, Three-dimensional (3D) tissues, Pluripotent stem cells (PSCs), Adult stem cells (ASCs), Embryonic stem cells (ESCs), Induced pluripotent stem cells (iPSC), Extracellular matrix (ECM), Light-sheet fluorescence microscopy (LSFM), Epi-illumination selective-plane illumination microscope (eSPIM), Open-top light-sheet (OTLS) microscopes, Single-cell RNA-sequencing (scRNA-seq), Multivariate-barcoded mass cytometry (MC), High-dimensional immunophenotyping, Imaging mass cytometry (IMC), Multiplexed error-robust fluorescence in situ hybridization (MERFISH), Sequential fluorescence in situ hybridization (seqFISH), Cyclic immunofluorescence (CycIF), Matrix-assisted laser desorption/ionization (MALDI), Time-of-flight secondary ion mass spectrometry (TOF-SIMS), Patient-derived xenograft (PDX), Regenerative medicine, Precision medicine

## 1 Introduction

Regenerative medicine has identified organoids as a promising alternative to traditional 2D culture systems with the ability to shorten the gap in translational research. Organoids allow studying how to control spatio-temporal self-renewal and differentiation in cells. The key promises in organoids are the growth of organs

Usha N. Kasid and Robert Clarke (eds.), *Cancer Systems and Integrative Biology*, Methods in Molecular Biology, vol. 2660, https://doi.org/10.1007/978-1-0716-3163-8_22,

in vitro and the ability to complement preclinical research as the organoids can be grown from either stem cells or patient-derived cells [1]. The first demonstration of organoids dates back to the late twentieth century when primary keratinocytes from human were co-cultured with 3T3 cells to proliferate stratified squamous epithelial colonies to recreate the epidermis; this demonstration links the study of the 3D organization of cells into tissues and then organs from a 2D surface [2]. The ability to grow hepatocytes from rat tail collagen while integrating the secretion of cytochrome P450 showed how organoids could retain tissues' typical characteristics [3]. Improvements in understanding extracellular matrix and cell suspension methods laid the foundation for 3D culture systems [4]. Alveolus-like structures from a 3D laminin matrix created mammary epithelial cells with the ability to secrete milk, demonstrating earlier examples of organoids [3–5].

After the initial establishment of the organoid platform, pluripotent stem cells guided fundamental advances in organoid research. Embryonic stem cell (ESC) retinal epithelium with epithelial vesicles formed the optic cup via intrinsic self-organization [6]. Additionally, intestinal stem cells created an epithelial structure similar to a gut [7]. Organ-specific stem cells and progenitor cells from the stomach, bladder, and lung can be grown as organoids and utilized in disease modeling and clinical translation. For example, organoids can guide treatment for patients with rare diseases such as cystic fibrosis via drug screening [3]. The initial focus was on growing stem cell-derived organoids and studying their ability to self-organize and differentiate, mimicking in vivo organs. The definition of an organoid was previously associated with a 3D complex structure, derived from stem cells, that self-organizes. Patient-derived cells have recently been used to produce organoids from cancerous and normal tissues [8]. Established protocols developed for patient-derived organoids, also referred to as tissue-derived organoids, can be applied for drug screening [9]. The current formation procedure of a patient-derived organoid is a tumor cell culture from a patient that can be expanded and retain morphological and genetic features of the tumor, yielding a predictive tool for treatment options [10].

Along with expanding organoid derivations, a growing body of research has utilized single-cell analysis for organoid screening. Single-cell details capture the heterogeneity of cell populations, previously not obtained in sequencing and spectrometry experiments. It can help understand the organoid-to-organoid variability in individual cells at high throughput. For example, single-cell RNA sequencing (scRNA-seq) and accessible chromatin profiling enable understanding of organoid development and explore how gene expression varies from one cell to another [11]. Organoids and organs typically contain multiple cell types and states in their microenvironments. Combining single-cell analysis and spatial single-cell profiling would locate different cell types and elucidate the structural heterogeneity, leading to advances in organoid research and benefiting disease modeling.

In this perspective, we discuss how organoids can be derived and successful organoid advancements from the significant classes of organoids. An overview of the preparation of organoids, emerging single-cell analysis techniques, and the potential applications of organoids is provided (Fig. 1). We highlight the technical challenges with organoid development, administrative concerns with organoids, along with the translational research potential to showcase future research areas. Notably, the diversity of culture media compositions for growing organoids can be attributed to a lack of lab-to-lab standardization. The lack of native tissue compartments such as vasculature in organoid systems hinders multicellular assemblies' formation [15, 16]. The shapes of diverse organoid types and the related cell types are presented to provide a visual representation of the differences within each organoid type. Emerging single-cell analysis, spatial single-cell profiling of the cell-cell interactions, and signaling of the heterogeneous cell types in organoids can improve organoid screening and bridge the gap between in vitro and in vivo screening [17].

## 2 Emergence and Construction of Organoids

Organoids serve as in vitro 3D tissue models, derived from stem cells and normal or diseased tissues. In organoids, cells are self-assembled during differentiation events under the influence of exogenous and endogenous signals. Simultaneously, cell types assemble at the micro-level based on the physical cell interactions and at the macro-level through the alterations in morphological architecture via contraction or expansion [4]. Organoids typically contain multiple cell types and are co-cultured with target or effector cells in unique growth conditions [18]. Typical 2D cell lines cannot capture 3D interactions of multiple cell types, but multicellular organoids can bridge that gap. Organoids preserve the native tissue specimens' microenvironment and recapitulate the varying cell states [19]. Multicellular organoids create a complex model to include 3D cellular components using ample cell types such as epithelial cells, immune cells, and cancer-associated fibroblasts. For example, the enteric nervous system can be incorporated into the human intestinal organoid for disease modeling of the gut, blood, and neural crosstalk [20].

### 2.1 Emergence of Organoids

Organoids are 3D cultures grown from pluripotent stem cells (PSCs), adult stem cells (ASCs), and patient-derived tissues. The PSCs mainly used in organoid research are ESCs and induced pluripotent stem cells (iPSC). PSCs create the three germ layers (endoderm, ectoderm, and mesoderm). Organoids are produced

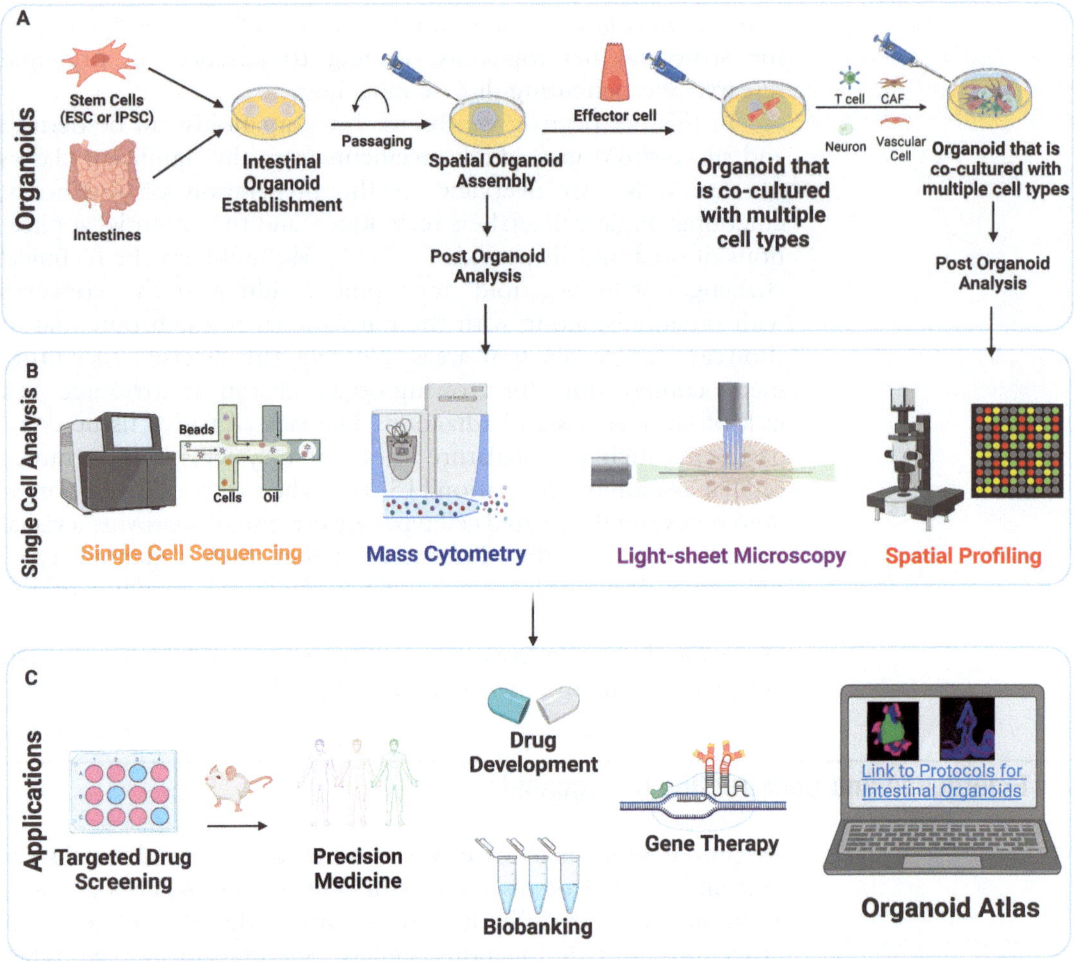


**Fig. 1** Building, deciphering, and leveraging organoids. This figure presents an overview of the process of harvesting and co-culturing organoids with different cell types, both patient-derived and stem cell-derived, for single-cell analysis. An intestinal organoid is illustrated here, but this process applies to all types of organoids. After the organoid is initially established and assembled in a spatial scaffold, it is prepared for single-cell analysis. These organoids can become multicellular with different cell types by using an effector cell. Alternatively, organoids can be co-cultured with cancer-associated fibroblasts (CAF), neurons, immune cells, or vascular cells to resemble their desired organ; post co-culture, the organoid is prepared for analysis [12, 13]. Popular single-cell analysis techniques involve single-cell sequencing, mass cytometry light-sheet microscopy, and spatial profiling. These imaging techniques allow for the different cell types present in an organoid to be studied and comprehend how the cellular level affects the organoid development. Some potential applications of organoids are highlighted here. Drugs can be developed and tested on organoids to ensure better translational research; drugs can be screened on the patient-derived organoid before administration and allow the personalization of the drug to the patient. Biobanking stores patients' diseased samples for future organoid use and treatment; it captures the disease's complexity. CRISPR-Cas9 technology used in conjunction with organoids allows the study of human genetics and tests the CRISPR-Cas9 gene therapy's applicability in organoids. A proposed solution to the lack of lab-to-lab standardization of protocols is creating an organoid atlas in which organoid pioneers can crowdsource their protocols and organoid shapes. Organoid atlas enables efficient testing protocols for each type of organoid. An online atlas platform can be generated as a list of the different types of organoids available to share the corresponding protocols and images to each shape. Experts can then judge the best protocols based on the available data or decide to lead more guided efforts in elucidating the best protocol [14]

either by manipulating the patterning or letting the cell differentiate independently with a scaffold, e.g., hydrogel, Matrigel, collagen, extracellular matrix (ECM), and other membrane matrices, which will be referred to in the subsequent organoid culture protocols [21, 22]. ASCs are isolated from human tissues and currently established via epithelial tissues, while using them in non-epithelial tissue platforms is possible [21, 23]. Patient-derived organoids can maintain the morphological and phenotypic characteristics of their original tissues, allowing them to capture the sensitivity to drugs and the tissue heterogeneity [24]. Patient-derived tissues can be both tumor and non-diseased tissues to aid in understanding organogenesis.

Organoids created from iPSCs can help model infectious diseases. For example, an iPSC-derived brain organoid demonstrated that the Zika virus effectively destroys early-stage organoids representing a first-trimester fetal brain [25]. IPSC-derived lung organoids are also used in studying the pathogenesis of COVID-19 and can serve as a model for candidates of COVID-19 inhibitors [26]. Furthermore, iPSCs can be beneficial at studying genetic diseases and disease modeling. For instance, CRISPR/Cas9 genome editing on iPSC-derived lung organoids showed that amending the most common mutation that causes cystic fibrosis can cure the disease [27]. Thyroids and lungs derived from ESCs capture the developmental landmarks and can help identify potential treatments for diseases commonly affecting these organs [28]. Differentiated iPSCs can create biobanks to screen for drugs' efficacy and toxicity to complement animal testing [29].

ASC-derived organoids are particularly useful when determining how tissue or tumor will respond to particular drugs [30], especially for testing whether specific tumors acquire resistance during treatment. A biobank of organoids derived from ASCs would have a wide range of genetic mutations to collect human-specific data for drugs. ASCs can help create a new class of drugs for diseases that have too small of a market to get tested in a clinical setting and can be used to test the efficacy and toxicity of treatment. ASCs can also be leveraged for genetic manipulations such as CRISPR/Cas9 or lentivirus treatments on human specimens to see the effects on tissue without having to experiment with a patient [23].

Similar to ASCs, patient-derived organoids are also harvested through the human body. Patient-derived organoids, specifically cancer cell-derived organoids, are used in research to elucidate the best cancer treatments. In particular, patient-derived organoids have a variety of analysis techniques. Genomic profiling allows the identification of drug targets and where resistance to treatment can occur. Survival assays, flow cytometry, staining, and optical imaging test the treatment efficacy [31]. For example, when ovarian cancer cell organoids were compared to 2D cell culture to show the drugs' effect on various patients' genomes, the genomic differences were only captured in organoids. This experiment demonstrated the potential of using patient-derived organoids to design cancer treatments with greater efficiency and in a personalized manner [32]. Growing normal (non-diseased) organoids is possible for studying development of human tissues. For example, comparative analysis of normal and tumor colon and rectal tissue-derived organoids showed the effect of vitamin D on these 3D assemblies [33]. This study showed that vitamin D upregulated genes related to stemness, such as LGR5 and LRIG1, and downregulated differentiation marker genes such as TFF2 and MUC2. A similar patient-derived organoid model of head and neck squamous cell carcinoma with its corresponding normal wild-type organoid (harvested from normal tissues next to the tumor tissues) provided comparisons of epidermal growth factor receptor (EGFR) expression levels between the normal and diseased organoids [34]. Comparative studies of normal organoids with their tumor counterparts have great potential to elucidate molecular regulation patterns. Thus, biobanks of patient-derived normal and tumor organoids would create a library to study cellular disease mechanisms in tissue biopsies that are typically convenient to retrieve, making them a suitable derivation method for organoids [31].

Stem cell-derived organoids and patient-derived organoids can be applied to multiple types of organoids. Table 1 highlights different organoid platforms, their origins, and the advantages and disadvantages associated with each platform. For example, an organoid platform with the brain that consists of a serum-free floating culture of embryoid body-like aggregates with quick reaggregation demonstrates that the central nervous system develops from neuroectoderm (the outermost layer in the embryo). For instance, the neuroectoderm develops to the spinal cord with stem cells that can divide both symmetrically to increase in number and asymmetrically to drive the differentiation in a specific direction [35]. Another use of organoids to model the brain's function is through mini-brain development, where stem cell aggregates are embedded into the laminin-rich extracellular matrix and can develop distinct brain regions upon activation of distinct growth factors [47]. A similar model developed other organoids such as the retina and the olfactory pulp using the neurons [36]. Besides, the same approach facilitated the growth of different organs in the human digestive tract, originating from the same stem cells of the stomach, liver, and salivary glands, upon the activation of the corresponding growth factors [37, 39, 42].

Organoids model crucial organs in living conditions outside the human body. Specifically, the study of the lungs is modeled through an air–liquid interface to develop a deeper understanding of rare diseases, including cystic fibrosis [38]. Previously, there was a gap in knowledge on the human kidney pathology and its in vivo function. ESCs induced ureteric buds and metanephric

**Table 1**
**3D culture system uses the stem cells' natural ability to self-organize through cell sorting and lineage commitment. Organoids are widely used in multiple applications, including understanding the fundamental physiology in organs, tracking disease progression, developing drugs, and predicting treatment response. Organoids are divided into two leading platforms, stem cell-derived organoids and patient-derived organoids. Patient-derived organoids use normal/healthy and diseased tissues from humans. Organoid platforms for each organ type have their advantages and disadvantages, as shown below**

| Cell origin | Biological details | Application | Advantages | Disadvantages |
|---|---|---|---|---|
| *Brain* | ESCs co-cultured with embryonic forebrain tissue [35] | Spatial and temporal development of the brain | Proved that the central nervous system develops from the neuroectoderm<br>Proof of neuroectoderm development to the spinal cord with the stem cells both symmetrically and asymmetrically | Lacks recapitulation of the six-layered neocortical tissues using in vitro culture conditions<br>Lacks ability to capture the native physiology of the brain |
| | iPSCs to produce forebrain, midbrain, and hypothalamic regions [25] | Mini brain | Simple and cost-effective system<br>Able to model several brain regions<br>Use the same model to understand the effect of Zika virus on the brain | Lacks modeling of the human brain beyond the second trimester due to size limitation, nutrition, and oxygen depletion |
| | ESCs to produce neural retinal epithelium [36] | Retinal-optic cup organoids | Regenerates retinal cells from neuroectoderm<br>Observed self-organization of the optic cup, rods, and cones differentiation<br>Effective to study the development of the human eye | Still needs ability to observe outer segment development which is essential for light sensitivity |
| *Stomach* | ESCs and iPSCs to produce endoderm [37] | Understanding the rule of adding different growth factor on cell fate | Changes the cell differentiation states multiple times upon the addition of different growth factors<br>Can be used to study stomach development and disease | Lacks ability to recapitulate the complex 3D architecture of the stomach with the different cell types that make up the organ |
| *Lung* | ESCs and iPSCs to produce airway epithelia [38] | Using lung organoids to understand the development of cystic fibrosis | Models the air–liquid interface of the lung<br>Uses the same model to study cystic fibrosis by using CFTR mutant iPSC | Observation of heterogeneous endoderm lineages could have affected their final conclusions in understanding CFTR pathway |

(continued)

**Table 1**
**(continued)**

| Cell origin | Biological details | Application | Advantages | Disadvantages |
|---|---|---|---|---|
| *Liver* | iPSC co-cultured with mesenchymal stem cells and human umbilical vein endothelial cells [39] | Developing liver organoids and transplanting them into in vivo animal models. | Mimics the native architecture of the human liver by co-culturing the hepatic cells with mesenchymal stem cells and endothelial cells<br>Observes the formation of vascularized liver-bud aggregates | Potential immune reactions and other side effects could complicate its application in human studies |
| Kidney | ESCs [40, 41] | Recapitulating the structural complexity of human kidneys into organoids | Manages to form ureteric buds and nephron cells from human PSCs | Lacks ability to capture the native physiology of the entire organ |
| Salivary glands | Salivary glands [42] | Restoring the function of radiation-damaged salivary glands | Expands ASCs into salivary gland cell<br>Implants the resulting salivary glands into irradiated mice and proved that they restored function and protein production | Potential immune reactions and other side effects could complicate its application in human studies |
| Breast | Primary breast cancer tissues [43] | Using breast cancer organoids as a model for breast cancer | Outlet to study human diseases, which are difficult to study in animal models<br>More comprehensive profile of disease compared to 2D cell lines<br>Faster reaction endpoints and quantitative data collection compared to animal studies<br>More practical for immunofluorescence<br>Derivable from various tissue sources. Fibrinogen and fibroblasts which can cross-link with PEG to form hydrogel can be used to study metastasis based on stiffness of product<br>Phenotype retained for up to 70 days | Loss of oncogene in rich growth factor conditions<br>Weak vascularization leading to hypoxia which triggers activation of genes to mediate cell survival<br>Cost and difficulty to scale up<br>Murine origin of Matrigel might not be compatible to be transplanted into humans<br>Rigidity of ECM component might dampen drug molecules introduced |

| | | | | |
|---|---|---|---|---|
| *Prostate* | Patient-derived organoids from circulating tumor cells [44] | Developing liver organoids and transplanting them into in vivo animal models | Retains parental tumor genomic characteristics<br>Adaptable to high-throughput drug screens and gene functional studies<br>Cost-effective | Lacks ability to generate from primary prostate cancer<br>Needs tumor microenvironment<br>Low rate of success for organoid propagation<br>Remains unable to demonstrate disease progression |
| *Intestinal* | 3D approximation of the intestinal epithelial cell layer [45] | Studying the pathology of human intestinal diseases | Provides a physiologically relevant model to study the intestinal response to stimuli<br>Helps understand the pathology of gastrointestinal diseases | Lacks incorporation of apical membrane in the structure of the intestinal organoids<br>Unable to capture the interaction among different cell and tissue types |
| *Lymphoid* | Human lymphatic micro-organoids [46] | Replacing animal models with human organoids to test novel pharmaceuticals | Provides a physiologically relevant system to test immune functionality and the efficacy of biopharmaceuticals<br>Overcomes the lack of ex vivo lymphoid biopsies | Lacks essential physiological characteristics that impacts immune cell infiltration (e.g., oxygen and nutrients transport, metabolites gradients, etc.) |

mesenchyme to create a nephron in vitro through self-organization [40]. The ability to create renal tissue via pluripotent resources exhibits various implications such as tissue-based nephrotoxicity screens, disease modeling, and bioengineering of transplantable organs. Analysis of the gastrointestinal epithelial organoid gave insight into the disease physiology of cystic fibrosis [48].

### 2.2 2D and 3D Culture Models

Traditional cell monocultures lack the ability to initiate critical cell signaling pathways. Tissue explants rapidly lose their phenotype and cannot last as long as organoids despite their cell organization. Various 2D and 3D culture models have decreased discrepancies between cell cultures and a patient's microenvironment. 2D cell culture methods involving sandwich cultures are composed of two layers of an extracellular matrix and collagen with a layer of cells between. This is especially useful when modeling liver cells because hepatocytes are similarly surrounded by extracellular matrices. This culture method also allows for longer lasting cultures that maintain their functionality better than traditional 2D cultures. 3D cultures, aside from organoids and spheroids, include the cell sheet method that utilizes multiple layers of 2D cell cultures that together create a 3D culture and are able to vascularize when implanted into mice [49]. In contrast to these 2D models, spheroids and organoids are commonly used 3D culture models [50]. Organoids, spheroids, and 3D culture are interchangeably used, but the organoids primarily indicate aggregation of cells that grow in 3D to create units and mimic organs in their structures and functions [3]. Both 3D models can model a tumor microenvironment with accompanying cell-to-cell interactions.

## 3 Structural and Molecular Analysis of Organoids

Stem cells in 2D and 3D structures, bioprinting of human cells, and organ-on-a-chip platforms are useful tools in modeling human organ development and mimicking the organs' ability to recapitulate the development and tissue organization [51]. To shed light on similarities of synthetic organoids and native organs, cellular imaging and molecular analysis technologies can be used to determine the biochemical, genetic, and metabolic features of organoids. Analyzing the structure and formation of organoids at the single-cell level has the potential to reveal the underlying biology of human organ development and organ regeneration, creating opportunities for potential therapeutic and diagnostic solutions [14]. To this end, emerging single-cell and spatial analysis methods help decipher morphology and cellular compositions.

### 3.1 Structural Heterogeneity of Organoids

Organoids exhibit intrinsic features, self-organization capabilities, and diverse structural composition. Therefore, organoids contain diverse cell types [22]. Diverse organoids are engineered based on three different signal types that include secreted soluble signals, modular ECM signals, and non-modular extracellular matrix signals. Using a membrane matrix as an extracellular matrix (ECM) supports cell growth and cell adhesion, which leads stem cells to form into structures with organized groups of cells. The composition of extracellular matrix, including the macromolecules and minerals, including collagen, glycoproteins, and hydroxyapatite, can heavily influence the final characteristics of the organoids. In particular, the composition of ECM can influence chemical screening and genetic screening of organoids. Therefore, it is important to fully define what materials are in the ECM, as they affect the organoid culture [14]. In these 3D assemblies, cells often convene into heterogeneous organoids regarding viability, size, and shape. Various organoid types, variations of shape patterns within each organoid type, and the markers associated with each organoid type are shown (Fig. 2 and Table 2). Sixteen different types of organoids with five different images of each type show key differences in their shapes. Primary cell types for each organoid and distinct marker sets utilized to express primary cells in each organoid are presented. Organoid types resemble targeted organs in their structures, shapes, and cell compositions [22].

### 3.2 Single-Cell and Spatial Analysis of Organoids

Single-cell studies reveal the specific features related to human organoids' development by identifying distinct cell types, cell states, and the heterogeneity of individual cells within the organoid model (Fig. 3) [143]. The traditional 2D imaging for analyzing sectioned tissues has limitations in characterizing the complex 3D architecture of the organoids [139]. 3D characterization reveals the cellular fate decisions, cell–cell interactions, and cellular compositions, and the high-throughput image-based profiling provides a window into cellular dynamics and morphologies in a multicellular microenvironment [8, 139]. For example, noninvasive optical sectioning microscopy methods, including confocal microscopy and light-sheet fluorescence microscopy (LSFM), make it possible to visualize the spatial complexity and 3D characterization of intact organoids at a single-cell level [139]. Both confocal microscopy and LSFM visualize the finer details of the cellular structure and composition. LSFM as a technique for imaging 3D structure of relatively transparent specimens has the advantages of high imaging speed and reduced photobleaching over other imaging methods [140]. However, conventional immunofluorescence is limited by the number of colors used per imaging cycle and the LSFM captures images at lower resolution, posing challenges for accurate data analysis of cell–cell interactions [139, 144]. Also, traditional LSFM uses capillary containing both the sample and light-sheet

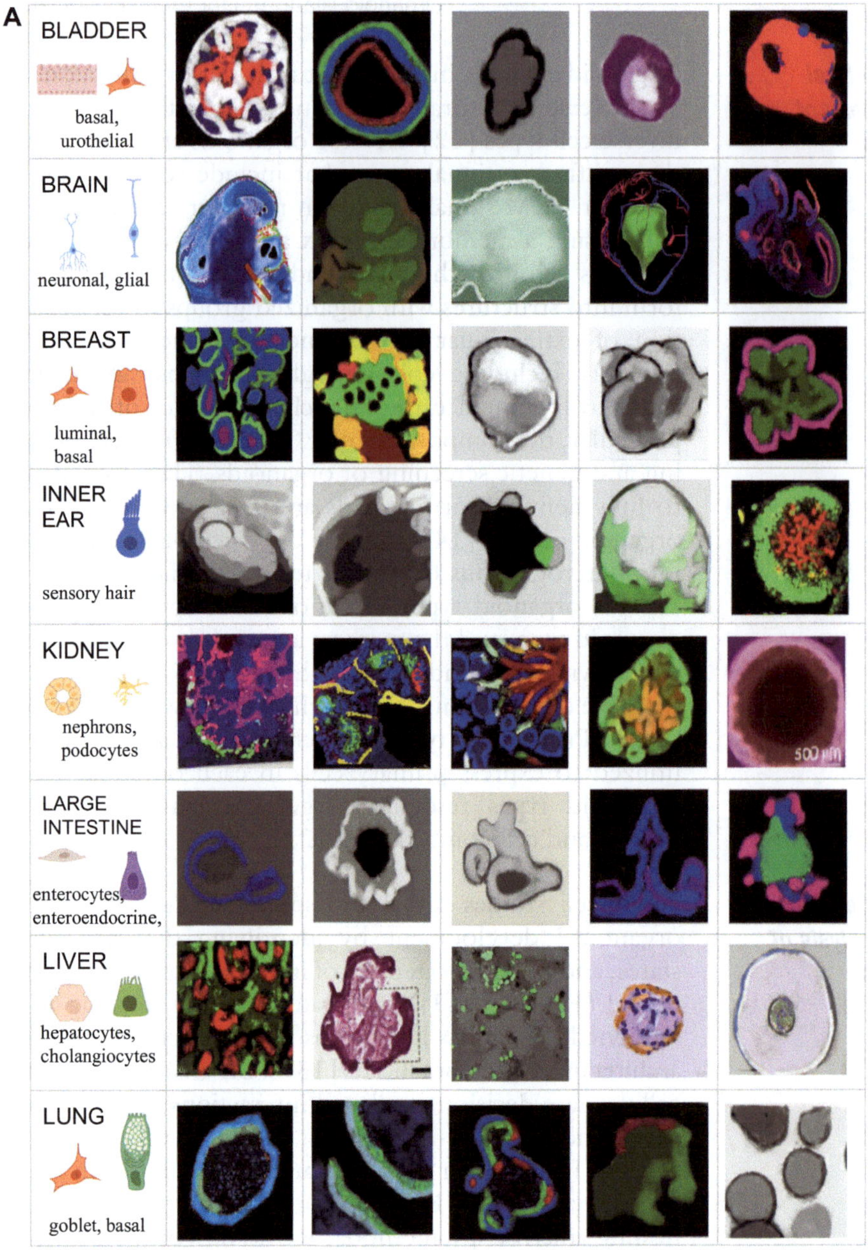


**Fig. 2** Morphological and cell-type survey of organoids. (**a**) Organoids: bladder, brain, breast, inner ear, kidney, large intestine, liver, and lung are shown. (**b**) Organoids: esophageal, retinal, ovarian, pancreas, prostate, small intestine, stomach, and thyroid are shown. Here 16 different types of organoids and 5 different images for each to observe the shapes, sizes, and structures are presented. In the first column, the type of organoid and a representative image of it is shown. The subsequent columns (2–6) include different images of each organoid. Most of these organoids generally range from 200 μm up to 500 μm, while the sizes of brain and kidney organoids can even go up to 1.5 m, 2 mm, and 2.5 mm. These structures resemble miniature organs and are composed of many different cell types. The first column includes major cell types of each corresponding organoid. (Bladder, basal and urothelial cells; brain, neuronal and glial cells; breast, luminal and basal cells; inner ear, sensory hair cells; intestinal, enterocytes, entero-endocrine, paneth, and goblet cells; kidney, nephrons and podocytes cells; liver, hepatocytes and cholangiocytes cells; lung, goblet and basal cells; esophageal, keratinocyte and basal cells; retinal, retinal ganglion, horizontal, and amacrine cells; ovarian, secretory and ciliated cells; pancreas, ductal and acinar cells; prostate, basal, luminal, and neuroendocrine cells; stomach, mucous and endocrine cells; thyroid, thyrocytes and follicular cells [47, 52–123]

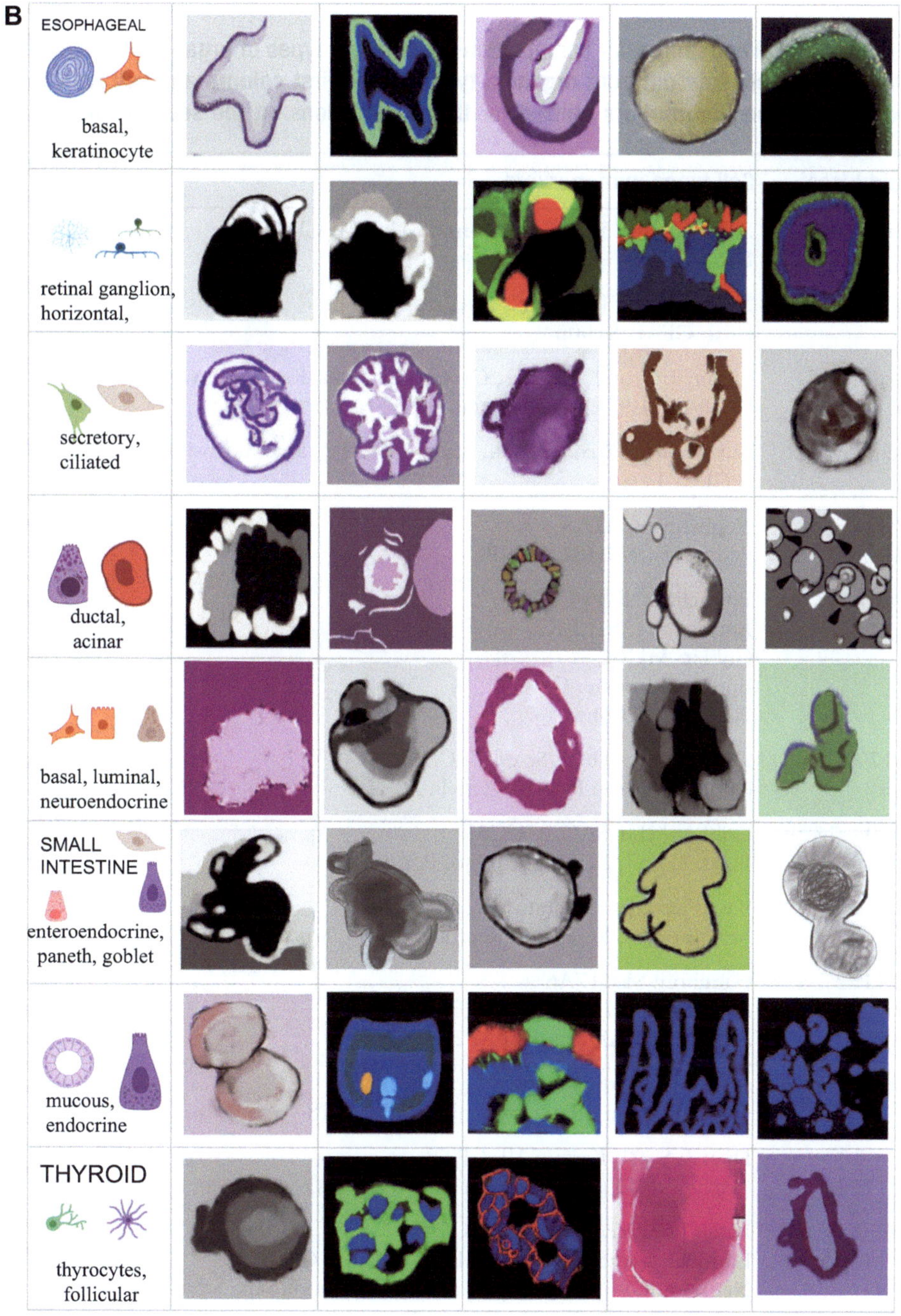


**Fig. 2** (continued)

**Table 2**
**Analysis of structural and molecular compositions of different types of organoids is summarized here in this table. The table lists different organoid types on the first column and cell types of each organoid with the specific markers utilized for their expressions on the second column**

| Organoid types | Cell types with markers |
|---|---|
| Bladder [116] | Basal: Keratin 5, p63, Keratin 20, UpkIII, CD44<br>Urothelial |
| Brain [124] | Neuronal: TUJ1, NeuN, MAP2<br>Glial: GFAP, S100β |
| Breast [88, 125] | Luminal: Trop2, CD271, CD49f, EpCAM, Keratin 8, GATA3<br>Basal: CD49f, EpCAM, Keratin 14, SMA |
| Inner ear [69, 126] | Sensory hair: Sox2, BMP4, Myo7a, CtBP2, F-actin-positive, Atoh1, EpCAM, CD271 |
| Kidney [127] | Nephrons: GATA3<br>Pedocytes: NPHS1, SYNPO |
| Intestinal [128] | Epithelial cell: EpCAM<br>Enterocytes: VILLIN<br>Goblet: MUC2<br>Enteroendocrine: CHGA<br>Paneth: Lysozyme |
| Liver [129] | Hepatocytes: Tbx3, Sox9, EpCAM<br>Cholangiocytes: FGF10, Activin A |
| Lung [108, 130] | Goblet: MUC5AC, MUC5B<br>Basal: Trp63, Keratin 5, Itga6, Pdpn |
| Esophageal [131] | Basal: Keratin 5, Keratin 14, p63, Sox2<br>Keratinocyte: Keratin 5, Keratin 14, p63, Sox2 |
| Retinal [132] | Ganglion: SLC17A6<br>Horizontal: LHX1, ONECUT1<br>Amacrine: GAD1 and TFAP2A |
| Ovarian [117] | Secretory: Pax8, Keratin 8<br>Ciliated: Dnali1, Foxj1, Cytokeratin 8, Pax8 |
| Pancreas [133, 134] | Ductal: Sox9, Keratin 19, CFTR<br>Acinar: AMY, Chymotrypsin C, Sox9, PDX1 |
| Prostate [135, 136] | Basal: P63, Keratin 5, ERβ, CD49f<br>Luminal: Keratin 8, Keratin 18, AR, ER β, CD26<br>Neuroendocrine: Chromogranin A, Synaptophysin |
| Stomach [37, 137] | Mucous: MUC5AC, MUC6, TFF2<br>Endocrine: GAST, GHRL, SST, 5-HT |
| Thyroid [138] | Thyrocytes: Nkx2-1, Pax8, Tg, Tsh-r, Nis, and Tpo<br>Follicular: Nkx2-1 |

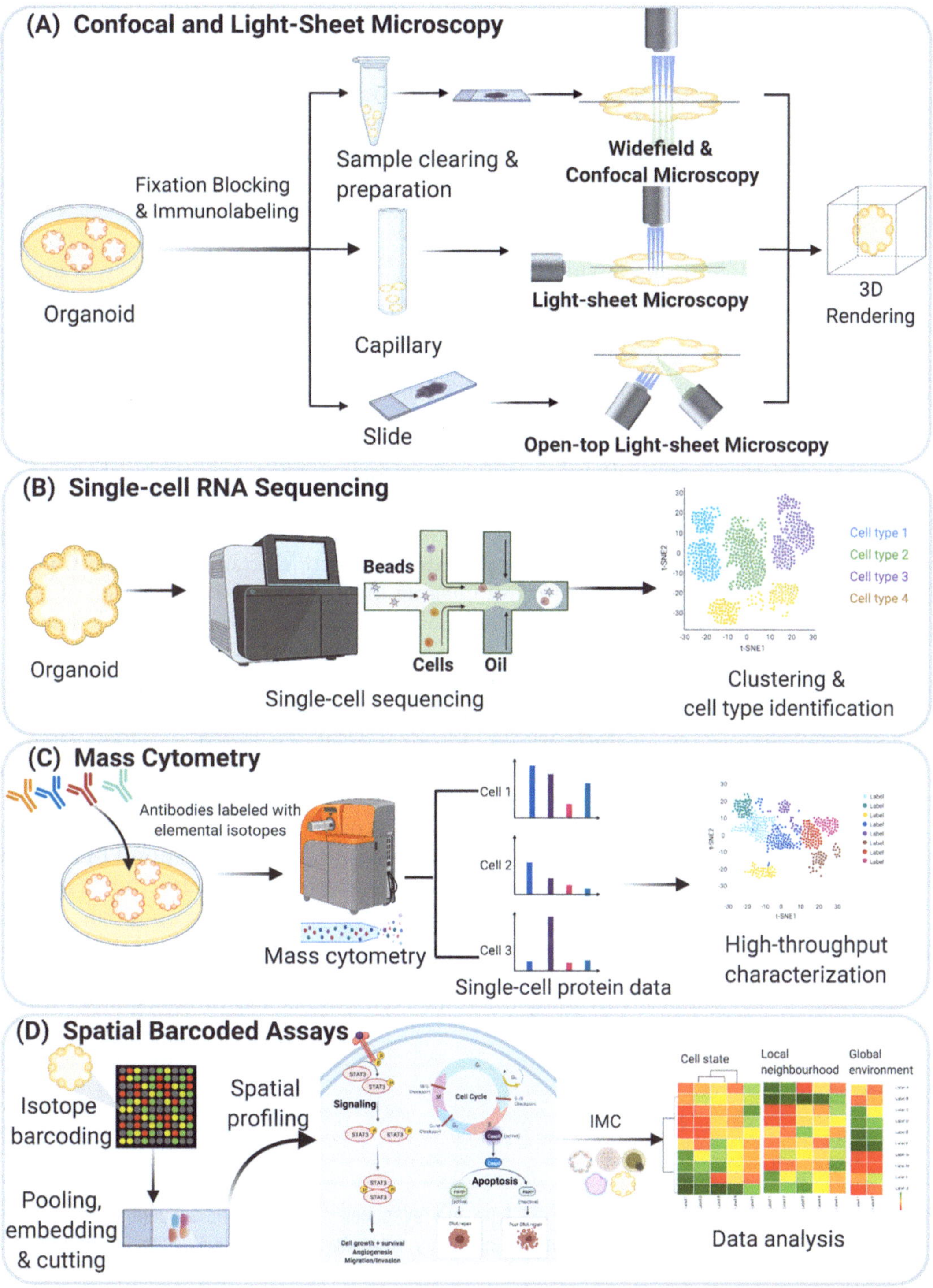


**Fig. 3** Single-cell analysis methods for organoids. (**b**) Schematic overview of widefield, confocal, light-sheet microscopy, and open-top light-sheet microscopy. Organoids are recovered from their 3D matrix, and the sample is fixed, blocked, and cleaned before immunolabeling with antibodies. The 3D rendering of images is performed with imaging analysis [139, 140]. (**b**) Schematic overview of scRNA-seq. scRNA-seq examines the

embedding solution, and the sample is exposed by pushing down the solution. However, this may result in specimen contamination and undesired dilution of reagents. To improve the specimen quality and ease-of-use of LSFM, open-top light-sheet microscopes (OTLS) have been developed with both illumination and collection objectives orientating 45 degrees with respect to the vertical axis. Also, the specimen is placed above the objective lenses. The design of OTLS makes it function as an inverted fluorescence microscope and eliminates the need of immersing the sample in medium by using a typical slide holder [140]. To solve the problem of limited resolution and sensitivity, people have designed a single-objective oblique epi-illumination selective-plane illumination microscope (eSPIM). eSPIM system utilizes a water-immersion objective with high numerical aperture and utilizes the index mismatch between the working media of an air objective and a second water-immersion objective. This microscope design builds on the existing inverted fluorescent microscopes, and it provides a platform for screening the sample with high spatial-temporal resolution [145]. The rise of live imaging enables both spatial and temporal analysis of the organoids. Some considerations for selecting the microscope for live imaging include phototoxicity, speed of imaging, sensitivity, and spatial/temporal resolution. Hof et al. designed the time-resolved live LSF, which allows the observations of the dynamics of the formation in a single organoid. The single-cell resolution pipeline achieves the tracking and extraction of single-cell features within individual organoids [146]. Confocal spinning disk has been used for live imaging to visualize the biological processes of organoids. The spinning disk with multiple pinholes enables the light illuminated at multiple points of the sample simultaneously, increasing the speed of image acquisition [147]. Also, the laser scanning confocal microscope can achieve time-lapse 3D live cell imaging of fluorescently labeled chromatin. The signals distinguish live and dead cells. The quantification of organoid 3D volume, surface area, and the number of live/dead cells provides means to study the tumor cell growth and drug responses within organoids [148].

**Fig. 3** (continued) individual cells to identify cell type and reveal cell differences via sequencing using Drop-seq [11, 141]. (**c**) Schematic overview of mass cytometry workflow. The organoids can be labeled with elemental isotopes staining. Mass cytometry allows cell-type identification and high-throughput characterization [17]. (**d**) Schematic overview of spatial barcoded assays. After labeling with a metal-based barcoding approach, the organoids are pooled for efficient embedding and cutting. Sections from the organoids can be imaged using IMC to study the spatial multiplexity, including the cell state, neighborhood, and global environment [142]

Another single-cell screening method, single-cell RNA-sequencing (scRNA-seq), achieves the genomic measurements of organoids without spatial details. The combination of scRNA-seq and accessible chromatin profiling reveals gene expression heterogeneity in human cerebral organoid development [149]. Although scRNA-sequencing could reveal the cell types within the organoids, it fails to describe the cell–cell interactions in the organoids' signaling networks. Also, most scRNA-seq methods achieve the sequencing of transcriptome by lysing the cells. To preserve the complex cell diversity in organoids, people have used the Drop-Seq method based on droplet microfluidics to enable sequencing over thousands of cells from whole-brain organoids by barcoding RNAs without physically isolating the cells. However, one of the drawbacks of scRNA-seq is the limitation of sequencing certain regions of the transcriptome, since only the 3′ or 5′ of the RNA is sequenced [141]. Smits et al. were able to use the Drop-seq method to sequence the different neuronal cells of human midbrain organoids. Although scRNA-seq enables high-throughput analysis of cells within the organoids, the difficulties of dissociating the 3D complexity of the organoids into single cells may hinder the identification and quantification of cell types [150]. Also, since most scRNA-seq methods require physically capturing a small volume of cells, the low number of cells pose another challenge in revealing the complexity of the organoids.

The development of the multivariate-barcoded mass cytometry (MC) method provides the window for demonstrating the single-cell signaling interactions in organoids [17]. MC is an approach used to study the high-dimensional immunophenotyping of post-transcriptional modifications, using heavy-metal conjugated antibodies to detect more than 35 proteins in single cells. The multivariate-barcoded mass cytometry method reveals the growth and characterization of cells [17]. To achieve high-throughput analysis of the cellular phenotypes under different microenvironment and cell states, an imaging mass cytometry (IMC) method was developed to leverage metal-barcoded antibodies for capturing the locations of cells in organoids. For example, IMC quantified the interplay of environment, neighborhood, and cell state on phenotypic variability by dozens of phenotypic features in up to 240 spheroids. This method provides a model for large-scale studies on any 3D microtissues [142]. The schematic workflow of the four single-cell analysis approaches, confocal and LSFM microscopy, scRNA-seq, MC, and IMC, is shown (Fig. 3), and the detailed examples of organoid analysis using different technologies are presented (Table 3). These high-throughput imaging-based approaches have contributed to the characterization of organoids' complex cellular compositions and the analysis of phenotypic features in cells. Analyzing the organoids' heterogeneity under different microenvironments and internal signals provides insights into the screening and identifying potential molecular candidates [8].

**Table 3**
**The applications of different single-cell analysis approaches in studying organoids are displayed in the table. These eight techniques apply to different types of organoids. The table also summarizes the number of markers studied using the techniques and whether the technique shows spatial multiplexity**

| Techniques | Brief description | Multiplexity | Organoid types | Spatial multiplexity |
|---|---|---|---|---|
| *Mass cytometry* | Study cell type-specific and cell state-specific signaling networks in organoids [17] | Measure 28 signaling post-translational modifications across six cell types in >1 million fixed single cells in small intestinal organoids | Small intestine | No |
| | Study the growth and characterization of multiple normal mammary epithelial cell lineages in a single culture and expression patterns of mammary markers [125] | 38 markers | Mammary epithelial cells | No |
| *Flow cytometry* | Analyze the proximal differentiated airway organoids to predict the infectivity of influenza viruses [151] | Four cell types of airway epithelial cells | Differentiated human airway organoids | No |
| | Analyze the stem cell surface markers of colorectal cancer liver metastases organoids [152] | 8 markers | Colorectal cancer liver metastases organoids | No |
| *Confocal IF* | Validate cell type markers and markers used to annotate specific cell clusters [127] | 3 markers/case, total 11 markers | Human iPSC-derived kidney organoids | No |
| | Validate different types of markers in brain organoids [153] | 3 markers/case, total 16 markers | Brain | No |
| *Light-sheet fluorescence microscopy* | Capture 3D organoids in full and quantify the expression of marker [139] | 2–4 markers | human airway, colon, kidney, liver, and breast tumor organoids | No |
| | Track cell fates using a image analysis pipeline [154] | Four features as possibly linked to tumor formation | Breast organoids | No |

(continued)

**Table 3 (continued)**

| Techniques | Brief description | Multiplexity | Organoid types | Spatial multiplexity |
|---|---|---|---|---|
| *IHC* | Validate pluripotency markers of human and chimpanzee cerebral organoids [149] | 4 markers | Human and chimpanzee cerebral organoids | No |
| | Identify the morphological types and sizes of organoid structures [155] | 4 markers and 4 morphological types | human colon cancer stem cell | No |
| *Hematoxylin and eosin (H&E) staining* | Identify the morphology of organoids [156] | Complex internal folding structures—morphology | Fallopian tube | No |
| | Identify organoids' morphology and organoid diameters [157] | Detect monolayer of polarized columnar epithelial cells—morphology | Gastrointestinal organoid | No |
| *scRNA-seq* | Identify the specific cell types and differentiation routes by comparing human cortical brain organoids to fetal brain [143] | Five clusters of cells across 190,022 single cells | Multiple brain organoids and fetal brain | No |
| | Identify cell types in initial tissue explants during organoid formation [11] | Three cell types across 4391 cells | Primary gastric organoid model | No |
| *Quantitative PCR* | Study intestinal drug metabolism and toxicity [158] | 26 genes | Intestinal crypt organoid | No |
| | Characterize the cerebral organoids [159] | 14 genes | Human Cerebral Organoids | No |

The emerging spatially resolved, multiplexed imaging technologies clarify organoids' complexity and heterogeneity while analyzing cell functions and cell types. Imaging-based transcriptomic approaches such as multiplexed error-robust fluorescence in situ hybridization (MERFISH) and sequential fluorescence in situ hybridization (seqFISH) have achieved the simultaneous quantification of RNA expression profiles and spatial distribution of RNAs in subcellular compartments [160, 161]. These approaches reveal the transcriptionally distinct cell states and subregions by analyzing many transcriptomic features relating to molecular factors, subcellular localization, and cell morphology. The integrated study in phenotypic states such as different cell-cycle phases and microenvironment could advance the understanding of the interplay between microenvironment and gene regulation [160–162]. For proteomics analysis, co-detection by indexing (CODEX) is an emerging multiplexed cytometric imaging technique that is used to visualize the fluorescence-labeled and DNA-barcoded antibodies for quantification of more than 50 markers at the single-cell level [163]. Cyclic immunofluorescence (CycIF) is another highly multiplexed imaging approach for deciphering the expression profiles and spatial patterning of proteins within cells [164]. By imaging the markers in an iterative process, CycIF can be used to quantify signaling cascades and to determine the cell phenotypes, morphology, and the cell states [164, 165]. To gather the information about the inorganic and organic molecules within the cells, matrix-assisted laser desorption/ionization (MALDI) and time-of-flight secondary ion mass spectrometry (TOF-SIMS) can achieve depth profiling of the molecular architecture in a high spatial resolution, including the subcellular localization of various lipid compounds. In general, MALDI and TOF-SIMS can be used to investigate the metabolite distribution in different cell stages within organoids [166, 167].

## 4 Applications of Organoid Platforms

Organoids ease the translational process from research to medicine and serve as the stepping stone between 3D cell culture and in vivo models. The translational drug development process with and without the use of organoids is illustrated (Fig. 4). Organoids can model the human body's response more effectively than 2D human cell cultures and give the ability to test potential drug candidates without in vivo research on mice. The 3D cell culture of organoids mimics the body more effectively because (1) they have biomimetic tissue models that are more physiologically relevant and more structurally complex and (2) they can maintain homeostasis for longer time. These characteristics make organoids more predictive than 2D cell cultures for drug screening [171]. Furthermore, 3D cell cultures allow for the creation of complex systems linked by microfluidics that can provide more accuracy regarding how different macromolecules interact using organs-on-a-chip platforms [172]. This ability to integrate fluid flow enables learning about how cells respond through differentiation and metabolic adaptation to facilitate testing of separate organ compartments by overcoming barrier tissues [172]. Since 2D cell cultures are inexpensive, they can still be used for translational drug development process as a preliminary step to organoid testing by initially screening drug candidates. Organoids fail to replace the in vivo models because the full physiological systems of a human body are lacking. In vivo testing in organoids identifies key drugs from a large library with in vivo models to reduce total cost as a decision-making step in

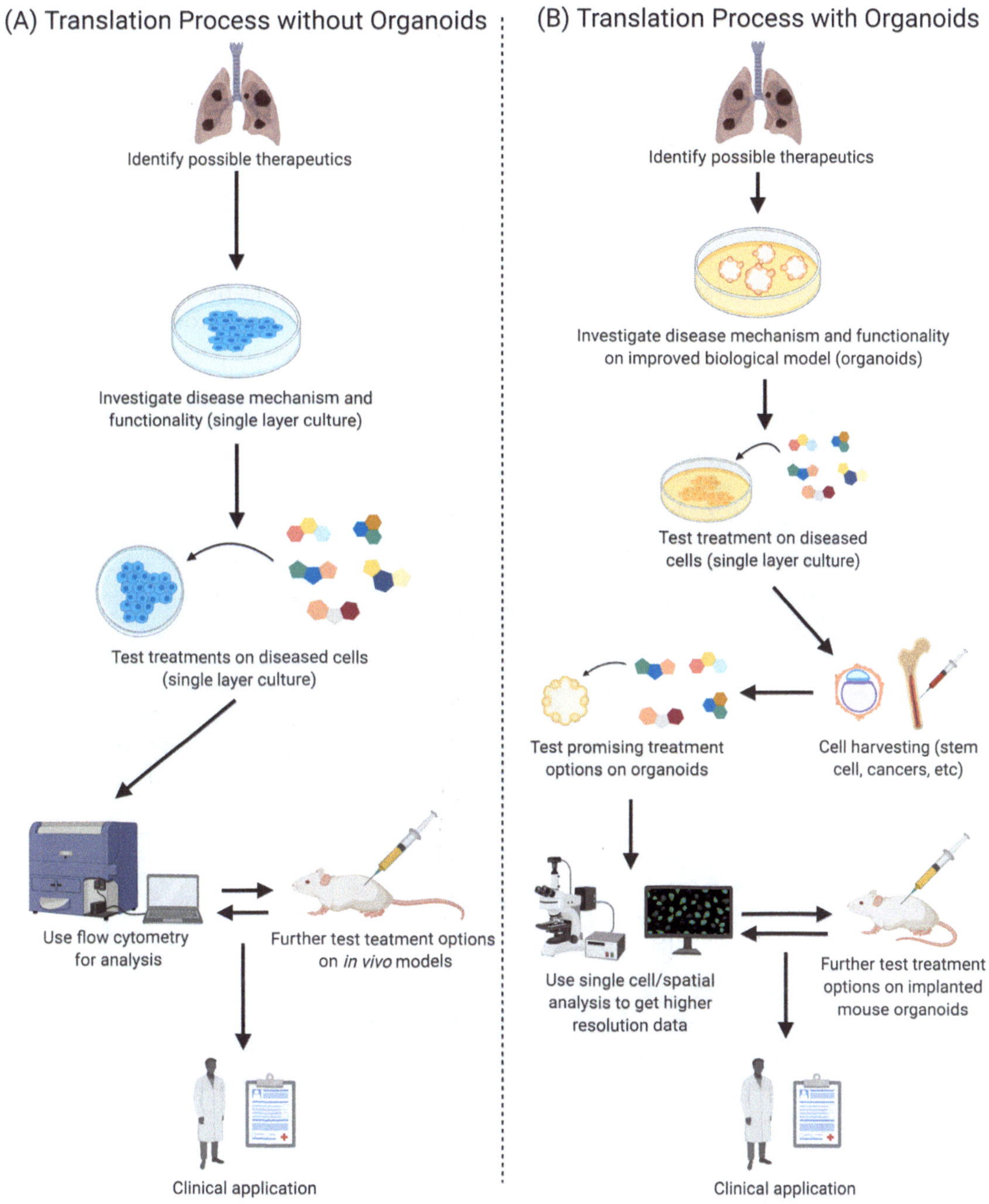


**Fig. 4** Translational path of organoid platforms. (**a**) The left panel shows a simplified model of drug development's translational process without organoids, using single-layer lung cell cultures (blue cells). In this model, research starts by identifying possible therapeutics for a disease, investigating the disease mechanism and functionality in a single layer culture, and testing promising treatment options in diseased cells of single layer culture [168]. Flow cytometry is used for analysis, after which translational drug development moves onto clinical applications or goes onto in vivo testing. This path then shifts to clinical applications [168, 169]. The process without organoids differs from (**b**) the translational process with pancreatic organoids (yellow cells). The process with organoids also starts with identifying possible

clinical trials. This combination of using (1) 2D cell cultures as a preliminary drug screening step, (2) moving onto organoid screening, and (3) then onto in vivo testing may increase the efficacy of drug candidates during clinical trials to potentially improve the current 86% failure rate of drug treatments [173].

A comparison of the translational process is demonstrated using single-layer cell cultures with conventional flow cytometry and the emergence of analysis methods for organoids using spatial single-cell analysis (Fig. 4). Flow cytometry is a useful method of rapidly profiling the cellular composition of a tissue sample to study diseases such as cancers. For example, flow cytometry quantified

(C)

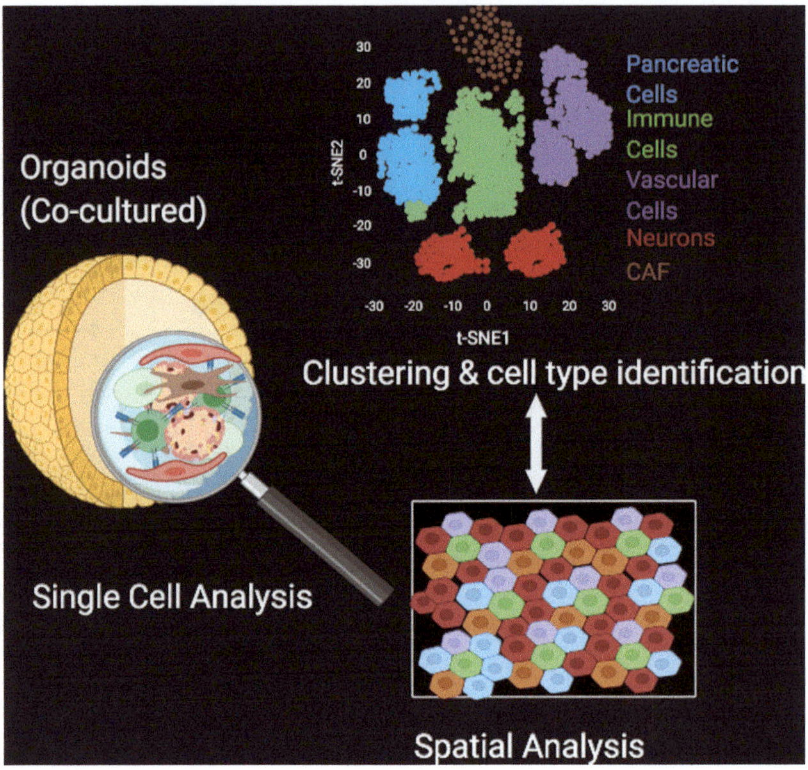


**Fig. 4** (continued)

**Fig. 4** (continued) therapeutics for a disease, investigating the disease mechanism and functionality with organoids, and testing a potential treatment on diseased cells in a single-layer culture [168]. After these steps, the translational process diverges, and the stem cells or patient-derived cells can be harvested for treatment testing on organoids [168]. In this process, single-cell spatial analysis and flow cytometry can be performed on the organoids after which drug screening mechanisms move onto in vivo testing and then clinical trials or just straight to clinical trials [168, 170]. (**c**) Organoids derived from stem cells and patient tissues have various potential applications, mainly in the pharmaceutical and regenerative medicine field. Through molecular and structural analysis, major technical challenges and practical issues can be addressed. Here, organoids are shown as a promising 3D culture with their biomedical applications and areas of growth measurable residual disease in biopsy samples of adults with acute myeloid leukemia before and after hematopoietic cell transplantation [174]. While flow cytometry is useful in understanding a patient's disease, spatial analysis is key to obtain cell–cell interactions in organoids for enhancing personalized treatments. Even a simple spatial analysis with confocal microscopy is advantageous when analyzing how a pathogen can affect multiple cell types. This approach analyzed the effect of Zika viruses on brain organoids [175]. In the meantime, single-cell RNA sequencing helped find treatment options specific to the genetic information that was measured in brain organoids [176]. The combination of flow cytometry and spatial analysis will be powerful for deciphering the tissue complexity of the organoids.

Organoids have played a role in expanding and improving the utility of personalized medicine using ASCs [23]. Personalized medicine benefits from the large-scale deployment of the patient-derived organoids [133, 177, 178]. For instance, iPSC-derived pancreatic organoids helped screen drugs with an mRNA-mediated gene supplementation with cystic fibrosis patients, showing that cystic fibrosis patients have normal pancreatic functions until the fetal stage [178]. Furthermore, the growth of pancreatic cancer organoids from patient-derived cells helped confirm cancer-associated fibroblast heterogeneity in pancreatic ductal adenocarcinoma tumor biology [177]. These applications provide insights into the field of personalized medicine abetted by organoid technology development.

One benefit to using organoids in medical research is the increased complexity of a self-organizing 3D culture, which allows for a microenvironment similar to that of the patient. Patient-derived xenograft (PDX) collects, cultures, and implants human-derived tissue samples into mice [179]. Despite complexity of implantation, the PDX model with organoid cultures is important for in vivo mechanistic studies of disease formation. The organoids can also be extensively genetically modified using a large target libraries, whereas PDX models are limited to genetic modifications of a few targets [44]. Thus, the use of a patient's samples in organoid-PDX models allows for a systematic approach to mechanistic exploration of treatment options with an increased accuracy.

Organoids have the potential to be used in the creation of synthetic organs. For instance, synthetic livers for patients were created due to the shortage of donor organs. One of the challenges is that organs have many different cell types and functions. To recapitulate the functionalities of an entire organ, the majority of cell functions must be replicated in organoid platforms. For example, an important factor in creating patient-derived liver organoids was to make sure harvested cells express the leucine-rich repeat-containing G coupled protein receptor 5 (Lgr5) [180]. Interestingly, although healthy liver cells lack this Lgr5 receptor, it is present near bile ducts in diseased and injured livers. The presence of Lgr5 indicates a differentiation possibility, and specifically, liver cells with Lgr5 have the potential to differentiate into hepatocytes and cholangiocytes in organoids, an important function of the liver organs. Additionally, liver organoids' cell cultures can divide rapidly, another key feature of organoids to give rise to the transplantable organs [180]. Therefore, patient-derived organoids are preferred in synthetic organ fabrication as the original tissue's functionality are better preserved compared to stem cell-derived organoids.

## 5 Challenges and Practical Issues of Organoids

### *5.1 Technical Challenges*

The development of organoids has provided a 3D tissue culture platform to study human organs' underlying biology. However, there are still challenges for growing organoids. As an ex vivo 3D tissue culture model, organoids can either self-assemble or be constructed under specific organogenesis cues [51]. Organoids use secreted growth factor signals, and ECM-instructed cues via membrane matrices to assemble [15, 51]. A decisive component for organoid development is the culture media for growing stem cells. The critical compositions of organoid culture media formulations are essential growth factors that include epidermal growth factor (EGF), R-spondins, and Noggin or Gremlin 1. The current approaches to the growth factor production are based on eukaryotic expression systems, which are essential for correct disulfide bond formation and protein folding. However, the culture media containing these essential growth factors are not widely available and present batch-to-batch variance. Also, producing growth factors from specific pathways may contaminate the media and affect organoid growth. People have developed a precise and cost-effective protocol for producing highly pure recombinant Gremlin 1 and R-spondins from the bacterial expression for organoid culture media [15]. An associated challenge with including growth factors in the media is that oncogenes often lose their function in a growth factor-rich environment. Using hydrogels, devoid of growth factors, post-successful differentiation to maintain the culture of the organoids addresses this loss of function [43].

Another challenge to current organoid systems is their inability to construct a vascular system [16]. The vasculature plays a critical role in organ interactions, soluble factors, nutrients, and cells. It can function as a transport barrier while regulating homeostasis and tissue regeneration [181]. However, in brain organoid systems, depletion of nutrients and oxygen inside the organoid constrains development due to the limited diffusion through dense tissue [16]. Organoid systems typically grow different organoid types by excluding vascular and immune cells, which explains the missing component of a vascular system within the organoid platform [19].

Current solutions involve generating flow in culture systems using tilting and rotary elements and developing microfluidic devices incorporated into 3D spheroid cultures for organoids. Generating flow decreases the variability in culture and increases the organoid's growth, while the microfluidic device intends to increase nutrient diffusion. To improve the organoid culture conditions for long-term live imaging, the reaction chamber facilitates the organoids' culturing, using microfluidics as inspiration; this ensures efficient nutrient diffusion [16]. A 3D bioprinting method fabricates engineered tissue-constructs, embedding vasculature, multiple types of cells, and ECM; hydrogel and polydimethylsiloxane (PDMS) build the 3D architecture. While further research needs to be done on optimizing scaffold build time, this method is scalable and can lead to the rapid manufacturing of tissues and organoids [182]. Additionally, people used a new 3D stamping technique to build a stable biodegradable scaffold with a built-in branching microchannel network to achieve vasculature in 3D culture [183]. However, integrating these techniques in organoids may interfere with the organoid's cell organization, which limits the incorporation of vasculature in organoids. Therefore, further research needs to determine the effects of endogenous vasculature generation on brain organoid growth and survival [52].

Another critical challenge in organoid use is the high variability between organoid samples. Since the organoids are cultured in vitro, the variation between samples is considerable due to the variation in biochemical media conditions and the sample's initial state. The variation in media composition from different protocols results in the heterogeneity of cultured organoids; the initial experimental condition across different protocols is poorly controlled. The effects of media composition need to be identified for the differentiation protocols, where the initial conditions, including the organoid size, shape, and composition on organoid assembly, would be used as optimization criteria [16]. Furthermore, the current methods for growing organoids lack feedback for development and network formation. In brain organoids, the biochemical inputs and feedback interact with each other to form the signaling network. However, this interaction failed to be recapitulated in organoid cultures. Therefore, advances in high-throughput imaging and techniques achieving synthetic vascular scaffolds are needed to understand organoid development further [16].

### 5.2 Practical Challenges

Another practical issue that comes with the use of organoids is animal testing. PDX that are derived from organoids are implanted into mice to test the efficacy of treatments and drugs. This implantation leads to the debate on animal testing, which includes the anatomical and physiological differences in humans and other animals [184]. Testing via traditional 2D cell cultures cannot account for microenvironmental factors or accurately represent the appropriate cellular responses, so 2D cultures are implanted into animals like mice to address these concerns [179]. To create an accurate environment, PDX, derived from organoids, uses its ability to self-organize into a 3D structure and implant into the mice [44]. Despite these solutions having greater accuracy, the differences in the physiology of and diseases' consequences on humans and mice lead to inaccurate preclinical results. Additionally, the synthetic laboratory environment can cause physiological differences in mice due to increased stress, causing further inaccuracies in test results [184]. The combination of PDX and organoids can increase accuracy in treatment testing; implanting this combination into mice can better demonstrate cell-to-cell communication with a more accurate microenvironment.

## 6 Conclusion

This chapter gives a biological overview of organoids, their applications, the need for molecular and structural analysis in this field, and the technical and practical challenges that arise with organoid technology. Organoids are 3D tissue models that self-organize. They can be derived from PSCs, ASCs, and patient-derived cells and tissues. Organoids can be made up of multiple cell types and are often co-cultured with different cell types or mixed with growth factors to mimic the tissue or tumor microenvironment. This paper highlights the necessity of structural and molecular analysis in organoids. Varying culture conditions, absence of vasculature, and other factors contribute to the lack of standardization in growing and maintaining organoids. This lack of standardization is shown in the variability of shape within the same organoid type. Along with shape variations, each organoid type has different markers that are best suited to observe their expression. Through molecular analysis, specifically via single-cell analysis and spatial profiling, the distinct cell types, cell states, and heterogeneity between individual cells can be observed through high-throughput screening. This information provides better insights into drug screenings, organogenesis, and gene therapies. Furthermore, growth in organoid technology can lead to better drug discovery methods, disease modeling, and precision medicine. Organoids can decrease the translational gap from in vitro to in vivo and replace animal research in the pharmaceutical field. Organoids derived from patient-derived cells and tissues, used in combination with PDX or by themselves, can be used to create personalized treatment plans and to understand disease progression. Additionally, synthetic organs can be potentially produced through the organoid platform and transplanted as organs.

Organoid variability remains a concern at each level, and it is crucial to identify techniques to alleviate it. Variability could be between diverse starting cell lines, organoids with the same origins, or several organoids within the same culture [22]. Single-cell analysis can reveal the heterogeneity in these organoid models. This variability leads to problems in translational research, such as drug screening. With high variability, it is not easy to ascertain the effects of the drug. Organoids also vary in how much of the organ's complexity and its structure are captured. Additionally, iPSCs do not always mature into ASCs, which decreases their use in adult tissue biology. Current research solutions attempt to increase nutrient availability via bioreactors and employment of vascular networks to address this problem of variability. Other solutions attempt to regulate self-organization or organoid communication through microstructured scaffolds, spatial positioning, and organ-on-chip technologies [4]. Another way to understand organoid-to-organoid variability would be to create a crowdsourced resource atlas to which pioneers in the organoid field contribute. As different organoids and protocols are generated and refined, the shapes of each organoid type and their accompanying protocols can be uploaded to a cloud server, creating an organoid atlas. With this information, progress can be made in increasing lab-to-lab standardization to decrease variability. Single-cell analysis can also support the development of an organoid atlas in understanding organoids' shortcomings by comparing organoid cell types to their in vivo counterparts [14]. Organoids hold the potential to (1) improve the translation from in vitro to in vivo, (2) reduce the use of animals, (3) with bioengineering tools, can better model diseases, (4) become transplantable organs, and (5) impact precision medicine.

## Acknowledgements

A.F.C. holds a Career Award at the Scientific Interface from Burroughs Wellcome Fund and a Bernie-Marcus Early-Career Professorship. A. F. C. was supported by start-up funds from the Georgia Institute of Technology and Emory University. A.K., S.C., and M.A. contributed equally to work and are equal first co-authors on this study. Figures were created with BioRender.com.

### References

1. Clevers H (2016) Modeling development and disease with organoids. Cell 165(7): 1586–1597
2. Rheinwald JG, Green H (1975) Serial cultivation of strains of human epidermal keratinocytes: the formation of keratinizing colonies from single cells. Cell 6(3):331–343
3. Muthuswamy SK (2017) Bringing together the organoid field: from early beginnings to the road ahead. Development 144:963–967

4. Rossi G, Manfrin A, Lutolf MP (2018) Progress and potential in organoid research. Nat Rev Genet 19(11):671–687
5. Petersen OW, Rønnov-Jessen L, Howlett AR, Bissell MJ (1992) Interaction with basement membrane serves to rapidly distinguish growth and differentiation pattern of normal and malignant human breast epithelial cells. Proc Natl Acad Sci U S A 89(19):9064–9068
6. Eiraku M, Takata N, Ishibashi H et al (2011) Self-organizing optic-cup morphogenesis in three-dimensional culture. Nature 472:51–56
7. Sato T, Vries RG, Snippert HJ et al (2009) Single Lgr5 stem cells build crypt-villus structures in vitro without a mesenchymal niche. Nature 459(7244):262–265
8. Betge J, Rindtorff N, Sauer J et al (2019) Multiparametric phenotyping of compound effects on patient derived organoids. bioRxiv. https://doi.org/10.1101/660993
9. Driehuis E, Kretzschmar K, Clevers H (2020) Establishment of patient-derived cancer organoids for drug-screening applications. Nat Protoc 15(10):3380–3409
10. Ooft SN, Weeber F, Dijkstra KK et al (2019) Patient-derived organoids can predict response to chemotherapy in metastatic colorectal cancer patients. Sci Transl Med 11. https://doi.org/10.1126/scitranslmed.aay2574
11. Chen J, Lau BT, Andor N et al (2019) Single-cell transcriptome analysis identifies distinct cell types and niche signaling in a primary gastric organoid model. Sci Rep 9(1):4536
12. Cattaneo CM, Dijkstra KK, Fanchi LF et al (2020) Tumor organoid-T-cell coculture systems. Nat Protoc 15(1):15–39
13. Nozaki K, Mochizuki W, Matsumoto Y et al (2016) Co-culture with intestinal epithelial organoids allows efficient expansion and motility analysis of intraepithelial lymphocytes. J Gastroenterol 51(3):206–213
14. Kim J, Koo BK, Knoblich JA (2020) Human organoids: model systems for human biology and medicine. Nat Rev Mol Cell Biol 21:571–584
15. Urbischek M, Rannikmae H, Foets T et al (2019) Organoid culture media formulated with growth factors of defined cellular activity. Sci Rep 9:6193
16. Karzbrun E, Reiner O (2019) Brain organoids-A bottom-up approach for studying human neurodevelopment. Bioengineering (Basel) 6(1):9
17. Qin X, Sufi J, Vlckova P et al (2020) Cell-type-specific signaling networks in heterocellular organoids. Nat Methods 17(3):335–342
18. Lancaster MA, Huch M (2019) Disease modelling in human organoids. Dis Model Mech 12(7):dmm039347
19. Camp JG, Treutlein B (2017) Human organomics: a fresh approach to understanding human development using single-cell transcriptomics. Development 144:1584–1587
20. Fiorini E, Veghini L, Corbo V (2020) Modeling cell communication in cancer with organoids: making the complex simple. Front Cell Dev Biol 8. https://doi.org/10.3389/fcell.2020.00166
21. Huch M, Koo BK (2015) Modeling mouse and human development using organoid cultures. Development 142(18):3113–3125
22. de Souza N (2018) Organoids. Nat Methods 15:23
23. Drost J, Clevers H (2017) Translational applications of adult stem cell-derived organoids. Development 144(6):968–975
24. Fujii M, Sato T (2021) Somatic cell-derived organoids as prototypes of human epithelial tissues and diseases. Nat Mater 20:156–169
25. Qian X, Nguyen HN, Song MM et al (2016) Brain-region-specific organoids using mini-bioreactors for modeling ZIKV exposure. Cell 165(5):1238–1254
26. Chugh RM, Bhanja P, Norris A, Saha S (2021) Experimental models to study COVID-19 effect in stem cells. Cell 10:91. https://doi.org/10.3390/cells10010091
27. Quiroz EJ, Ryan AL (Firth) (2019) In: Stem Cell-Based Ther Lung Dis (Eds: JK Burgess, IH Heijink), Springer International Publishing, Cham, pp. 153–178
28. Longmire TA, Ikonomou L, Hawkins F et al (2012) Efficient derivation of purified lung and thyroid progenitors from embryonic stem cells. Cell Stem Cell 10(4):398–411
29. Liang P, Lan F, Lee AS et al (2013) Drug screening using a library of human induced pluripotent stem cell-derived cardiomyocytes reveals disease-specific patterns of cardiotoxicity. Circulation 127(16):1677–1691
30. Bartfeld S, Clevers H (2017) Stem cell-derived organoids and their application for medical research and patient treatment. J Mol Med (Berl) 95(7):729–738
31. Naglea PW, Plukkerc JTM, Muijsb CT, van Luijkb P, Coppes RP (2018) Patient-derived tumor organoids for prediction of cancer treatment response. Semin Cancer Biol 53: 258–264
32. Jabs J, Zickgraf FM, Park J et al (2017) Screening drug effects in patient-derived cancer cells links organoid responses to genome alterations. Mol Syst Biol 13(11):955

33. Costales-Carrera A, Fernández-Barral A, Bustamante-Madrid P et al (2020) Comparative study of organoids from patient-derived normal and tumor colon and rectal tissue. Cancers (Basel) 12(8):2302
34. Driehuis E, Spelier S, Beltrán Hernández I et al (2019) Patient-derived head and neck cancer organoids recapitulate EGFR expression levels of respective tissues and are responsive to EGFR-targeted photodynamic therapy. J Clin Med 8:1880
35. Eiraku M, Watanabe K, Matsuo-Takasaki M et al (2008) Self-organized formation of polarized cortical tissues from ESCs and its active manipulation by extrinsic signals. Cell Stem Cell 3(5):519–532
36. Nakano T, Ando S, Takata N et al (2012) Self-formation of optic cups and storable stratified neural retina from human ESCs. Cell Stem Cell 10(6):771–785
37. Sinagoga KL, Schumacher M, Rockich BE et al (2014) Modelling human development and disease in pluripotent stem-cell-derived gastric organoids. Nature 516(7531):400–404
38. Wong AP, Bear CE, Chin S et al (2012) Directed differentiation of human pluripotent stem cells into mature airway epithelia expressing functional CFTR protein. Nat Biotechnol 30(9):876–882
39. Takebe T, Sekine K, Enomura M et al (2013) Vascularized and functional human liver from an iPSC-derived organ bud transplant. Nature 499(7459):481–484
40. Takasato M, Er PX, Becroft M et al (2014) Directing human embryonic stem cell differentiation towards a renal lineage generates a self-organizing kidney. Nat Cell Biol 16(1):118–126
41. Takasato M, Er PX, Chiu HS et al (2015) Kidney organoids from human iPS cells contain multiple lineages and model human nephrogenesis. Nature 526(7574):564–568
42. Maimets M, Rocchi C, Bron R et al (2016) Long-term in vitro expansion of salivary gland stem cells driven by Wnt signals. Stem Cell Reports 6(1):150–162
43. Octavian SA, Pei WY (2018) Organoids as reliable breast cancer study models: an update. Int J Oncol Res 1. https://doi.org/10.23937/ijor-2017/1710008
44. Gleave AM, Ci X, Lin D, Wang Y (2020) A synopsis of prostate organoid methodologies, applications, and limitations. Prostate 80(6):518–526
45. Wallach TE, Bayrer JR (2017) Intestinal organoids: new frontiers in the study of intestinal disease and physiology. J Pediatr Gastroenterol Nutr 64(2):180–185
46. Giese C, Lubitz A, Demmler CD et al (2010) Immunological substance testing on human lymphatic micro-organoids in vitro. J Biotechnol 148(1):38–45
47. Lancaster MA, Renner M, Martin CA et al (2013) Cerebral organoids model human brain development and microcephaly. Nature 501(7467):373–379
48. Dedhia PH, Bertaux-Skeirik N, Zavros Y et al (2016) Organoid models of human gastrointestinal development and disease. Gastroenterology 150(5):1098–1112
49. Duval K, Grover H, Han LH et al (2017) Modeling physiological events in 2D vs. 3D cell culture. Physiology (Bethesda) 32(4):266–277
50. Yin X, Mead BE, Safaee H et al (2016) Engineering stem cell organoids. Cell Stem Cell 18(1):25–38
51. Shah SB, Singh A (2017) Cellular self-assembly and biomaterials-based organoid models of development and diseases. Acta Biomater 53:29–45
52. Pham MT, Pollock KM, Rose MD et al (2018) Generation of human vascularized brain organoids. Neuroreport 29(7):588–593
53. Bagley JA, Reumann D, Bian S, Lévi-Strauss J, Knoblich JA (2017) Fused cerebral organoids model interactions between brain regions. Nat Methods 14(7):743–751
54. Wang Y, Wang L, Zhu Y, Qin J (2018) Human brain organoid-on-a-chip to model prenatal nicotine exposure. Lab Chip 18:851
55. Czerniecki SM, Cruz NM, Harder JL et al (2018) High-throughput screening enhances kidney organoid differentiation from human pluripotent stem cells and enables automated multidimensional phenotyping. Cell Stem Cell 22(6):929–940
56. Lancaster MA, Corsini NS, Wolfinger S et al (2017) Guided self-organization and cortical plate formation in human brain organoids. Nat Biotechnol 35(7):659–666
57. Broutier L, Andersson-Rolf A, Hindley CJ et al (2016) Culture and establishment of self-renewing human and mouse adult liver and pancreas 3D organoids and their genetic manipulation. Nat Protoc 11(9):1724–1743
58. Broutier L, Mastrogiovanni G, Verstegen MM et al (2017) Human primary liver cancer-derived organoid cultures for disease modeling and drug screening. Nat Med 23(12):1424–1435

59. Low JH, Li P, Chew EGY et al (2019) Generation of human PSC-derived kidney organoids with patterned nephron segments and a de novo vascular network. Cell Stem Cell 25(3):373–387
60. Morizane R, Lam AQ, Freedman BS et al (2015) Nephron organoids derived from human pluripotent stem cells model kidney development and injury. Nat Biotechnol 33(11):1193–1200
61. van den Berg CW, Ritsma L, Avramut MC et al (2018) Renal subcapsular transplantation of PSC-derived kidney organoids induces neo-vasculogenesis and significant glomerular and tubular maturation in vivo. Stem Cell Reports 10(3):751–765
62. Asai A, Aihara E, Watson C et al (2017) Paracrine signals regulate human liver organoid maturation from induced pluripotent stem cells. Development 144(6):1056–1064
63. Rennert K, Steinborn S, Gröger M et al (2015) A microfluidically perfused three dimensional human liver model. Biomaterials 71:119–131
64. Leite SB, Roosens T, El Taghdouini A et al (2016) Novel human hepatic organoid model enables testing of drug-induced liver fibrosis in vitro. Biomaterials 78:1–10
65. Koehler K, Hashino E (2014) 3D mouse embryonic stem cell culture for generating inner ear organoids. Nat Protoc 9:1229–1244
66. DeJonge RE, Liu XP, Deig CR (2016) Modulation of Wnt signaling enhances inner ear organoid development in 3D culture. PLOS ONE 11(9):e0162508
67. Koehler KR, Nie J, Longworth-Mills E et al (2017) Generation of inner ear organoids containing functional hair cells from human pluripotent stem cells. Nat Biotechnol 35(6): 583–589
68. Liu XP, Koehler K, Mikosz A et al (2016) Functional development of mechanosensitive hair cells in stem cell-derived organoids parallels native vestibular hair cells. Nat Commun 7:11508
69. Jeong M, O'Reilly M, Kirkwood NK et al (2018) Generating inner ear organoids containing putative cochlear hair cells from human pluripotent stem cells. Cell Death Dis 9:1
70. Seino T, Kawasaki S, Shimokawa M et al (2018) Human pancreatic tumor organoids reveal loss of stem cell niche factor dependence during disease progression. Cell Stem Cell 22(3):454–467
71. Boj SF, Hwang CI, Baker LA et al (2015) Organoid models of human and mouse ductal pancreatic cancer. Cell 160:324–338
72. Huang L, Holtzinger A, Jagan I et al (2015) Ductal pancreatic cancer modeling and drug screening using human pluripotent stem cell- and patient-derived tumor organoids. Nat Med 21(11):1364–1371
73. Greggio C, De Franceschi F, Figueiredo-Larsen M et al (2013) Artificial three-dimensional niches deconstruct pancreas development in vitro. Development 140(21):4452–4462
74. Chen HY, Kaya KD, Dong L, Swaroop A (2016) Three-dimensional retinal organoids from mouse pluripotent stem cells mimic in vivo development with enhanced stratification and rod photoreceptor differentiation. Mol Vis 22:1077–1094
75. DiStefano T, Chen HY, Panebianco C et al (2018) Accelerated and improved differentiation of retinal organoids from pluripotent stem cells in rotating-wall vessel bioreactors. Stem Cell Reports 10(1):300–313
76. Völkner M, Zschätzsch M, Rostovskaya M et al (2016) Retinal organoids from pluripotent stem cells efficiently recapitulate retinogenesis. Stem Cell Reports 6(4):525–538
77. Deng WL, Gao ML, Lei XL et al (2018) Gene correction reverses ciliopathy and photoreceptor loss in iPSC-derived retinal organoids from retinitis pigmentosa patients. Stem Cell Reports 10(4):1267–1281
78. Reichman S, Slembrouck A, Gagliardi G et al (2017) Generation of storable retinal organoids and retinal pigmented epithelium from adherent human iPS cells in xeno-free and feeder-free conditions. Stem Cells 35(5): 1176–1188
79. Gracz AD, Ramalingam S, Magness ST (2010) Sox9 expression marks a subset of CD24-expressing small intestine epithelial stem cells that form organoids in vitro. Am J Physiol Gastrointest Liver Physiol 298(5): G590–G600
80. Petersen N, Reimann F, Bartfeld S et al (2014) Generation of L-cells in mouse and human small intestine organoids. Diabetes 63:410–420
81. Dekkers JF, Wiegerinck CL, de Jonge HR et al (2013) A functional CFTR assay using primary cystic fibrosis intestinal organoids. Nat Med 19(7):939–945
82. Wilson SS, Tocchi A, Holly MK, Parks WC, Smith JG (2015) A small intestinal organoid model of non-invasive enteric pathogen-epithelial cell interactions. Mucosal Immunol 8(2):352–361
83. Tsuruta T, Saito S, Osaki Y et al (2016) Organoids as an ex vivo model for studying the serotonin system in the murine small intestine

and colon epithelium. Biochem Biophys Res Commun 474(1):161–167
84. Sachs N, de Ligt J, Kopper O et al (2018) A living biobank of breast cancer organoids captures disease heterogeneity. Cell 172(1–2): 373–386
85. Srivastava V, Huycke TR, Phong KT, Gartner ZJ (2020) Organoid models for mammary gland dynamics and breast cancer. Curr Opin Cell Biol 66:51–58
86. Chandhoke AS, Chanda A, Karve K et al (2017) The PIAS3-Smurf2 sumoylation pathway suppresses breast cancer organoid invasiveness. Oncotarget 8(13):21001–21014
87. Walsh AJ, Cook RS, Sanders ME et al (2014) Quantitative optical imaging of primary tumor organoid metabolism predicts drug response in breast cancer. Cancer Res 74(18):5184–5194
88. Goldhammer N, Kim J, Timmermans-Wielenga V, Petersen OW (2019) Characterization of organoid cultured human breast cancer. Breast Cancer Res 21(1):141
89. Gao D, Vela I, Sboner A et al (2014) Organoid cultures derived from patients with advanced prostate cancer. Cell 159(1): 176–187
90. Karthaus WR, Iaquinta PJ, Drost J et al (2014) Identification of multipotent luminal progenitor cells in human prostate organoid cultures. Cell 159(1):163–175
91. Chua CW, Shibata M, Lei M et al (2014) Single luminal epithelial progenitors can generate prostate organoids in culture. Nat Cell Biol 16(10):951–954
92. Calderon-Gierszal EL, Prins GS (2015) Directed differentiation of human embryonic stem cells into prostate organoids in vitro and its perturbation by low-dose bisphenol A exposure. PLoS One 10(7):e0133238
93. Richards Z, McCray T, Marsili J et al (2019) Prostate stroma increases the viability and maintains the branching phenotype of human prostate organoids. iScience 12:304–317
94. Saito Y, Onishi N, Takami H et al (2018) Development of a functional thyroid model based on an organoid culture system. Biochem Biophy Res Comm 497:783–789
95. Bartfeld S, Clevers H (2015) Organoids as model for infectious diseases: culture of human and murine stomach organoids and microinjection of Helicobacter Pylori. J Vis Exp 105:53359
96. Yan HHN, Siu HC, Law S et al (2018) A comprehensive human gastric cancer organoid biobank captures tumor subtype heterogeneity and enables therapeutic screening. Cell Stem Cell 23(6):882–897
97. McCracken KW, Catá EM, Crawford CM et al (2014) Modelling human development and disease in pluripotent stem-cell-derived gastric organoids. Nature 516(7531):400–404
98. Bouchi R, Foo KS, Hua H et al (2014) FOXO1 inhibition yields functional insulin-producing cells in human gut organoid cultures. Nat Commun 5:4242
99. Steele NG, Chakrabarti J, Wang J et al (2019) An organoid-based preclinical model of human gastric cancer. Cell Mol Gastroenterol Hepatol 7(1):161–184
100. Sato T, Stange DE, Ferrante M et al (2011) Long-term expansion of epithelial organoids from human colon, adenoma, adenocarcinoma, and Barrett's epithelium. Gastroenterology 141(5):1762–1772
101. Lukovac S, Roeselers G (2015) In Impact food bioact health vitro ex vivo models (Eds: K Verhoeckx P, Cotter I et al), Springer International Publishing, Cham, pp. 245–253
102. Spence JR, Mayhew CN, Rankin SA et al (2011) Directed differentiation of human pluripotent stem cells into intestinal tissue in vitro. Nature 470(7332):105–109
103. DeWard AD, Cramer J, Lagasse E (2014) Cellular heterogeneity in the mouse esophagus implicates the presence of a nonquiescent epithelial stem cell population. Cell Rep 9(2): 701–711
104. Trisno SL, Philo KED, McCracken KW et al (2018) Esophageal organoids from human pluripotent stem cells delineate Sox2 functions during esophageal specification. Cell Stem Cell 23(4):501–515
105. Dye BR, Hill DR, Ferguson MAH et al (2015) In vitro generation of human pluripotent stem cell derived lung organoids. elife 4: e05098
106. Zhang Y, Yang Y, Jiang M et al (2018) 3D modeling of esophageal development using human PSC-derived basal progenitors reveals a critical role for notch signaling. Cell Stem Cell 23(4):516–529
107. Shacham-Silverberg V, Wells JM (2020) In Methods Cell Biol (Ed JR Spence), Academic Press, pp. 1–22
108. Miller AJ, Dye BR, Ferrer-Torres D et al (2019) Generation of lung organoids from human pluripotent stem cells in vitro. Nat Protoc 14(2):518–540
109. Gjorevski N, Sachs N, Manfrin A et al (2016) Designer matrices for intestinal stem cell and organoid culture. Nature 539(7630): 560–564

110. Chen YW, Huang SX, de Carvalho ALRT et al (2017) A three-dimensional model of human lung development and disease from pluripotent stem cells. Nat Cell Biol 19(5):542–549
111. Kim M, Mun H, Sung CO et al (2019) Patient-derived lung cancer organoids as in vitro cancer models for therapeutic screening. Nat Commun 10(1):3991
112. Lehmann R, Lee CM, Shugart EC et al (2019) Human organoids: a new dimension in cell biology. Mol Biol Cell 30(10): 1129–1137
113. Angus HCK, Butt AG, Schultz M, Kemp RA (2020) Intestinal organoids as a tool for inflammatory bowel disease research. Front Med 6. https://doi.org/10.3389/fmed.2019.00334
114. Takebe T, Wells JM, Helmrath MA, Zorn AM (2018) Organoid center strategies for accelerating clinical translation. Cell Stem Cell 22(6):806–809
115. Ma R, Morshed SA, Latif R, Davies TF (2015) Thyroid cell differentiation from murine induced pluripotent stem cells. Front Endocrinol 6. https://doi.org/10.3389/fendo.2015.00056
116. Mullenders J, de Jongh E, Brousali A et al (2019) Mouse and human urothelial cancer organoids: a tool for bladder cancer research. Proc Natl Acad Sci USA 116(10):4567–4574
117. Lõhmussaar K, Kopper O, Korving J et al (2020) Assessing the origin of high-grade serous ovarian cancer using CRISPR-modification of mouse organoids. Nat Commun 11:2660
118. Nanki Y, Chiyoda T, Hirasawa A et al (2020) Patient-derived ovarian cancer organoids capture the genomic profiles of primary tumours applicable for drug sensitivity and resistance testing. Sci Rep 10(1):12581
119. Maenhoudt N, Defraye C, Boretto M et al (2020) Developing organoids from ovarian cancer as experimental and preclinical models. Stem Cell Reports 14(4):717–729
120. Rothschild D, Srinivasan T, Aponte-Santiago L et al (2016) The ex vivo culture and pattern recognition receptor stimulation of mouse intestinal organoids. J Vis Exp. https://doi.org/10.3791/54033
121. Lee SH, Hu W, Matulay JT et al (2018) Tumor evolution and drug response in patient-derived organoid models of bladder cancer. Cell 173(2):515–528
122. Eldred KC, Hadyniak S E, Hussey KA et al (2018) Thyroid hormone signaling specifies cone subtypes in human retinal organoids. bioRxiv. https://doi.org/10.1101/359950
123. Drost J, Karthaus WR, Gao D et al (2016) Organoid culture systems for prostate epithelial and cancer tissue. Nat Protoc 11(2): 347–358
124. Yakoub AM (2019) Cerebral organoids exhibit mature neurons and astrocytes and recapitulate electrophysiological activity of the human brain. Neural Regen Res 14(5): 757–761
125. Rosenbluth JM, Schackmann RCJ, Gray GK et al (2020) Organoid cultures from normal and cancer-prone human breast tissues preserve complex epithelial lineages. Nat Commun 11(1):1711
126. Roccio M, Edge ASB (2019) Inner ear organoids: new tools to understand neurosensory cell development, degeneration and regeneration. Development 146(17):dev177188
127. Subramanian A, Sidhom EH, Emani M et al (2019) Single cell census of human kidney organoids shows reproducibility and diminished off-target cells after transplantation. Nat Commun 10:5462
128. Yoshida S, Miwa H, Kawachi T et al (2020) Generation of intestinal organoids derived from human pluripotent stem cells for drug testing. Sci Rep 10(1):5989
129. Akbari S, Arslan N, Senturk S, Erdal E (2019) Next-generation liver medicine using organoid models. Front Cell Dev Biol 7. https://doi.org/10.3389/fcell.2019.00345
130. Barkauskas CE, Chung MI, Fioret B et al (2017) Lung organoids: current uses and future promise. Development 144(6): 986–997
131. Whelan KA, Muir AB, Nakagawa H (2018) Esophageal 3D culture systems as modeling tools in esophageal epithelial pathobiology and personalized medicine. Cell Mol Gastroenterol Hepatol 5(4):461–478
132. Cowan CS, Renner M, De Gennaro M et al (2020) Cell types of the human retina and its organoids at single-cell resolution. Cell 182(6):1623–1640
133. Hohwieler M, Illing A, Hermann PC et al (2017) Human pluripotent stem cell-derived acinar/ductal organoids generate human pancreas upon orthotopic transplantation and allow disease modelling. Gut 66(3): 473–486
134. Moreira L, Bakir B, Chatterji P et al (2017) Pancreas 3D organoids: current and future aspects as a research platform for personalized medicine in pancreatic cancer. Cell Mol Gastroenterol Hepatol 5(3):289–298
135. Elbadawy M, Abugomaa A, Yamawaki H et al (2020) Development of prostate cancer

organoid culture models in basic medicine and translational research. Cancers 12:777

136. McCray T, Richards Z, Marsili J et al (2019) Handling and assessment of human primary prostate organoid culture. Vis Exp JoVE. https://doi.org/10.3791/59051
137. Seidlitz T, Merker SR, Rothe A et al (2019) Human gastric cancer modelling using organoids. Gut 68(2):207–217
138. Kurmann AA, Serra M, Hawkins F et al (2015) Regeneration of thyroid function by transplantation of differentiated pluripotent stem cells. Cell Stem Cell 17(5):527–542
139. Dekkers JF, Alieva M, Wellens LM et al (2019) High-resolution 3D imaging of fixed and cleared organoids. Nat Protoc 14(6): 1756–1771
140. Glaser AK, Reder NP, Chen Y et al (2019) Multi-immersion open-top light-sheet microscope for high-throughput imaging of cleared tissues. Nat Commun 10(1):2781
141. Brazovskaja A, Treutlein B, Camp JG (2019) High-throughput single-cell transcriptomics on organoids. Curr Opin Biotechnol 55: 167–171
142. Zanotelli VR, Leutenegger M, Lun XK et al (2020) A quantitative analysis of the interplay of environment, neighborhood, and cell state in 3D spheroids. Mol Syst Biol 16(12):e9798
143. Tanaka Y, Cakir B, Xiang Y, Sullivan GJ, Park IH (2020) Synthetic analyses of single-cell transcriptomes from multiple brain organoids and fetal brain. Cell Rep 30(6):1682–1689
144. Andilla J, Jorand R, Olarte O et al (2017) Imaging tissue-mimic with light sheet microscopy: a comparative guideline. Sci Rep 7: 44939
145. Yang B, Chen X, Wang Y et al (2019) Epi-illumination SPIM for volumetric imaging with high spatial-temporal resolution. Nat Methods 16(6):501–504
146. Hof L, Moreth T, Koch M et al (2020) Long-term live imaging of epithelial organoids and corresponding multiscale analysis reveal high heterogeneity and identifies core regulatory principles. bioRxiv 2020.07.12.199463
147. Bolhaqueiro ACF, van Jaarsveld RH, Ponsioen B et al (2018) In Methods Cell Biol (Eds: H Maiato, M Schuh), Academic Press, pp. 91–106
148. Kim S, Choung S, Sun RX et al (2020) Comparison of cell and organoid-level analysis of patient-derived 3D organoids to evaluate tumor cell growth dynamics and drug response. SLAS Discov Adv Sci Drug Discov 25:744
149. Kanton S, Boyle MJ, He Z et al (2019) Organoid single-cell genomic atlas uncovers human-specific features of brain development. Nature 574(7778):418–422
150. Smits LM, Magni S, Kinugawa K et al (2020) Single-cell transcriptomics reveals multiple neuronal cell types in human midbrain-specific organoids. Cell Tissue Res 382(3): 463–476
151. Zhou J, Li C, Sachs N et al (2018) Differentiated human airway organoids to assess infectivity of emerging influenza virus. Proc Natl Acad Sci U S A 115:6822
152. Buzzelli JN, Ouaret D, Brown G et al (2018) Colorectal cancer liver metastases organoids retain characteristics of original tumor and acquire chemotherapy resistance. Stem Cell Res 27:109–120
153. Velasco S, Kedaigle AJ, Simmons SK et al (2019) Individual brain organoids reproducibly form cell diversity of the human cerebral cortex. Nature 570:523–527
154. Alladin A, Chaible L, Reither S et al (2019) Tracking the cells of tumor origin in breast organoids by light sheet microscopy. bioRxiv. https://doi.org/10.1101/617837
155. Fujii E, Yamazaki M, Kawai S et al (2018) A simple method for histopathological evaluation of organoids. J Toxicol Pathol 31(1): 81–85
156. Xie Y, Park ES, Xiang D, Li Z (2018) Long-term organoid culture reveals enrichment of organoid-forming epithelial cells in the fimbrial portion of mouse fallopian tube. Stem Cell Res 32:51–60
157. Pastuła A, Middelhoff M, Brandtner A et al (2016) Three-dimensional gastrointestinal organoid culture in combination with nerves or fibroblasts: a method to characterize the gastrointestinal stem cell niche. Stem Cells Int:3710836
158. Lu W, Rettenmeier E, Paszek M et al (2017) Crypt organoid culture as an in vitro model in drug metabolism and cytotoxicity studies. Drug Metab Dispos 45(7):748–754
159. Yakoub AM, Sadek M (2018) Development and characterization of human cerebral organoids: an optimized protocol. Cell Transplant 27(3):393–406
160. Xia C, Fana J, Emanuel G et al (2019) Spatial transcriptome profiling by MERFISH reveals subcellular RNA compartmentalization and cell cycle-dependent gene expression. Proc Natl Acad Sci 116:19490
161. Shah S, Lubeck E, Zhou W, Cai L (2016) In situ transcription profiling of single cells

reveals spatial organization of cells in the mouse hippocampus. Neuron 92(2): 342–357

162. Mayr U, Serra D, Liberali P (2019) Exploring single cells in space and time during tissue development, homeostasis and regeneration. Development 146(12):dev176727

163. Schürch CM, Bhate SS, Barlow GL et al (2020) Coordinated cellular neighborhoods orchestrate antitumoral immunity at the colorectal cancer invasive front. Cell 182(5): 1341–1359

164. Lin JR, Fallahi-Sichani M, Chen JY, Sorger PK (2016) Cyclic Immunofluorescence (CycIF), A highly multiplexed method for single-cell imaging. Curr Protoc Chem Biol 8(4):251–264

165. Lin JR, Izar B, Wang C et al (2018) Highly multiplexed immunofluorescence imaging of human tissues and tumors using t-CyCIF and conventional optical microscopes. elife 7: e31657

166. Dueñas ME, Essner JJ, Lee YJ (2017) 3D MALDI Mass spectrometry imaging of a single cell: spatial mapping of lipids in the embryonic development of zebrafish. Sci Rep 7:14946

167. Fouquet T, Mertz G, Desbenoit N et al (2014) TOF-SIMS/MALDI-TOF combination for the molecular weight depth profiling of polymeric bilayer. Mater Lett 128:23–26

168. Kaitin KI (2010) Deconstructing the drug development process: the new face of innovation. Clin Pharmacol Ther 87(3):356–361

169. Brehm-Stecher BF (2014) In Encycl Food Microbiol Second Ed. (Eds: CA Batt, ML Tortorello), Academic Press, Oxford, pp. 943–953.

170. Allam M, Cai S, Coskun AF (2020) Multiplex bioimaging of single-cell spatial profiles for precision cancer diagnostics and therapeutics. NPJ Precis Oncol 4:11

171. Saglam-Metiner P, Gulce-Iz S, Biray-Avci C (2019) Bioengineering-inspired three-dimensional culture systems: organoids to create tumor microenvironment. Gene 686:203–212

172. Yu F, Hunziker W, Choudhury D (2019) Engineering microfluidic organoid-on-a-chip platforms. Micromachines 10:165

173. Wong CH, Siah KW, Lo AW (2019) Estimation of clinical trial success rates and related parameters. Biostatistics 20(2):273–286

174. Zhou Y, Othus M, Araki D et al (2016) Pre- and post-transplant quantification of measurable ('minimal') residual disease via multiparameter flow cytometry in adult acute myeloid leukemia. Leukemia 30(7):1456–1464

175. Garcez PP, Loiola EC, Madeiro da Costa R et al (2016) Zika virus impairs growth in human neurospheres and brain organoids. Science 352(6287):816–818

176. Quadrato G, Nguyen T, Macosko EZ et al (2017) Cell diversity and network dynamics in photosensitive human brain organoids. Nature 545(7652):48–53

177. Öhlund D, Handly-Santana A, Biffi G et al (2017) Distinct populations of inflammatory fibroblasts and myofibroblasts in pancreatic cancer. J Exp Med 214(3):579–596

178. Bartucci M, Ferrari AC, Kim IY et al (2016) Personalized medicine approaches in prostate cancer employing patient derived 3D organoids and humanized mice. Front Cell Dev Biol 4. https://doi.org/10.3389/fcell.2016.00064

179. Al-Lamki RS, Bradley JR, Pober JS (2017) Human organ culture: updating the approach to bridge the gap from in vitro to in vivo in inflammation, cancer, and stem cell biology. Front Med 4. https://doi.org/10.3389/fmed.2017.00148

180. Willemse J, Lieshout R, van der Laan LJW, Verstegen MMA (2017) From organoids to organs: bioengineering liver grafts from hepatic stem cells and matrix. Best Pract Res Clin Gastroenterol 31(2):151–159

181. Herron LA, Hansen CS, Abaci HE (2019) Engineering tissue-specific blood vessels. Bioeng Transl Med 4:e10139. https://doi.org/10.1002/btm2.10139

182. Kolesky DB, Truby RL, Gladman AS et al (2014) 3D bioprinting of vascularized, heterogeneous cell-laden tissue constructs. Adv Mater 26(19):3124–3130

183. Zhang B, Montgomery M, Chamberlain MD et al (2016) Biodegradable scaffold with built-in vasculature for organ-on-a-chip engineering and direct surgical anastomosis. Nat Mater 15(6):669–678

184. Akhtar A (2015) The flaws and human harms of animal experimentation. Camb Q Healthc Ethics 24(4):407–419

# Chapter 23

# Analysis of Phase-Separated Biomolecular Condensates in Cancer

**Wei Li and Hao Jiang**

## Abstract

Phase-separated biomolecular condensates play important roles in virtually all cellular processes, and their dysregulation is associated with many pathological processes including cancer. Here we concisely review some basic methodologies and strategies to analyze the phase-separated biomolecular condensates in cancer, including physical characterization of phase separation for the protein of interest, functional demonstration of this property in cancer regulation, as well as mechanistic studies on how phase separation regulates the protein's function in cancer.

**Key words** Liquid–liquid phase separation (LLPS), Protein of interest (POI), Biomolecular condensates, Intrinsically disordered regions (IDRs), Cancer

## 1 Introduction

Many macromolecules in cells, especially proteins, contain sequence features that allow them to engage in weak and multivalent intermolecular interactions, which effectively separate these molecules from the solvent phase, a process called liquid–liquid phase separation (LLPS) [1]. This process drives the formation of membraneless micro-compartments, also termed biomolecular condensates [2], which enrich specific molecules. This property plays an important role in the spatiotemporal regulation of these molecules in cellular space and biochemistry. Homotypic and heterotypic interactions of both structured domains and intrinsically disordered regions (IDRs) on proteins as well as other biomolecules (RNA and DNA) can all contribute to the formation of the phase-separated biomolecular condensates. Moreover, biomolecular condensates can adopt very wide ranges of different material and compositional properties, which can profoundly impact the biological processes that these condensates are associated with [2].

Usha N. Kasid and Robert Clarke (eds.), *Cancer Systems and Integrative Biology*, Methods in Molecular Biology, vol. 2660, https://doi.org/10.1007/978-1-0716-3163-8_23,

Alterations in biomolecular phase separation can lead to dysregulated biological processes and disease [3–5]. Reports over the past few years have started to experimentally demonstrate the role of biomolecular condensates in cancer [6–19]. Instead of providing detailed experimental protocols, here we briefly review some of the basic methodologies used in the published studies for the analysis of protein condensates in cancer. We start from the basic characterization of LLPS of a protein of interest (POI) at different levels, and direct readers to excellent and detailed reviews on methods and guidelines for characterization of biomolecular condensates and their physicochemical properties using specialized assays [20, 21]. We then summarize some of the key strategies used to establish the functional importance of LLPS in cancer regulation and then use a couple of published examples to discuss the mechanistic studies of how LLPS regulates cancer.

## 2 Does it Happen? Analysis of Phase Separation for the Protein of Interest

### 2.1 In Silico

Over the past few years, a number of computational programs have been developed that aim to predict the proteins and/or protein regions with phase separation capacities. Many of these programs are based on protein sequence features thought to drive phase separation [22]. Some programs use algorithms trained on datasets containing experimentally validated phase-separating protein sequences [23–25]. While these programs can be helpful in the initial stages of studying phase separation in a specific biological setting (such as cancer), it is important to recognize that the in silico approaches are limited by our incomplete understanding of the molecular interactions underlying LLPS, the biased datasets, and the context-dependent nature of this process. In addition, computational simulations can help predict and understand molecular mechanisms of LLPS, especially IDR-driven LLPS [26, 27]. The in silico approaches are complementary to the experimental approaches discussed below.

### 2.2 In Vitro

#### 2.2.1 Why In Vitro?

By in vitro, we mean biomolecules outside cellular structures. Cellular factors and structures can have profound effects on whether and how a protein from phase-separated condensates. Without in vitro studies, it is difficult to determine if a protein forms condensates in cells because of its own LLPS capacity driven by its own sequence features or because it mere gets passively incorporated into other pre-formed condensates. In vitro assays allow perturbation of LLPS by changes of experimental conditions, including protein and salt concentrations, temperature, etc., as well as many other factors that may regulate LLPS of POI. Reconstitution of the in vitro system that can recapitulate the in vivo condensation is highly valuable in revealing what are sufficient in mediating

LLPS of POI, and the differences seen between the in vitro system and in vivo behavior would suggest future research directions to identify cellular factors that regulate POI LLPS. Moreover, in vitro assays allow the identification of key sequence features in driving LLPS, which then allows perturbation of LLPS through mutagenesis and functional assays to show the role of LLPS in a specific biological process, such as cancer, as discussed in sections below.

#### 2.2.2 Reconstituted Systems with Purified Molecules

In vitro protein condensates can be shown without purification when they are produced from cell-free expression systems or in cell extracts. However, this approach has a number of drawbacks, including unclear protein concentrations, difficulty in changing protein concentrations, and difficulty in distinguishing the intrinsic ability versus effects from other cellular factors. Therefore, whenever possible, it is highly recommended to purify the POI before proceeding to LLPS assays in vitro.

The POI can be recombinantly expressed and purified from bacterial, insect, or mammalian cells. One potential challenge is that many LLPS-prone proteins tend to aggregate and have low solubility. Readers are directed to guidelines and protocols from experts in the LLPS field for purification of LLPS-prone proteins [28]. Once purified to a certain level of homogeneity, many different methods can be used to characterize the protein phase behavior. As these different methods all have their unique advantages and disadvantages, a combination of multiple methods should be used. In all these methods, the purified proteins need to be stored in non-LLPS conditions and then switched to conditions that allow LLPS to occur. This can be achieved by storing the protein at a high salt concentration or non-native pH level that prevents LLPS and then switching to the physiological salt concentration and pH level, or expressing the protein with soluble tags (e.g., Maltose-binding protein) which can be cleaved off by specific proteases.

Proteins undergoing LLPS often exhibit visible turbidity that can be quantitatively measured for the optical density (typically at wavelengths of 600 nm or 340 nm). However, this often requires a relatively high concentration of the protein that may be higher than the endogenous concentration and provides little information of the biophysical status of the condensates.

Physical centrifugation can be used to show LLPS. Proteins in the condensed phase can be spun down into a pellet while the proteins in the diffuse phase remain in the supernatant. The resultant pellet and supernatant fractions as well as the total proteins before centrifugation can be resolved by SDS-PAGE and detected by Coomassie staining or immunoblotting, thus allowing quantitatively determination of the partition percentage of the protein. As we do not know for sure the sedimentation limit of the condensates in a specific centrifugation process, one can always argue whether the proteins in the supernatant are truly not in any forms of condensates (similar arguments are made for the imaging-based methods). However, it is useful in comparing the relative LLPS capacity of different protein samples. In addition, this method can be used to test the in vitro activity of the molecules in the separated phases by recovering each fraction after centrifugation.

Light microscopy-based imaging is the primary method to characterize protein LLPS [28]. It provides multi-aspect information of LLPS, including whether the protein forms condensates, the size, number, morphology, mobility, fusion of the condensates, and the quantitative partition coefficient based on the signal intensities in and out of the condensates. We note that while the morphological features (e.g., roundness of the droplets) can provide the first line of information on the material properties of the condensates, in our experience they can be sometimes misleading and not consistent with the molecular dynamics and liquidity of the condensates as more rigorously measured by other biophysical assays [20].

#### 2.2.3 How to Demonstrate that the POI Forms Condensates Instead of Just "Aggregates"?

Condensates are a type of aggregates in a broad sense. But in this question, the term "aggregates" usually refers to solid assemblies formed in an irreversible process by misfolded proteins that have lost their biological activities and thus no longer subjects of meaningful studies. On the contrary, biomolecular condensates usually have certain levels of liquidity, can be dissolved and reformed, and most importantly, are biologically active—and often need to be in condensates to be optimally active. These descriptions may not apply to certain non-dynamic condensates that serve specific biological purposes [29], although in those cases the inactivity can be viewed as their required biological activity. Therefore, it is important to perform assays to determine fluidity (through fluorescence recovery after photobleaching [FRAP], observing droplet fusion, etc.), reversibility (through change experimental conditions including salt concentration, pH, temperature), and biological activity (sections below), of the condensates.

### 2.3 In Cells

Protein condensation in live cells can be studied by expression of the POI fused with a fluorescent protein. As the easiest method in cells, transient transfection leads to overexpression in certain cells, and its high protein concentration facilitates formation of protein condensates. Such non-physiological protein concentration is, of course, also the major caveat of the method. However, when combined with other methods described below to show intracellular condensation of proteins near endogenous levels, it remains a useful and convenient method in demonstrating the relative differences in condensation to bring out the importance of sequence or structural elements in the protein.

Stable and inducible cell lines can be established to allow the expression of the transgene encoding POI fused with a fluorescent protein to be near its endogenous level. The most rigorous way to demonstrate endogenous protein condensation in live cells requires knock-in of a fluorescent protein tag to a terminus of the coding sequence of the endogenous gene. Genomic editing to delete or mutate the key region or residues for endogenous protein condensation can be performed. The resultant cells can be used to demonstrate the effects on condensation of the endogenous protein and functional studies for LLPS (below). For example, UTX core IDR was deleted from the endogenous locus by CRISPR/Cas9-mediated approach in mouse embryonic stem cells, and the mutant UTX that lacks core IDR but has mEGFP tag exhibits reduced condensation [7]. Moreover, the stem cells harboring such mutation show impaired differentiation capacity similar to UTX knockout cells [7]. These genomic engineering assays are thus very helpful in studying both the condensation properties and the functional role of condensation of the endogenous protein.

A key question here is how to demonstrate that the POI undergoes LLPS in live cells? This is very difficult, as we have very limited ways to manipulate the experimental conditions in cells without eliciting indirect effects. Our opinion is that one should try the best to rigorously demonstrate LLPS of the POI in vitro, identify key residues for LLPS in vitro, and then demonstrate that the in vivo condensation of the protein is equally dependent on the same residues. This will suggest that the in vivo and in vitro condensation are likely to be governed by the same nature of molecular underpinnings.

## 3 Is it Important? Functional Analysis of LLPS in Cancer

After showing that the POI indeed undergoes LLPS in physiological conditions, the next important question would be: is its LLPS required or functionally important for its role in cancer regulation? This is a crucial question most researchers face, yet not sufficiently discussed in many review articles.

As in any approach to addressing the “requirement” question, one will need to specifically perturb LLPS and examine the biological outcomes. In cancer research, this can include cancer-related activity assays involving cultured cells or live animals and in vitro functional assays on a molecular activity that is important for cancer. The central challenge here is how to SPECIFICALLY perturb LLPS without (or minimally) affecting other molecular properties, as most proteins have multiple molecular properties that together confer the biological function.

### 3.1 LLPS Perturbation by Protein-Extrinsic Factors and Conditions

For in vitro functional assays, it is possible to use specific reagents and conditions to perturb LLPS and show effects on the activities. Salt concentration, temperature, as well as crowding agents can be changed to affect LLPS. 1,6-hexanediol is the most commonly used reagent [30]. Sometimes, other biomolecules that are known to bind to the POI and also have LLPS capacity may be used to show the effects. However, all of these treatments can potentially affect other known and unknown molecular activities of the POI, making it difficult to conclude specifically on the role of LLPS on the POI function measured in the in vitro assays.

For functional assays based on cultured cells or animal models, it is more difficult to use these treatments for the purpose of specifically perturbing LLPS, although certain treatments may be performed on cultured cells as a supplementary approach. Intracellular condensation of engineered proteins can be controlled by light or chemical activation methods [31] and employed to show the functional capability of the induced condensation of the POI. These methods can be used in combination with the methods discussed below to better demonstrate the functional relevance of the POI LLPS in a biological or pathological process.

### 3.2 LLPS Perturbation by Changing Protein Sequences

The most important approach to the functional demonstration of LLPS in a biological process is through specific alterations of the POI sequences that control its LLPS properties. This would need to start by mapping the region(s) on the POI that are important for its LLPS in vitro and in cells. Careful examination of the sequence of the required region can be helpful to reveal features known to promote LLPS, including aromatic residues, oppositely charged clusters, as well as highly enriched residues. These residues can be mutated. The mutants are then subject to assays for condensation in vitro and in vivo, biochemical activities, and biological functions. Importantly, these mutants need to be tested for other known properties, which often include binding to other molecules that are involved in the function of the POI. Ideally, the mutants should not affect other known molecular properties. If this is not true, further mutagenesis of narrower residues should be pursued to allow separation of effects on condensation from effects on other properties. This may be challenging for certain proteins whose condensation properties are functionally associated with other known properties. For example, LLPS of certain proteins may be driven by heterotypic interactions with other proteins or may be important for the other properties. In these scenarios, deeper mechanistic dissections can aid to enhance the confidence level of the role of LLPS in the specific biological processes.

IDR-driven LLPS is often not sensitive to mutation of a single or just a few amino acid residues, due to the weak and multivalency nature of the molecular interactions underlying this process. Most published studies on IDR-driven LLPS thus mutate multiple residues (of the same or different amino acids) to show effects on LLPS. One would rightly question whether the mutation of the multiple residues would affect known or unknown properties other than LLPS. This concern can never be formally excluded as one can never demonstrate an unknown (but possible) property and is exacerbated by the often large number of residues mutated.

To alleviate this concern, one can replace the LLPS-responsible IDR in the POI with LLPS-promoting IDRs from a number of unrelated proteins and test the function of the resultant chimeric proteins. The rationale behind this approach is that, if these randomly selected foreign IDRs can maintain the function of the POI (while the IDR-deleted mutant cannot), they should share some key properties, in this case, LLPS. The chance that these unrelated protein IDRs may also have a specific non-LLPS property (such as binding to a protein that the POI binds to) that drives the functional readout is small. Therefore, the more unrelated the proteins are to the POI, and the larger number of such proteins can be shown to maintain function, the higher confidence one has in stating that the primary role of the IDR of POI is to mediate LLPS and thus the effects of its mutations are mainly through LLPS. It is important to note that failure of certain foreign IDR to maintain function does not necessarily mean LLPS is not involved, as these unrelated protein IDRs may harbor different properties that could interfere with the POI function. Moreover, different physicochemical properties of condensates formed from different IDRs can affect the function of the chimeric proteins. To build an even stronger case, one could mutate the key residues in the foreign IDRs and show the effects on the function of the chimeric proteins [12, 14].

### 3.3 Analysis of Cancer-Associated Mutations and Variations

Cancer-associated mutations are highly meaningful in functional demonstration if these mutations alter LLPS or the physicochemical properties of condensates. Most reports involve larger genetic alterations, including fusion proteins that acquire aberrant LLPS [8, 9, 13–15, 32–34] or impair LLPS [10] and truncating mutations that alter LLPS properties [7]. Because these alterations often involve change of the large regions with different domains, further and more focused mutations (discussed in the section above) within the altered regions are usually necessary in order to functionally link the altered LLPS properties to cancer. Missense mutations or in-frame alterations have also been shown to affect LLPS and cancer [11, 16, 18], but these are mostly limited to structured domains instead of IDRs. While missense mutations in IDRs have been hypothesized to regulate cancer through LLPS [35], and some cancer-associated IDR missense mutations have been shown to alter LLPS or condensate properties [7], little experimental data have been shown to demonstrate their causal effect in cancer development. This is consistent with the relatively low conservation of

IDR at the sequence level. Another likely reason is that the cancer models used to show the effects of missense mutations are likely insufficient to recapitulate the complex and long process of cancer development in human patients.

Natural variations of proteins may also link altered LLPS to pathologies including cancer. Mainly due to the sequence feature variations in the IDR, the chromatin modulator UTY (on Y chromosome) has a higher propensity to form condensates with reduced liquidity and molecular diffusion rate than its X chromosomal homolog, UTX. The altered condensate properties likely account for the impaired tumor-suppressive activity of UTY and may contribute to the higher cancer incidence in men [7]. Similarly, the IDR of the RNA helicase DDX3Y also has a stronger LLPS capacity than that of its X chromosomal homolog, DDX3X, leading to enhanced repression of mRNA translation, and promotes stronger aggregation of FUS and TDP-43 [36]. These sexually dimorphic properties in LLPS may contribute to sex bias in human disease.

It is challenging to rigorously demonstrate the causality of LLPS in cancer (or other biological or pathological processes), but we think that the collective evidence of the approaches discussed above can provide a reasonably strong support. Moreover, revelation of mechanisms, as discussed below, also greatly boosts the confidence for causality.

## 4 How Does it Work? Mechanistic Studies on how Phase-Separated Biomolecular Condensates Regulate Cancer

From the perspective of LLPS biology, the roadmap laid by Rosen and colleagues should be helpful for understanding the functions of biomolecular condensates across different size scales and organization levels [37]. From the perspective of cancer biology, one needs to connect the condensates functions at the molecular scale (e.g., enhancing biochemical reaction rates) to the pathways in cancer. Cancer can result from dysregulation of many different pathways and molecular mechanisms, and dysregulated LLPS may contribute to the aberrance of many of these mechanisms [6]. It is thus not realistic to discuss in details all the different mechanistic studies on how LLPS dysregulation causes cancer. Rather, we will dissect mechanistic studies into two major cancer-related pathways, signal transduction and gene regulation, to exemplify some strategies to address a central question: "How does the mechanism generate unique insight that would not be possible without the concept of LLPS?" Another way to ask the question is, "How does unique biology arise from the phase-separated biomolecular condensates that drives the process in the study"?

A prime example is the discovery that the disease (including cancer)-associated SHP2 mutations drive aberrant LLPS to promote MAPK activation [11]. SHP2 is a non-receptor protein tyrosine phosphatase that positively regulates the Ras-MAPK signaling pathway. A longstanding perplex regarding SHP2 is why its enzymatically activating and inactivating mutations cause disease with overlapping clinical manifestations. The authors show that these mutations promote SHP2 LLPS driven by electrostatic interactions mediated by the folded phosphatase domain through transitioning to more open conformations. The authors engineered a set of mutations on charged residues to demonstrate the functional importance of the electrostatic interaction-driven LLPS in stimulating the SHP2 enzymatic activity and the RAS-MAPK signaling. LLPS of the SHP2 disease mutants locally concentrate both the enzyme itself and the substrate into the condensates, thereby enhancing the reaction rates. However, this simplistic mechanism cannot support the elevated MAPK signaling by those catalytically inactivated SHP2 mutants. To understand the mechanistic basis for how LLPS can solve this perplex, the authors studied the effect of LLPS on the wild-type copy of SHP2 given the heterozygous nature of the mutations. They found that, while the catalytically inactivated SHP2 mutants still cannot be sufficiently activated by LLPS, they recruit the wild-type SHP2 into the condensates to promote their enzymatic activity, resulting in the overall enhanced MAPK signaling in the cell.

This work is inspirational on how methods can be designed to demonstrate the unique importance and mechanistic insight of LLPS in the involved biological process that goes awry. The enigma of the SHP2 mutations would not be solved without introducing the concept of LLPS here. LLPS can almost be viewed as another, unknown, functional dimension of SHP2, as opposed to the enzymatic activity dimension of the affected protein itself we are normally fixated on. An important bonus from this work is the finding that allosteric inhibitors of SHP2 can inhibit its LLPS and enzymatic activity and may be used for treating SHP2-associated disease including cancer.

In another example, LLPS was found to be a missing property that solves a puzzle of an important tumor suppressor. UTX/KDM6A, a histone H3K27 demethylase, is known to suppress cancer often in a demethylase-independent manner. It was recently shown that UTX undergoes IDR-dependent LLPS, which underlies its activity in suppressing cancer and regulating embryonic stem cell differentiation [7]. The LLPS property of UTX is lost in the most frequent UTX cancer mutation, as it truncates the core IDR important for LLPS. To show the functional importance of the IDR-mediated LLPS for the tumor-suppressive activity of UTX, the authors mutated key residues in the core IDR and showed that those mutations that abolished UTX LLPS also abolished the tumor-suppressive activity and those that hardened the condensates greatly impaired the tumor-suppressive activity. Moreover, replacing its IDR with IDRs from unrelated proteins effectively maintained the tumor-suppressive activity. The authors then used a number of different experimental approaches including biochemical, biophysical, and cell-based assays to study the mechanisms by which UTX LLPS regulates tumor suppression. Both biochemical reconstitution and optogenetic assays show that UTX forms co-condensates with and enrich the enzymatic activities of KMT2D/MLL3 and p300, two important histone modifiers that function mainly at enhancers. Genomic assays in cells show that UTX condensation orchestrates a tumor-suppressive transcriptional program through regulating genomic histone modifications and higher-order chromatin interactions. Certain cancer-associated IDR mutations in UTX as well as IDR sequence alterations in UTY enhance the LLPS activity but reduce the molecular dynamics in the condensates and impairs the tumor-suppressive activity. These data suggest that UTX IDR appears to have evolved to adopt sequence features to support an optimal LLPS ability and material state of the condensates [7]. In this study, the concept of LLPS here is crucial in unveiling the underlying mechanisms for the tumor-suppressive activity of a chromatin modulator that is mainly viewed as an epigenetic enzyme.

## 5 Concluding Remarks

Over the past few years, physical characterization of phase-separated condensates has been extensively reported. Yet we are just in the very beginning of studying the complex nature, properties, regulation, function, and molecular mechanisms of the native cellular condensates in biological and pathological processes including cancer. New methods are needed to specifically probe and control (induce, dissolve, or tune) the endogenous condensates. These methods will allow better understanding of how the cellular condensates regulate cancer and make it possible to develop novel strategies to fight cancer based on these condensates. While we hope that the basic approaches and methodologies (mainly based on published works) sketched here are helpful to readers, it is rather not the intention of the authors that this essay restrict the creativity of the researchers in developing novel methods for this very young field that has enormous potential for unexpected findings.

## Acknowledgements

Work in the authors' laboratory is supported by the National Institute of Health (1 R01 CA259573-01, 1 R21 CA257936-01), Department of Defense (BC190343), American Cancer Society Research Scholar Award (128609-RSG-15-166-01-DMC), and the Leukemia and Lymphoma Society Scholar Award.

### References

1. Hyman AA, Weber CA, Julicher F (2014) Liquid-liquid phase separation in biology. Annu Rev Cell Dev Biol 30:39–58
2. Banani SF, Lee HO, Hyman AA, Rosen MK (2017) Biomolecular condensates: organizers of cellular biochemistry. Nat Rev Mol Cell Biol 18(5):285–298
3. Jiang S, Fagman JB, Chen C, Alberti S, Liu B (2020) Protein phase separation and its role in tumorigenesis. eLife 9. https://doi.org/10.7554/eLife.60264
4. Alberti S, Dormann D (2019) Liquid-liquid phase separation in disease. Annu Rev Genet 53:171–194
5. Boija A, Klein IA, Young RA (2021) Biomolecular condensates and cancer. Cancer Cell 39(2):174–192
6. Mehta S, Zhang J (2022) Liquid-liquid phase separation drives cellular function and dysfunction in cancer. Nat Rev Cancer 22(4):239–252
7. Shi B, Li W, Song Y et al (2021) UTX condensation underlies its tumour-suppressive activity. Nature 597(7878):726–731
8. Qin Z, Sun H, Yue M et al (2021) Phase separation of EML4-ALK in firing downstream signaling and promoting lung tumorigenesis. Cell Discov 7:33. https://doi.org/10.1038/s41421-021-00270-5
9. Tulpule A, Guan J, Neel DS et al (2021) Kinase-mediated RAS signaling via membraneless cytoplasmic protein granules. Cell 184(10):2649–2664.e18
10. Zhang JZ, Lu TW, Stolerman LM et al (2020) Phase separation of a PKA regulatory subunit controls cAMP compartmentation and oncogenic signaling. Cell 182(6):1531–1544.e15
11. Zhu G, Xie J, Kong W et al (2020) Phase separation of disease-associated SHP2 mutants underlies MAPK hyperactivation. Cell 183(2):490–502.e18
12. Esposito M, Fang C, Cook KC et al (2021) TGF-β-induced DACT1 biomolecular condensates repress Wnt signalling to promote bone metastasis. Nat Cell Biol 23(3):257–267
13. Boulay G, Sandoval GJ, Riggi N et al (2017) Cancer-specific retargeting of BAF complexes by a prion-like domain. Cell 171(1):163–178.e19
14. Ahn JH, Davis ES, Daugird TA et al (2021) Phase separation drives aberrant chromatin looping and cancer development. Nature 595(7868):591–595
15. Chandra B, Michmerhuizen NL, Shirnekhi HK et al (2021) Phase separation mediates NUP98 fusion oncoprotein leukemic transformation. Cancer Discov 12:1152. https://doi.org/10.1158/2159-8290.cd-21-0674
16. Wan L, Chong S, Xuan F et al (2020) Impaired cell fate through gain-of-function mutations in a chromatin reader. Nature 577(7788):121–126
17. Li W, Hu J, Shi B, Palomba F et al (2020) Biophysical properties of AKAP95 protein condensates regulate splicing and tumorigenesis. Nat Cell Biol 22(8):960–972
18. Bouchard JJ, Otero JH, Scott DC et al (2018) Cancer mutations of the tumor suppressor SPOP disrupt the formation of active, phase-separated compartments. Mol Cell 72(1):19–36.e8
19. Cheng Y, Xie W, Pickering BF et al (2021) N(6)-Methyladenosine on mRNA facilitates a phase-separated nuclear body that suppresses myeloid leukemic differentiation. Cancer Cell 39(7):958–972.e8
20. Mitrea DM, Chandra B, Ferrolino MC et al (2018) Methods for physical characterization of phase-separated bodies and membrane-less organelles. J Mol Biol 430(23):4773–4805
21. Alberti S, Gladfelter A, Mittag T (2019) Considerations and challenges in studying liquid-liquid phase separation and biomolecular condensates. Cell 176(3):419–434
22. Vernon RM, Forman-Kay JD (2019) First-generation predictors of biological protein phase separation. Curr Opin Struct Biol 58:88–96

23. van Mierlo G, Jansen JRG, Wang J, Poser I, van Heeringen SJ, Vermeulen M (2021) Predicting protein condensate formation using machine learning. Cell Rep 34(5):108705
24. Chu X, Sun T, Li Q et al (2022) Prediction of liquid-liquid phase separating proteins using machine learning. BMC Bioinform 23(1):72. https://doi.org/10.1186/s12859-022-04599-w
25. Hardenberg M, Horvath A, Ambrus V, Fuxreiter M, Vendruscolo M (2020) Widespread occurrence of the droplet state of proteins in the human proteome. Proc Natl Acad Sci USA 117(52):33254–33262
26. Dignon GL, Zheng W, Mittal J (2019) Simulation methods for liquid-liquid phase separation of disordered proteins. Curr Opin Chem Eng 23:92–98
27. Shea JE, Best RB, Mittal J (2021) Physics-based computational and theoretical approaches to intrinsically disordered proteins. Curr Opin Struct Biol 67:219–225
28. Alberti S, Saha S, Woodruff JB, Franzmann TM, Wang J, Hyman AA (2018) A user's guide for phase separation assays with purified proteins. J Mol Biol 430(23):4806–4820
29. Woodruff JB, Hyman AA, Boke E (2018) Organization and function of non-dynamic biomolecular condensates. Trends Biochem Sci 43(2):81–94
30. Kroschwald S, Maharana S, Simon A (2017) Hexanediol: a chemical probe to investigate the material properties of membrane-less compartments. Matters. https://doi.org/10.19185/matters.201702000010
31. Bracha D, Walls MT, Brangwynne CP (2019) Probing and engineering liquid-phase organelles. Nat Biotechnol 37(12):1435–1445
32. Owen I, Yee D, Wyne H et al (2021) The oncogenic transcription factor FUS-CHOP can undergo nuclear liquid-liquid phase separation. J Cell Sci 134(17). https://doi.org/10.1242/jcs.258578
33. Davis RB, Kaur T, Moosa MM, Banerjee PR (2021) FUS oncofusion protein condensates recruit mSWI/SNF chromatin remodeler via heterotypic interactions between prion-like domains. Protein Sci 30(7):1454–1466
34. Davis RB, Moosa MM, Banerjee PR (2022) Ectopic biomolecular phase transitions: fusion proteins in cancer pathologies. Trends Cell Biol 32:681. https://doi.org/10.1016/j.tcb.2022.03.005
35. Tsang B, Pritišanac I, Scherer SW, Moses AM, Forman-Kay JD (2020) Phase separation as a missing mechanism for interpretation of disease mutations. Cell 183(7):1742–1756
36. Shen H, Yanas A, Owens MC et al (2022) Sexually dimorphic RNA helicases DDX3X and DDX3Y differentially regulate RNA metabolism through phase separation. Mol Cell 82:2588. https://doi.org/10.1016/j.molcel.2022.04.022
37. Lyon AS, Peeples WB, Rosen MK (2021) A framework for understanding the functions of biomolecular condensates across scales. Nat Rev Mol Cell Biol 22(3):215–235

# Chapter 24

## Gain-of-Function Variomics and Multi-omics Network Biology for Precision Medicine

**Mark M. Li, Sharad Awasthi, Sumanta Ghosh, Deepa Bisht, Zeynep H. Coban Akdemir, Gloria M. Sheynkman, Nidhi Sahni, and S. Stephen Yi**

### Abstract

Traditionally, disease causal mutations were thought to disrupt gene function. However, it becomes more clear that many deleterious mutations could exhibit a "gain-of-function" (GOF) behavior. Systematic investigation of such mutations has been lacking and largely overlooked. Advances in next-generation sequencing have identified thousands of genomic variants that perturb the normal functions of proteins, further contributing to diverse phenotypic consequences in disease. Elucidating the functional pathways rewired by GOF mutations will be crucial for prioritizing disease-causing variants and their resultant therapeutic liabilities. In distinct cell types (with varying genotypes), precise signal transduction controls cell decision, including gene regulation and phenotypic output. When signal transduction goes awry due to GOF mutations, it would give rise to various disease types. Quantitative and molecular understanding of network perturbations by GOF mutations may provide explanations for 'missing heritability" in previous genome-wide association studies. We envision that it will be instrumental to push current paradigm toward a thorough functional and quantitative modeling of all GOF mutations and their mechanistic molecular events involved in disease development and progression. Many fundamental questions pertaining to genotype–phenotype relationships remain unresolved. For example, which GOF mutations are key for gene regulation and cellular decisions? What are the GOF mechanisms at various regulation levels? How do interaction networks undergo rewiring upon GOF mutations? Is it possible to leverage GOF mutations to reprogram signal transduction in cells, aiming to cure disease? To begin to address these questions, we will cover a wide range of topics regarding GOF disease mutations and their characterization by multi-omic networks. We highlight the fundamental function of GOF mutations and discuss the potential mechanistic effects in the context of signaling networks. We also discuss advances in bioinformatic and computational resources, which will dramatically help with studies on the functional and phenotypic consequences of GOF mutations.

**Key words** Gain-of-Function mutations, Protein-protein interactions, Computational mutation annotation, Coding and noncoding variants, Post-translational regulation, Liquid-liquid phase separation

Usha N. Kasid and Robert Clarke (eds.), *Cancer Systems and Integrative Biology*, Methods in Molecular Biology, vol. 2660, https://doi.org/10.1007/978-1-0716-3163-8_24,

## 1 Gain-of-Function Mutations in Cancer

How variants of a genomic sequence alter biological functions and molecular activities has long been a key question of molecular biology [1, 2]. Past research in genomics has emphasized the importance of loss-of-function (LOF) and gain-of-function (GOF) mutations in understanding mutational effects on gene expression, protein activity, and phenotypic plasticity [3]. For instance, in protein-coding genes, LOF mutations disrupt or reduce protein functions compared to their wild-type counterparts [4]. On the other hand, GOF mutations produce mutant proteins that exhibit either new or overactive functions that would not usually exist [4, 5].

The field of oncology has continuously focused on identifying LOF and GOF mutations that may lead to many different types of cancers. Significantly, these mutations are usually identified in cancer patients, reflecting their clinical relevance. LOF mutations, usually involving tumor suppressor genes, can cause loss of protein function essential for inhibiting cancer-causing factors. These proteins typically suppressing cell growth or promoting cell death are now unable to function at the level they normally would, thereby promoting cancer growth. Meanwhile, GOF mutations in proto-oncogenes have the capacity to become neomorphic [4, 6], or produce new protein functions, and create other changes to cellular systems that favor overactive, uncontrolled cell growth. Compared to research on LOF cancer mutations, research regarding GOF mutations on cancer is largely limited as LOF mutations are more easily identifiable [7]. This review will predominantly serve to inform on recent research advances highlighting GOF mutations in cancer genomics with a focus on covering this gap in knowledge. Meanwhile, GOF mutant effects in neurodegenerative diseases will be briefly discussed as well.

While all generally serve the same effect, GOF mutations often come in different forms. Such mutations may be categorized into two main areas of interest: protein-coding mutations and noncoding mutations. The majority of GOF mutations studied are limited to the coding regions of the genome and directly affect the makeup of proteins. The mutant proteins may further impact biological features, such as protein–protein interactions and enzymatic activity, to affect cell growth or other influences on the phenotype. Furthermore, recent literature has brought attention to significant GOF mutation changes in epigenetics and noncoding regions of the genome affecting tumor growth as well. These notable mutations affect cancer cell hallmarks through a variety of distinct molecular mechanisms. Identifying all types of mutations implicated in cancer progression is imperative for developing new ways of early detection and therapeutic interventions in cancer. In this review, the latest functional implications of both coding and noncoding mutations in cancer will be highlighted.

## 2 Epigenetic Regulation

The relationship between cancer-associated GOF mutations and epigenetic regulation has not been thoroughly studied, introducing a rather unexplored branch in cancer research. Without affecting DNA sequences, epigenetic GOF changes may create tumorigenic activity by influencing the physical structure of DNA or associated co-factors/modifiers instead (Fig. 1a). These changes may be induced by a diverse range of signals yet all similarly function to activate oncogenic transcriptional programs.

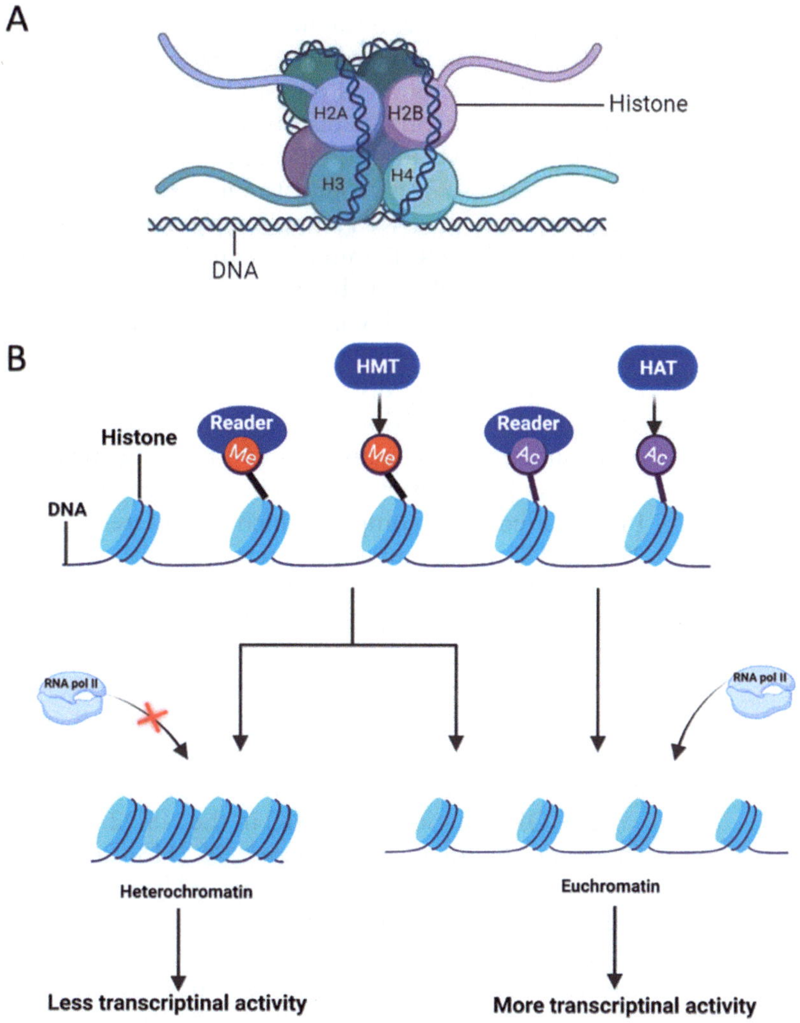


**Fig. 1** (**a**) Nucleosome consisting of histones H2A, H2B, H3, and H4 wrapped by DNA; (**b**) process of epigenetic regulation via histone acetylation and methylation. Histone acetyltransferases (HATs) act by acetylating histones, causing them to lose positive charge which loosens their interaction with negatively charged DNA. This enables RNA polymerase to better access to DNA, enabling more transcriptional activity. Histone methylation may either increase or decrease the association of histones and DNA. Depending on context, histone methylation may therefore increase or decrease transcriptional activity

### 2.1 Histone Acetylation and Methylation

A major area of research on epigenetic regulation looks at the GOF changes toward histone acetylation and methylation. These biological processes are catalyzed by three subcategories of tools: writers, readers, and erasers. Writers are enzymes capable of modifying base pairs in DNA and histone proteins, erasers are enzymes capable of reversing the work of writers, and readers are protein domains capable of identifying epigenetic changes and mediating them [8]. Catalyzed by the writer, histone acetyltransferase, histone acetylation loosens the interaction of DNA with histones and subsequently increases transcriptional activity. On the other hand, histone methylation, carried out by the writer histone methyltransferases, can either repress or increase transcriptional activity depending on where in the histone they methylate (Fig. 1b).

Research has revealed the role of GOF mutations in histone methyltransferase and histone acetyltransferase genes, modifying their binding potential to histone tails and changing levels of methylation and acetylation. For example, GOF variants in G9a histone methyltransferase were found to increase WNT signaling through the inhibition of DKK1, implicating it as a possible cause for melanoma [9]. GOF mutations in CREB-binding protein and p300, two homologous lysine acetyltransferases, were also found responsible for amplifying DNA replication through multiple non-transcriptional and transcriptional processes [10]. Interestingly, another study discovered mutant-enhanced self-association of a histone acetylation reader, leading to increased chromatin occupancy and gene activation contributing toward pediatric kidney cancer [11]. Several studies have also found GOF mutations in PRC2, affecting a complex which methylates histone 3 on lysine 27 [12, 13]. Such mutations interfering with the role and function of PR2C have been implicated in multiple cancers, such as myeloid malignancies and malignant peripheral nerve sheath tumors among many more aggressive cancers [14, 15]. Another study found histone deacetylase inhibition to reduce GOF effects of p53, highlighting how an eraser too may be implicated in tumorigenesis [16]. A separate finding determined a GOF mutation in IDH1 to depend heavily on HDAC in order to promote glioma [17].

A GOF mutation (E1099K) in the NSD2 histone methyltransferase specific for histone 3 lysine 36 (H3K36), results in altered enzyme substrate binding and increased level of H3K36 dimethylation (transcription activation mark) and decreased level of H3K27 trimethylation (repressive mark) specifically at H3.1. H3K36me2 is normally found at the 5′ end of genes near transcriptional start sites, and mutant NSD2 plays an important role in proliferation and progression of acute lymphocytic leukemia. Further, in vivo studies suggest that the mice which are xenografted with $NSD2^{\mathrm{WT/E1099K}}$ survive shorter in comparison to mice xenografted with wild-type *NSD2* [18]. Chromosomal translocation is another example of GOF events which occur in 10–20% multiple myeloma cases and places *NSD2* under the control of immunoglobulin heavy chain promoter [19]. This results in overexpression of *NSD* which leads to an increase in H3K36me2 and a reduced H3K27me2 mark. The increased H3K36me2 level results in aberrant gene activation, promoting the cancer cell growth and tumor progression [20–22]. Additionally, Y641 mutation in *EZH2* increases the level of H3K27me3 and reduces the level of H3K27 monomethylation and dimethylation in B-cell lymphomas, follicular lymphomas, and metastatic skin melanoma. The cells with GOF mutation Y641 undergo enhanced motility and significant growth advantage compared to wild-type cells [23–25].

Together, these mutations reflect important areas of consideration for drug targeting. Herein lies evidence of epigenetic GOF changes promoting tumorigenesis. Showcasing the importance of uncovering GOF effects in histone modifying proteins, these examples highlight a significant area of interest for future research in cancer biology.

## 3 Transcription Factors

GOF mutations can also influence transcription factor binding sites as another way outside of coding regions of the genome to produce oncogenic effects. Recent research highlights tumor suppressor gene *TP53* for exhibiting multiple GOF mutant properties, including those affecting transcription factors. In recognizing genes encoding cell cycle-dependent proteins cyclin A and CHKI, GOF p53 mutants are able to localize to their regulatory regions, accelerating their transcription by inducing origin firing, protecting replication forks, and promoting micronuclei formation [26]. Other research highlights separate novel mutations affecting transcriptional control through neomorphic, GOF transcription factor interactions. A mutation resulting in an N-terminally truncated variant of a transcription factor named C/EBPα, or p30, was discovered to be implicated in leukemogenesis by binding to upstream enhancers of *NT5E*. *NT5E* regulates the expression of *CD73*, promoting cell proliferation and stopping apoptosis in leukemia cells [27]. Taken together, changes to and interactions among transcription factors, impacting noncoding regions of the genome, are shown to play a significant role in gene expression.

## 4 Noncoding Elements

Elements of the noncoding genome, such as promoters and enhancers, also play important roles in oncogenesis and cancer progression.

### 4.1 Enhancers and Promoters

The role of enhancers in gene regulation came into picture in early 80s, before that binding of regulatory proteins with DNA was known as only a factor involved in gene regulation. Genome-wide association studies show that, many mutations in cis-regulatory regions are also associated with disease (PMID: 25261935). A single nucleotide variation (SNV) in the α-globin cluster creates an entirely new promoter region which causes decreased α-globin expression leading to α-thalassaemia. The hinderance in gene expression is caused by the newly formed regulatory region which lies in between the α-globin gene and their associated super enhancers in an orientation dependent manner (PMID: 34155213). Another study shows that the mutation in the promoter region of the telomerase reverse transcriptase (TERT), encoding a catalytic subunit of the telomerase, as the most abundant (71%) type of mutations occurring in melanomas examined in the study [30]. This mutation, when studied with reporter assays, exhibited upregulated transcription of TERT (PMID: 23348506). Genome-wide analysis of mutations in noncoding regions showed that mutations in regulatory regions such as promoters and 5' UTR are more frequent than mutations in 3' UTR and distal enhancer regions. The study reported that these mutations are recurrent in the promoter regions of PLEKHS1, WDR74 and SDHD as well as previously reported TERT which are further associated with different types of cancers (PMID: 25261935).

## 5 Protein–Protein Interactions

Protein–protein interactions (PPIs) are characterized by the physical association of proteins to mediate cellular processes and are also influenced by GOF mutations. Depending on the protein functions and the type of interactions, such acquired interactions may produce oncogenic effects in several unique ways [32].

Recent discoveries have shown various new GOF PPIs as a result of mutant protein p53. One mechanism in which p53 was shown to be implicated in cancer progression was through facilitating DNA replication. A recent study investigating GOF effects in mutant protein p53 found that p53 has the capacity to recruit MCM and PARP1 proteins on replicating DNA to promote replication upon DNA damage. This, in turn, causes tumorigenesis by allowing uncontrolled replication [33]. p53 was also discovered to mediate other PPIs enabling cancer progression as well. p53 mutants enhancing STAT3 activation by binding to STAT3 and displacing SHP2 were found to promote the growth of tumor cells in colorectal cancer [34]. Furthermore, another recent study elucidated the role of mutant p53 PPIs in suppressing the immune system to promote cancer cell proliferation. According to the study, p53 is able to bind to TANK-binding protein kinase 1, preventing the formation of a protein complex required for the activation of the innate immune response [35]. Taken together, these recent studies reveal a large number of neomorphic gain-of-PPI mutations in one gene alone. Further studies will elucidate the diverse role of p53 in creating neomorphic, oncogenic PPIs [36].

## 6 Post-translational Regulation

Enzymatic activity takes on a critical role in the activation and inactivation of proteins. This becomes immediately apparent when reviewing the significance of signaling pathways and kinase activation on cancer. Kinase activation works by a phosphorylation cascade creating eventual cellular response through a chain of events. GOF mutations can play a role in changing these pathways, creating phenotypic behaviors through altered steps in the chain of reactions.

Notably, tyrosine kinases are important for a cell to grow, move, differentiate, and undergo metabolism [37]. As such, changes to tyrosine kinases are frequent triggers for the onset of cancer, proving to be an active area of GOF research. Focal adhesion kinase (FAK) is a cytoplasmic non-receptor tyrosine kinase that has been linked to driving invasion and metastasis in many cancer types. A recent study interrogated a GOF mutation in the GTPase RHOA and its biochemical relation toward inducing the activity of FAK and promoting diffuse gastric cancer [38]. Other GOF mutations promote separate kinase pathways, some of which related to FAK, such as one involving mutant PI3K, a kinase involved in the PI3K/AKT/mTOR pathway which is important in cell growth and proliferation. Another study determined that *WWP1* was found to inactivate *PTEN*, a tumor suppressor gene, and contribute toward overactive PI3K signaling, causing heightened cell growth [39]. A different study interrogated the ineffectiveness of a HER2 tyrosine kinase inhibitor due to high HER2/HER3 heterodimer catalytic activity. It was shown that high HER2/HER3 heterodimer activity leads to strong activation of the PI3K/AKT/mTOR pathway and eventually promotes cell growth, invasiveness, and drug resistance [40]. As demonstrated by these recent discoveries, multiple modes of action may be at play upon hyperactivation of tyrosine kinase signaling pathways.

Other work has recently explored areas of inhibiting oncogenic KRAS involved in an effector pathway which promotes cell survival and proliferation. KRAS mutants are implicated in approximately 30% of cancers. A frequent GOF *KRAS* mutation in lung cancer impairs the intrinsic GTPase function of KRAS, causing constitutive downstream signaling of multiple kinase pathways [41, 42]. Efforts in the past three decades have been made to develop therapies for patients carrying *KRAS* mutations but to no avail [43]. However, new research highlights the significant discovery of the KRAS-G12C variant with a druggable pocket. Current research is evaluating inhibitors targeting this variant for treatment of non-small cell lung cancer [44].

A majority of other oncogenic GOF mutations described in current literature further investigates other kinase pathways, highlighting the significance toward understanding GOF kinase activity for drug targeting. GOF mutations in the *RET* (rearranged during transfection) gene which encodes a receptor tyrosine kinase important for several physiological functions drive tumor growth and proliferation in several types of cancers [45]. GOF mutations in *BRAF*, one of the most frequently found oncogenes in multiple cancer types, lead to overactivation of the MEK/ERK signaling pathway, promoting metastasis [46]. GOF mutations in *JAK2*, *CALR*, and *MPL* constitutively activate the JAK/STAT pathway, giving rise to myeloproliferative neoplasms, polycythemia vera, essential thrombocytosis, and primary myelofibrosis from pluripotent hematopoietic stem cells [47–49]. GOF mutations constitutively activating RAS and RAC1 proteins involved in the Ras signal transduction pathway promote cell proliferation, giving rise to cancers such as juvenile myelomonocytic leukemia and malignant melanoma [50–52]. GOF *KIT* mutations lead to kinase activation found in systemic mastocytosis, which is a myeloproliferative neoplasm that expands abnormal mast cells in many types of tissues. For example, gastrointestinal stromal tumors were demonstrated to have GOF mutations in *KIT*, exemplifying the apparent role of KIT in promoting cancer [53, 54]. Altogether, these diverse and abundant kinase GOF mutations showcase the wide array of kinase pathways which may be implicated in causing cancer. As such, further study on GOF mutations in proteins affecting kinase activity is significant and holds promising potential for innovative drug discovery.

## 7 Gain-of-Function Mutations in Other Diseases

Study of GOF mutations is not just limited to understanding cancer, but also to the development of other diseases as well. Notably, recent research has also highlighted neurodegenerative and inflammatory diseases influenced by GOF mutations, although not as well studied as in cancer.

### 7.1 Neurological Diseases

Neurodegenerative diseases, affecting nerve cells that cause debilitation, have been shown to be influenced in part by GOF mutations. GOF in kinase activity has been identified to contribute toward neurodegeneration. A recent study related GOF activity in protein kinase C (PKC) isozymes to unique neurodegenerative diseases. GOF mutations in PKCα were linked to Alzheimer's disease, and GOF mutations in PKCγ were linked to spinocerebellar ataxia type 14 [55]. Another study related GOF mutations in pro-inflammatory gene *TBK1*, or TANK binding kinase 1, to an increased risk in developing normal tension glaucoma [56]. Apart from kinase activity, recent study on Huntington's disease has shown that expanded CAG repeats may produce toxic truncated polyQ-containing huntingtin proteins, illuminating a new GOF mechanism [57]. Further, mutations in several different kinds of ion channels have been reported to be associated with neurodevelopmental, neurological, and/or psychiatric disorders. The patients with GOF mutation in KCNA2, which encodes for the potassium channel Kv1.2 exhibit severe phenotypic conditions in terms of epilepsy, ataxia, and intellectual disability (PMID: 25751627). Cav1.2 and Cav1.3 are two isoforms encoding calcium channels present in the brain. Mutations in these lead to several neurological abnormalities. A GOF mutation in Cav1.2 causes a rare multiorgan disorder Timothy syndrome and autism (PMID: 15454078). On the other hand, a GOF mutation in CACNA1D forming α1-subunit of Cav1.3 causes aldosteronism with seizures, neurologic abnormalities, and intellectual disability and might be the cause of autism spectrum disorder (ASD) in the patients (PMID: 25620733). In their next study, they found that the GOF mutation in CACNA1D is recurrent in patients and epilepsy (PMID: 28472301). Further, GOF mutations in CACNA1G, CACNA1H and CACNA1I encoding different subunits of Cav3 calcium channel, are associated with cerebellar atrophy, primary aldosteronism and epilepsy, respectively (PMIDs: 24277868; 17397049; 33704440). A GOF mutation in a Transient receptor potential gene, TRPA1 leads to a neuropathy disorder known as familial episodic pain syndrome (PMID: 20547126).

Current efforts at therapeutic strategies look to lowering huntingtin levels in order to counteract such GOF mechanisms [58]. As new research continues to investigate the significance of GOF effects in neurodegenerative diseases, they prove to be an invaluable endeavor for more effective therapeutics.

### 7.2 Inflammatory Diseases

Inflammatory diseases, characterized by harm from one's own over-inflammatory immune system, have also been linked to GOF mutations. Cryopyrin-associated periodic syndrome, an autoinflammatory disorder characterized by inflammation throughout multiple bodily tissues, was recently linked to GOF in *NLRP3*, resulting in an over-release of inflammatory cytokines [59]. *NLRP3* GOF has also been implicated in contributing toward several other inflammatory and autoimmune diseases such as inflammatory bowel disease [60]. Another autoinflammatory disease, familial Mediterranean fever (FMF), was recently linked to GOF mutations in the *MEFV* gene which encodes pyrin, a protein involved in regulating the innate immune system

[61]. Also notable, GOF mutations in the cGAS-STING signaling pathway, key to inflammation in infection, cellular stress, and tissue damage, has been shown to contribute toward a severe pediatric autoinflammatory syndrome called STING-associated vasculopathy (SAVI) [62]. GOF mutations in STING, a signaling molecule in cytosolic DNA sensing pathway, is a novel cause of familial chilblain lupus, which is a monogenic form of cutaneous lupus erythematosus (PMID: 27566796). Together, GOF mutations play a significant role in dysregulating the immune system as well as in inflammatory disease pathogenesis.

## 8 Protein Aggregation and Liquid–Liquid Phase Separation

Protein misfolding and aggregation lead to severe human diseases, such as Parkinson's disease, Alzheimer's disease, and amyotrophic lateral sclerosis. However, not only the neurodegenerative diseases but also cancer has been recently shown to be affected by aggregation, particularly by the aggregation of mutant variants of the tumor suppressor protein p53, which are present in more than 50% of malignant tumors. Mostly the effective therapeutic strategies for all these diseases are focused on the prevention of these aggregation processes. The aggregation of large amorphous or amyloid fibrils terminates a pathway and includes the formation of intermediate oligomeric species and protofibrils. Prior to the formation of more stable aggregate species, several proteins involved in neurodegenerative diseases tend to undergo phase separation and form biomolecular condensates usually by transiting from liquid-like to gel-like and solid-like states. This transition is particularly applicable for several nuclear proteins that associate with RNAs and DNAs to generate membraneless organelles, such as the nucleolus.

Mutant p53 tends to form aggregates with amyloid properties, especially amyloid oligomers inside the nucleus, which are believed to cause oncogenic GOFs. DNA-binding domain of p53 (p53C) undergoes liquid–liquid phase separation on the pathway to aggregation under various conditions. Similarly, mutant p53C (M237I and R249S) undergoes phase separation, evolved to solid-like biomolecular condensates than that in the case of wild-type p53C (PMID: 34163823). Live cell microscopic data indicated that transfection of mutant full-length p53 into the cells results in phase separation in the nuclear compartments, which is likely the cause of the GOF effects (Fig. 2).

Considering phase separated conformers as a crucial missing link of many fundamental biological processes, these transient biomolecular condensates may be assigned to have their own function to contribute. Just like the previous example, there are a few examples on involvement of mutant proteins in different phase separated

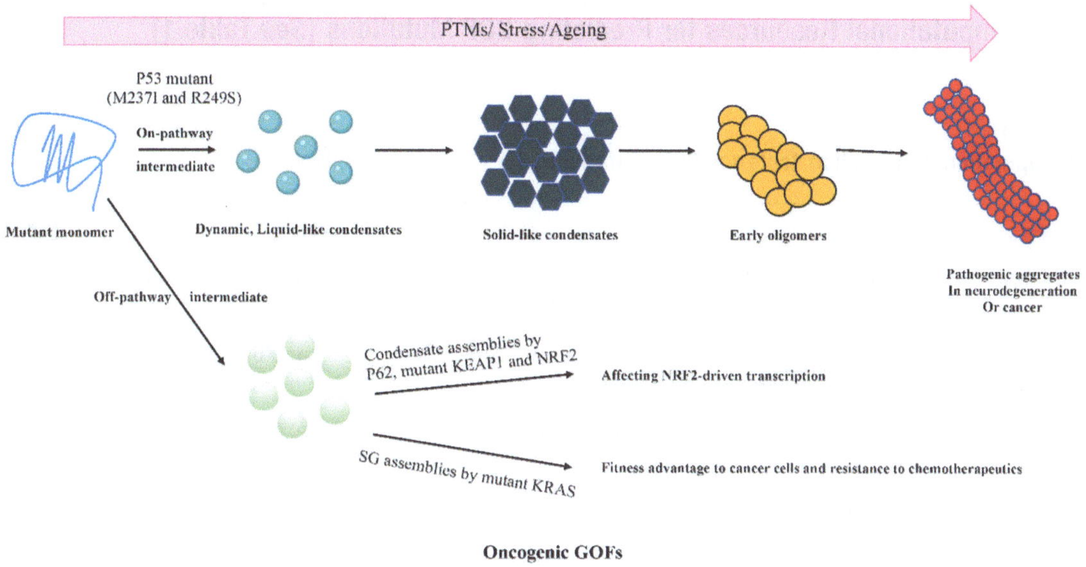


**Fig. 2** (*Scenario 1*) Mutant p53 (M237I and R249S) forms on-pathway liquid–liquid phase separation, followed by solid-like condensates which eventually leads to formation of pathological aggregates, an example of direct oncogenic GOF. (*Scenario 2*) Mutant KEAP1 assembles with p62 and NRF2 into off-pathway phase-separated condensates, which affects the function of NRF2 driven transcription. Cancer cells expressing mutant KRAS induce the formation of SG assembly which confers fitness advantage to the cells. Here both mutants are involved in phase separated assembly to accomplish an off-pathway function. So, those can be considered as example of indirect GOFs

condensates. p62 assembles into condensates together with mutant KEAP1 proteins and the transcription factor NRF2, thereby affecting NRF2-driven transcription (PMID: 30126895). Although not determined, it is speculated that p62 condensates are involved in the formation and autophagy-mediated disposal of various cellular condensates that promote or inhibit tumorigenesis. Stress granule (SG) formation was found to be induced in cancerous cells expressing mutant *KRAS*, and this was shown to be dependent on the secretion of a lipid that inactivates the eukaryotic initiation factor eIF4A (PMID: 27984728). Thus, mutant *KRAS* establishes a more stress-resistant cellular condition through SG assembly, and this confers a fitness advantage to cancer cells and presumably also resistance to chemotherapeutics. Taken together, more systematic profiling of GOF mutants in a large set of disease-associated genes is needed to correlate the current theme based on phase separation and protein aggregation.

## 9 Computational Resources for Predicting GOF Mutations (*See* Table 1)

**Table 1**
**Computational prediction tools for GOF mutations.**

| Tools | Description | Web | Ref |
|---|---|---|---|
| OncoLnc | Explores survival correlations using clinical data and expression data for mRNAs, miRNAs, or lncRNAs | http://www.oncolnc.org/ | [63] |
| OncodriveCLUST | Identifies genes with a bias toward mutations which cluster within a protein sequence | http://bg.upf.edu/oncodriveclust | [64] |
| Kin-driver | Human protein kinase database for driver mutations | http://kin-driver.leloir.org.ar/ | [65] |
| KinMutRF | Identifies pathogenic variants in human kinases | https://github.com/Rbbt-Workflows/KinMut2 | [66] |
| OncoVar | Database and analysis platform for oncogenic driver mutations | https://oncovar.org/ | [67] |
| OncoKB | Annotates oncogenic effects and the prognostic and predictive importance of mutations | http://oncokb.org/ | [68] |
| CHASM | Computational method which prioritizes missense mutations likely to cause functional changes that enhance cell proliferation | https://karchinlab.org/apps/appChasm.html | [69] |
| MEME Suite | Portal for online analysis and identification of sequence motifs representing transcription factors | http://meme.nbcr.net/ | [70] |
| RegulomeDB | Analyzes regulatory variants in the human genome | https://regulomedb.org/regulome-search/ | [71] |
| VEST | Scoring tool that predicts functional significance of missense mutations based on possibility of pathogenicity | http://www.cravat.us/CRAVAT/ | [72] |

## 10 Conclusion

Together, this review leads to an emerging area in molecular biology and is becoming an important area of research in the future. The area is innovative because it will provide unique insights in prioritizing functional GOF disease mutations and uncovering individualized molecular mechanisms. It is also significant because it will greatly facilitate the functional annotation of a large number of GOF mutations, providing a fundamental link between genotype and phenotype in human disease.

Since the introduction of next-generation sequencing methods, genomics research has taken a large step forward and revealed the importance of GOF mutations in contributing toward disease. Increasingly, more GOF mutations affecting cancer in the coding regions of the genome, changing PPIs and enzymatic activity, have been identified. Additionally, other novel GOF mutations in the noncoding regions of the genome, affecting transcriptional control, have also increasingly been identified and explored. Continuing forward, these GOF mutations prove to be important toward understanding both the pathogenesis of cancer and other (e.g., neurodegenerative and inflammatory) diseases. Furthermore, genomics study of GOF mutations helps elucidate potential targets for drug therapy, aiding in the combat against lethal diseases. Looking ahead, further research in GOF mutations seems promising and may bring large potential for effective treatment of various lethal diseases.

## Acknowledgements

This work was supported by the National Institutes of Health grant R35GM133658 (to S.Y.), Komen Foundation grant CCR19609287 (to S.Y.), Early Career Investigator Grant by Ovarian Cancer Research Alliance Grant# 649968 (to N.S.), R35GM137836 (to N.S.). S.Y. was also supported by Scialog grant #28418, sponsored jointly by Research Corporation for Science Advancement (RCSA) and the Gordon and Betty Moore Foundation; Scialog grant #28706, by RCSA, Chan Zuckerberg Initiative, and the Frederick Gardner Cottrell Foundation; and a pilot project program jointly by Oden Institute, MD Anderson and Texas Advanced Computing Center. N.S. is a CPRIT Scholar in Cancer Research with funding from the Cancer Prevention and Research Institute of Texas (CPRIT) New Investigator Grant RR160021 and Research Grant RP220292. We would also like to acknowledge the support from BioRender in figure preparation.

### References

1. Yi S, Liu NN, Hu L, Wang H, Sahni N (2017) Base-resolution stratification of cancer mutations using functional variomics. Nat Protoc 12(11):2323–2341
2. Yi S, Lin S, Li Y, Zhao W, Mills NB, Sahni N (2017) Functional variomics and network perturbation: connecting genotype to phenotype in cancer. Nat Rev Genet 18(7):395–410
3. Li Y, McGrail DJ, Xu J, Mills GB, Sahni N, Yi S (2018) Gene regulatory network perturbation by genetic and epigenetic variation. Trends Biochem Sci 43(8):576–592
4. Sahni N, Yi S, Taipale M et al (2015) Widespread macromolecular interaction perturbations in human genetic disorders. Cell 161(3):647–660
5. Fuxman Bass JI, Sahni N, Shrestha S et al (2015) Human gene-centered transcription factor networks for enhancers and disease variants. Cell 161(3):661–673

6. Ng PK, Li J, Jeong KJ et al (2018) Systematic functional annotation of somatic mutations in cancer. Cancer Cell 33(3):450–462. e10
7. Li Y, Zhang Y, Li X, Yi S, Xu J (2019) Gain-of-Function mutations: an emerging advantage for cancer biology. Trends Biochem Sci 44(8): 659–674
8. Biswas S, Rao CM (2018) Epigenetic tools (The Writers, The Readers and The Erasers) and their implications in cancer therapy. Eur J Pharmacol 837:8–24
9. Kato S, Weng QY, Insco ML et al (2020) Gain-of-Function genetic alterations of G9a drive oncogenesis. Cancer Discov 10(7):980–997
10. Attar N, Kurdistani SK (2017) Exploitation of EP300 and CREBBP lysine acetyltransferases by cancer. Cold Spring Harb Perspect Med 7(3):a026534
11. Wan L, Chong S, Xuan F et al (2020) Impaired cell fate through gain-of-function mutations in a chromatin reader. Nature 577(7788): 121–126
12. Cao L, Li R, Wu X (2021) The functions and mechanisms of PR-DUB in malignancy. Front Mol Biosci 8:657150
13. Kim KH, Roberts CW (2016) Targeting EZH2 in cancer. Nat Med 22(2):128–134
14. Asada S, Fujino T, Goyama S, Kitamura T (2019) The role of ASXL1 in hematopoiesis and myeloid malignancies. Cell Mol Life Sci 76(13):2511–2523
15. Korfhage J, Lombard DB (2019) Malignant peripheral nerve sheath tumors: from epigenome to bedside. Mol Cancer Res 17(7): 1417–1428
16. Meng X, Yang S, Li Y et al (2018) Combination of proteasome and histone deacetylase inhibitors overcomes the impact of gain-of-function p53 mutations. Dis Markers, p. 3810108
17. Sears TK, Horbinski CM, Woolard KD (2021) IDH1 mutant glioma is preferentially sensitive to the HDAC inhibitor panobinostat. J Neuro-Oncol 154(2):159–170
18. Swaroop A, Oyer JA, Will CM et al (2019) An activating mutation of the NSD2 histone methyltransferase drives oncogenic reprogramming in acute lymphocytic leukemia. Oncogene 38(5):671–686
19. Keats JJ, Maxwell CA, Taylor BJ et al (2005) Overexpression of transcripts originating from the MMSET locus characterizes all t(4;14) (p16;q32)-positive multiple myeloma patients. Blood 105(10):4060–4069
20. Popovic RE, Martinez-Garcia EG, Giannopoulou Q et al (2014) Histone methyltransferase MMSET/NSD2 alters EZH2 binding and reprograms the myeloma epigenome through global and focal changes in H3K36 and H3K27 methylation. PLoS Genet 10(9): e1004566
21. Lauring J, Abukhdeir AM, Konishi H et al (2008) The multiple myeloma associated MMSET gene contributes to cellular adhesion, clonogenic growth, and tumorigenicity. Blood 111(2):856–864
22. Kuo AJ, Cheung P, Chen K et al (2011) NSD2 links dimethylation of histone H3 at lysine 36 to oncogenic programming. Mol Cell 44(4):609–620
23. Zhou Z, Gao J, Popovic R et al (2015) Strong expression of EZH2 and accumulation of trimethylated H3K27 in diffuse large B-cell lymphoma independent of cell of origin and EZH2 codon 641 mutation. Leuk Lymphoma 56(10):2895–28901
24. Bodor C, Grossmann V, Popov N et al (2013) EZH2 mutations are frequent and represent an early event in follicular lymphoma. Blood 122(18):3165–3168
25. McCabe MT, Ott HM, Ganji G et al (2012) EZH2 inhibition as a therapeutic strategy for lymphoma with EZH2-activating mutations. Nature 492(7427):108–112
26. Singh S, Vaughan CA, Frum RA (2017) Mutant p53 establishes targetable tumor dependency by promoting unscheduled replication. J Clin Invest 127(5):1839–1855
27. Schmidt L, Heyes E, Grebien F (2020) Gain-of-Function effects of N-terminal CEBPA mutations in acute myeloid leukemia. BioEssays 42(2):e1900178
28. Statello L, Guo CJ, Chen LL, Huarte M (2021) Gene regulation by long non-coding RNAs and its biological functions. Nat Rev Mol Cell Biol 22(2):96–118
29. Qian X, Yang J, Qiu Q et al (2021) LCAT3, a novel m6A-regulated long non-coding RNA, plays an oncogenic role in lung cancer via binding with FUBP1 to activate c-MYC. J Hematol Oncol 14(1):112
30. Liu X, Yang Q, Yan J, Zhang X, Zheng M (2019) LncRNA MNX1-AS1 promotes the progression of cervical cancer through activating MAPK pathway. J Cell Biochem 120(3): 4268–4277
31. Zhang Z. Shi Z (2019) The pseudogene PTTG3P promotes cell migration and invasion in esophageal squamous cell carcinoma. Open Med (Wars) 14:516–522
32. Latysheva NS, Oates ME, Maddox L et al (2016) Molecular principles of gene fusion mediated rewiring of protein interaction networks in cancer. Mol Cell 63(4):579–592

33. Xiao G, Lundine D, Annor GK et al (2020) Gain-of-Function mutant p53 R273H interacts with replicating DNA and PARP1 in breast cancer. Cancer Res 80(3):394–405
34. Schulz-Heddergott R, Stark N, Edmunds SJ et al (2018) Therapeutic ablation of gain-of-function mutant p53 in colorectal cancer inhibits Stat3-mediated tumor growth and invasion. Cancer Cell 34(2):298–314. e7
35. Ghosh M, Saha S, Bettke J (2021) Mutant p53 suppresses innate immune signaling to promote tumorigenesis. Cancer Cell 39(4): 494–508. e5
36. Kim MP, Lozano G (2018) Mutant p53 partners in crime. Cell Death Differ 25(1): 161–168
37. Du Z, Lovly CM (2018) Mechanisms of receptor tyrosine kinase activation in cancer. Mol Cancer 17(1):58
38. Zhang H, Schaefer A, Wang Y et al (2020) Gain-of-Function RHOA mutations promote focal adhesion kinase activation and dependency in diffuse gastric cancer. Cancer Discov 10(2):288–305
39. Lee YR, Yehia L, Kishikawa T et al (2020) WWP1 Gain-of-Function inactivation of PTEN in cancer predisposition. N Engl J Med 382(22):2103–2116
40. Hanker AB, Brown BP, Meiler J et al (2021) Co-occurring gain-of-function mutations in HER2 and HER3 modulate HER2/HER3 activation, oncogenesis, and HER2 inhibitor sensitivity. Cancer Cell 39(8):1099–1114. e8
41. Friedlaender A, Drilon A, Weiss GJ et al (2020) KRAS as a druggable target in NSCLC: rising like a phoenix after decades of development failures. Cancer Treat Rev 85:101978
42. Ghimessy A, Radeczky P, Laszlo V et al (2020) Current therapy of KRAS-mutant lung cancer. Cancer Metastasis Rev 39(4):1159–1177
43. Liu P, Wang Y, Li X (2019) Targeting the untargetable KRAS in cancer therapy. Acta Pharm Sin B 9(5):871–879
44. Tang D, Kroemer G, Kang R (2021) Oncogenic KRAS blockade therapy: renewed enthusiasm and persistent challenges. Mol Cancer 20(1):128
45. Belli C, Anand S, Gainor JF et al (2020) Progresses toward precision medicine in RET-altered solid tumors. Clin Cancer Res 26(23):6102–6111
46. Pisapia P, Pepe F, Laccarino A (2020) BRAF: a two-faced Janus. Cell 9(12):E2549
47. Campanelli R, Massa M, Rosti V, Barosi G (2021) New markers of disease progression in myelofibrosis. Cancers (Basel) 13(21):5324
48. Misaka T, Kimishima Y, Yokokawa T, Ikeda K, Takeishi Y (2022) Clonal hematopoiesis and cardiovascular diseases: role of JAK2V617F. J Cardiol S0914-5087(22):00024–00027
49. Spivak JL (2021) Advances in polycythemia vera and lessons for acute leukemia. Best Pract Res Clin Haematol 34(4):101330
50. Baltanas FC, Zarich N, Rojas-Cabaneros JM, Santos E (2020) SOS GEFs in health and disease. Biochim Biophys Acta Rev Cancer 1874(2):188445
51. Cannon AC, Uribe-Alvarez C, Chernoff J (2020) RAC1 as a therapeutic target in malignant melanoma. Trends Cancer 6(6):478–488
52. Niemeyer CM, Flotho C (2019) Juvenile myelomonocytic leukemia: who's the driver at the wheel? Blood 133(10):1060–1070
53. Arima J, Hiramatsu M, Taniguchi K et al (2020) Multiple gastrointestinal stromal tumors caused by a novel germline KIT gene mutation (Asp820Gly): a case report and literature review. Gastric Cancer 23(4):760–764
54. Gilreath JA, Tchertanov L, Deininger MW (2019) Novel approaches to treating advanced systemic mastocytosis. Clin Pharmacol 11:77–92
55. Lorden G, Newton AC (2021) Conventional protein kinase C in the brain: repurposing cancer drugs for neurodegenerative treatment? Neuronal Signal 5(4):NS20210036
56. Wei X, Cho KS, Thee EF et al (2019) Neuroinflammation and microglia in glaucoma: time for a paradigm shift. J Neurosci Res 97(1): 70–76
57. Malik I, Kelley CP, Wang ET, Todd PK (2021) Molecular mechanisms underlying nucleotide repeat expansion disorders. Nat Rev Mol Cell Biol 22(9):589–607
58. Barker RA, Fujimaki M, Rogers P, Rubinsztein DC (2020) Huntingtin-lowering strategies for Huntington's disease. Expert Opin Investig Drugsw 29(10):1125–1132
59. Booshehri LM, Hoffman HM (2019) CAPS and NLRP3. J Clin Immunol 39(3):277–286
60. Chen QL, Yin HR, He QY, Wang Y (2021) Targeting the NLRP3 inflammasome as new therapeutic avenue for inflammatory bowel disease. Biomed Pharmacother 138:111442
61. Tufan A, Lachmann HJ (2020) Familial mediterranean fever, from pathogenesis to treatment: a contemporary review. Turk J Med Sci 50(SI-2):1591–1610
62. Decout A, Katz JD, Venkatraman S, Ablasser A (2021) The cGAS-STING pathway as a therapeutic target in inflammatory diseases. Nat Rev Immunol 21(9):548–569

63. Anaya J (2016) OncoLnc: linking TCGA survival data to mRNAs, miRNAs, and lncRNAs. PeerJ Comput Sci 2:e67
64. Tamborero D, Gonzalez-Perez A, Lopez-Bigas N (2013) OncodriveCLUST: exploiting the positional clustering of somatic mutations to identify cancer genes. Bioinformatics 29(18): 2238–2244
65. Simonetti FL, Tornador C, Nabau-Moreto N et al (2014) Kin-Driver: a database of driver mutations in protein kinases. Database (Oxford):bau104
66. Pons T, Vazquez M, Matey-Hernandez ML (2016) KinMutRF: a random forest classifier of sequence variants in the human protein kinase superfamily. BMC Genomics 17 Suppl 2:396
67. Wang T, Ruan S, Zhao X et al (2021) OncoVar: an integrated database and analysis platform for oncogenic driver variants in cancers. Nucleic Acids Res 49(D1):D1289–D1301
68. Chakravarty D, Gao J, Phillips SM et al (2017) OncoKB: a precision oncology knowledge base. JCO Precis Oncol 2017:1
69. Carter H, Chen S, Isik L et al (2009) Cancer-specific high-throughput annotation of somatic mutations: computational prediction of driver missense mutations. Cancer Res 69(16):6660–6667
70. Bailey TL, Boden M, Buske FA et al (2009) MEME SUITE: tools for motif discovery and searching. Nucleic Acids Res 37(Web Server issue):W202–8.
71. Boyle AP, Hong EL, Hariharan M et al (2012) Annotation of functional variation in personal genomes using RegulomeDB. Genome Res 22(9):1790–1797
72. Carter H, Douville C, Stenson PD et al (2013) Identifying mendelian disease genes with the variant effect scoring tool. BMC Genomics 14 Suppl 3:S3.

# Index



Usha N. Kasid and Robert Clarke (eds.), *Cancer Systems and Integrative Biology*, Methods in Molecular Biology, vol. 2660, https://doi.org/10.1007/978-1-0716-3163-8,

## Q



## R



## S



## T



## V



## W

MIX
Papier aus verantwortungsvollen Quellen
Paper from responsible sources
FSC® C105338

If you have any concerns about our products,
you can contact us on
**ProductSafety@springernature.com**

In case Publisher is established outside the EU,
the EU authorized representative is:
**Springer Nature Customer Service Center GmbH**
**Europaplatz 3, 69115 Heidelberg, Germany**

Printed by Libri Plureos GmbH
in Hamburg, Germany